Foundations of Quantitative Finance

Chapman & Hall/CRC Financial Mathematics Series

Series Editors

M.A.H. Dempster
Centre for Financial Research
Department of Pure Mathematics and Statistics
University of Cambridge, UK

Dilip B. Madan
Robert H. Smith School of Business
University of Maryland, USA

Rama Cont
Mathematical Institute
University of Oxford, UK

Robert A. Jarrow
Ronald P. & Susan E. Lynch Professor of Investment ManagementSamuel Curtis Johnson Graduate School of Management Cornell University

Recently Published Titles

Machine Learning for Factor Investing: Python Version
Guillaume Coqueret ad Tony Guida

Introduction to Stochastic Finance with Market Examples, Second Edition
Nicolas Privault

Commodities: Fundamental Theory of Futures, Forwards, and Derivatives Pricing, Second Edition
Edited by M.A.H. Dempster, Ke Tang

Introducing Financial Mathematics: Theory, Binomial Models, and Applications
Mladen Victor Wickerhauser

Financial Mathematics: From Discrete to Continuous Time
Kevin J. Hastings

Financial Mathematics: A Comprehensive Treatment in Discrete Time
Giuseppe Campolieti and Roman N. Makarov

Introduction to Financial Derivatives with Python
Elisa Alòs, Raúl Merino

The Handbook of Price Impact Modeling
Dr. Kevin Thomas Webster

Sustainable Life Insurance: Managing Risk Appetite for Insurance Savings & Retirement Products
Aymeric Kalife with Saad Mouti, Ludovic Goudenege, Xiaolu Tan, and Mounir Bellmane

Geometry of Derivation with Applications
Norman L. Johnson

Foundations of Quantitative Finance
Book I: Measure Spaces and Measurable Functions
Robert R. Reitano

Foundations of Quantitative Finance
Book II: Probability Spaces and Random Variables
Robert R. Reitano

Foundations of Quantitative Finance
Book III: The Integrals of Riemann, Lebesgue and (Riemann-)Stieltjes
Robert R. Reitano

Foundations of Quantitative Finance
Book IV: Distribution Functions and Expectations
Robert R. Reitano

Foundations of Quantitative Finance
Book V: General Measure and Integration Theory
Robert R. Reitano

For more information about this series please visit: https://www.routledge.com/Chapman-and-HallCRC-Financial-Mathematics-Series/book-series/CHFINANCMTH

Foundations of Quantitative Finance

Book V: General Measure and Integration Theory

Robert R. Reitano
Brandeis International Business School
Waltham, MA

CRC Press
Taylor & Francis Group
Boca Raton London New York

CRC Press is an imprint of the
Taylor & Francis Group, an **informa** business
A CHAPMAN & HALL BOOK

First edition published 2024
by CRC Press
2385 Executive Center Drive, Suite 320, Boca Raton, FL 33431

and by CRC Press
4 Park Square, Milton Park, Abingdon, Oxon, OX14 4RN

CRC Press is an imprint of Taylor & Francis Group, LLC

Library of Congress Cataloging-in-Publication Data

Names: Reitano, Robert R., 1950- author.
Title: Foundations of quantitative finance. Book V, General measure and integration theory / Robert R. Reitano.
Other titles: General measure and integration theory
Description: First edition. | Boca Raton, FL : CRC Press, 2024. | Series: Chapman & Hall/CRC financial mathematics series | Includes bibliographical references and index.
Identifiers: LCCN 2023038707 | ISBN 9781032206516 (hardback ; vol. 5) | ISBN 9781032206509 (paperback ; vol. 5) | ISBN 9781003264576 (ebook ; vol. 5)
Subjects: LCSH: Finance--Mathematical models. | Integration, Functional.
Classification: LCC HG106 .R4485 2023 | DDC 332.01/5195--dc23/eng/20230907
LC record available at https://lccn.loc.gov/2023038707

ISBN: 978-1-032-20651-6 (hbk)
ISBN: 978-1-032-20650-9 (pbk)
ISBN: 978-1-003-26457-6 (ebk)

DOI: 10.1201/9781003264576

Typeset in CMR10 font
by KnowledgeWorks Global Ltd.

to Michael, David, and Jeffrey

Contents

Preface

The idea for a reference book on the mathematical foundations of quantitative finance has been with me throughout my professional and academic careers in this field, but the commitment to finally write it didn't materialize until completing my introductory quantitative finance book in 2010.

My original academic studies were in pure mathematics in a field of mathematical analysis, and neither applications generally nor finance in particular were then even on my mind. But on completion of my degree, I decided to temporarily investigate a career in applied math, becoming an actuary, and in short order became enamored with mathematical applications in finance.

One of my first inquiries was into better understanding yield curve risk management, ultimately introducing the notion of partial durations and related immunization strategies. This experience led me to recognize the power of greater precision in the mathematical specification and solution of even an age-old problem. From there my commitment to mathematical finance was complete, and my temporary investigation into this field became permanent.

In my personal studies, I found that there were a great many books in finance that focused on markets, instruments, models, and strategies, which typically provided an informal acknowledgment of the background mathematics. There were also many books on mathematical finance focusing on more advanced mathematical models and methods, and typically written at a level of mathematical sophistication requiring a reader to have significant formal training and the time and motivation to derive omitted details.

The challenge of acquiring expertise is compounded by the fact that the field of quantitative finance utilizes advanced mathematical theories and models from a number of fields. While there are many good references on any of these topics, most are again written at a level beyond many students, practitioners and even researchers of quantitative finance. Such books develop materials with an eye to comprehensiveness in the given subject matter, rather than with an eye toward efficiently curating and developing the theories needed for applications in quantitative finance.

Thus the overriding goal I have for this collection of books is to provide a complete and detailed development of the many foundational mathematical theories and results one finds referenced in popular resources in finance and quantitative finance. The included topics have been curated from a vast mathematics and finance literature for the express purpose of supporting applications in quantitative finance.

I originally budgeted 700 pages per book, in two volumes. It soon became obvious this was too limiting, and two volumes ultimately turned into ten. In the end, each book was dedicated to a specific area of mathematics or probability theory, with a variety of applications to finance that are relevant to the needs of financial mathematicians.

My target readers are students, practitioners, and researchers in finance who are quantitatively literate and recognize the need for the materials and formal developments presented. My hope is that the approach taken in these books will motivate readers to navigate these details and master these materials.

Most importantly for a reference work, all 10 volumes are extensively self-referenced. The reader can enter the collection at any point of interest, and then using the references cited, work backward to prior books to fill in needed details. This approach also works for a course on a given volume's subject matter, with earlier books used for reference, and for both course-based and self-study approaches to sequential studies.

The reader will find that the developments herein are presented at a much greater level of detail than most advanced quantitative finance books. Such developments are of necessity typically longer, more meticulously reasoned, and therefore can be more demanding on the reader. Thus before committing to a detailed line-by-line study of a given result, it can be more efficient to first scan the derivation once or twice to better understand the overall logic flow.

I hope the scope of the materials, and the additional details presented, will support your journey to better understanding.

I am grateful for the support of my family: Lisa, Michael, David, and Jeffrey, as well as the support of friends and colleagues at Brandeis International Business School.

Robert R. Reitano

Brandeis International Business School

Author

Robert R. Reitano is Professor of the Practice of Finance at the Brandeis International Business School where he specializes in risk management and quantitative nance. He previously served as MSF Program Director, and Senior Academic Director. He has a PhD in mathematics from MIT, is a fellow of the Society of Actuaries, and a Chartered Enterprise Risk Analyst. Dr. Reitano consults in investment strategy and asset/liability risk management, and previously had a 29-year career at John Hancock/Manulife in investment strategy and asset/liability management, advancing to Executive Vice President & Chief Investment Strategist. His research papers have appeared in a number of journals and have won an Annual Prize of the Society of Actuaries and two F.M. Redington Prizes of the Investment Section of the Society of the Actuaries. Dr. Reitano serves on various not-for-prot boards and investment committees.

Introduction

Foundations of Quantitative Finance is structured as follows:

Book I: *Measure Spaces and Measurable Functions*
Book II: *Probability Spaces and Random Variables*
Book III: *The Integrals of Riemann, Lebesgue, and (Riemann-)Stieltjes*
Book IV: *Distribution Functions and Expectations*
Book V: *General Measure and Integration Theory*
Book VI: *Densities, Transformed Distributions, and Limit Theorems*
Book VII: *Brownian Motion and Other Stochastic Processes*
Book VIII: *Itô Integration and Stochastic Calculus 1*
Book IX: *Stochastic Calculus 2 and Stochastic Differential Equations*
Book X: *Classical Models and Applications in Finance*

The series is logically sequential. Books I, III, and V develop foundational mathematical results needed for the probability theory and finance applications of Books II, IV, and VI, respectively. Books VII, VIII, and IX then develop results in the theory of stochastic processes. While these latter three books introduce ideas from finance as appropriate, the final realization of the applications of these stochastic and other models to finance is deferred to Book X.

This Book V, *General Measure and Integration Theory*, generalizes the results of Books I and III on these respective topics. Because the Book I development is so foundational to the materials of this book, Chapter 1 sets out to review the key ideas and results that will be needed. Beginning with the definitional framework of measure spaces and the Lebesgue and Borel measure constructions, the general extension theory is then summarized and applied to the constructions of finite product measure spaces, general Borel measure spaces on $\mathbb{R}^n$, and finally to infinite dimensional product probability spaces.

The study of measurable functions is undertaken in Chapter 2. Following the development of a number of basic properties, the measurability of various limits of measurable functions is established, as is the approximation of measurable functions with simple function sequences. These approximations will be essential in the development of an integration theory. As in Book III, deeper properties can be derived from simpler results and appropriate limiting processes. An important tool is developed next, known as the monotone class theorem. After properties of monotone classes and their relationships to sigma algebras are developed, the important functional monotone class theorem is derived. This theorem provides a powerful approach to verify, through a sequence of simpler steps, that a given statement or identity applies to all bounded measurable functions.

Lebesgue-Steiltjes integration theory is developed in Chapter 3, largely following the approach and results of the Lebesgue development of Book III in this more abstract setting. Starting with simple functions, the approach is axiomatic in that the integrals of such functions are defined and shown to be consistent. The sequential approach of the Lebesgue theory is streamlined somewhat, omitting the development of integrals of bounded measurable functions and instead initiating the generalization from simple functions to nonnegative measurable functions. Key results for bounded functions are instead developed in the exercises and later in the chapter. Important "integration to the limit" results are derived for the integrals of nonnegative functions, including Fatou's lemma and Lebesgue's monotone convergence theorem, which are then instrumental in the development of the properties of such integrals. As an application of this theory, a Chapter I.7 result is generalized, addressing countable additivity of the set function defined on measurable rectangles of product measure spaces.

This integration theory is then applied to derive integrals of general measurable functions and their properties, along with the integration to the limit results of Beppo-Levi's theorem and Lebesgue's dominated convergence theorem. The bounded convergence theorem is then derived, as is an integration to the limit result not seen in Book III, called the uniform integrability convergence theorem. The Leibniz integral rule on the derivatives of parametrically defined integrals is then studied, generalizing the result of the Riemann theory of Book III. The chapter ends with a discussion of when the Lebesgue-Stieltjes integrals of this chapter agree with the Riemann-Stieltjes integrals of Book III.

Chapter 4 develops various results on change of variables in Lebesgue-Stieltjes integrals. The first investigation is fundamental in probability theory and addresses integrals whose measures are defined by integrals of measurable functions with respect to other measures. In probability theory, for example, this measure would be the Borel measure on $\mathbb{R}$ induced by the density function of a given random variable.

A more general realization of this idea is studied next, related to measurable transformations and their induced measures on range spaces. Integrals on the domain and range spaces can then be related using this transformation and related measures. Special cases of transformations are then studied in detail, beginning with linear transformations on $\mathbb{R}^n$, and then turning to continuously differentiable transformations.

Integrals on product spaces are studied in Chapter 5. The key question is, echoing a Riemann result of Book III, when can a multiple integral be evaluated using so-called iterated integrals, which integrate one variable at a time? Such results prove to be related to the product space sigma-algebra used, whereby the choices contrasted are the complete sigma algebra and the smallest sigma algebra that contains the defining algebra. In large measure, the results look quite similar, but there are details related to the need to qualify certain statements with "almost everywhere." The fundamental results here are then Fubini's theorem, applicable to integrable functions, and Tonelli's theorem, applicable to nonnegative measurable functions. Examples from earlier books are then used to illustrate the theory, though the next two chapters find deeper applications.

In Chapter 6, the first application of the Chapter 5 theory is to Lebesgue-Stieltjes integration by parts. For this, the notion of a "signed measure" is introduced and seen to possess all the properties of a measure other than nonnegativity. The integration theory is then developed for such measures when induced by functions of bounded variation, echoing the Riemann-Stieltjes work of Book III. The second application of the Chapter V theory is to the study of the integrability of the convolution of integrable functions, proving that such functions are indeed integrable.

The theory of Fourier transforms is developed in Chapter 7, a theory that will have its most important applications in Book VI in the guise of characteristic functions. After introducing the notion of the integral of complex-valued functions, the Fourier transform of integrable functions and finite Borel measures is defined and various properties developed. In particular, there are key connections between the decay at infinity of a function (or measure) and the differentiability of its Fourier transform. Conversely, there are connections between the existence and integrability of the derivatives of a function and the rate of decay at infinity of its Fourier transform.

Fourier inversion is then studied, whereby one recovers the measure from its Fourier transform, or, recovers the integrable function in the case where this transform is integrable. Finally, a continuity theory is studied, which addresses the connection between the weak convergence of probability measures and the pointwise convergence of their Fourier transforms. An application to Poisson's limit theorem is made, while the more powerful applications are deferred to Book VI.

Chapter 8 investigates general measure relationships and various decompositions of measures vis-a-vis other measures. The chapter begins with an example of the decomposition of a Borel measure on $\mathbb{R}$ into measures that have characteristic properties relative to Lebesgue measure. These characteristic properties are then generalized, whereby given a measure μ we identify what it means for other measures to be absolutely continuous, or, mutually singular, relative to μ. To generalize the Borel decomposition, the signed measures introduced in Chapter 6 are now studied in some detail. Various results are developed, such as the Hahn decomposition, which address the positive and negative sets underlying a signed measure, and the Jordan decomposition, which decomposes a signed measure into a difference of measures.

Perhaps the most famous and useful result on relationships between measures is the Radon-Nikodým theorem. It states that if a measure υ is absolutely continuous with respect to a measure μ, then υ is given by the μ-integral of a measurable function. This key result will have an important application in Chapter 9, and then in the stochastic process studies of Books VII–IX. The chapter ends with the Lebesgue decomposition theorem, which generalizes the chapter's Borel example to σ-finite measures.

The final Chapter 9 investigates Banach spaces, adding to their introduction in Book III by studying the so-called L_p spaces of variously integrable or bounded measurable functions, and their properties. Bounded linear functionals on L_p spaces are investigated and characterized by the Riesz representation theorem, which is proved with the aid of the Radon-Nikodým theorem. The special space of L_2, which is a Hilbert space, is addressed.

I hope this book and the other books in the collection serve you well.

Notation 0.1 (Referencing within FQF Series) *To simplify the referencing of results from other books in this series, we use the following convention.*

A reference to "Proposition I.3.33" is a reference to Proposition 3.33 of Book I, while "Chapter III.4" is a reference to Chapter 4 of Book III, and "II.(8.5)" is a reference to formula (8.5) of Book II, and so forth.

1

Measure Spaces

This chapter summarizes some of the more important results from Book I on the various constructions of a measure space $(X, \sigma(X), \mu)$. The importance of such constructions is twofold:

- The requirements for a measure space $(X, \sigma(X), \mu)$ are quite demanding and, in fact, so demanding that it is not immediately clear that such objects even exist.
- Given general existence, a measure space with particular properties will often be required. What if any restrictions are needed on these properties to ensure existence? For example, can we create a measure on $\mathbb{R}$ so that the measure of any interval $[a, b]$ is given by $f(b) - f(a)$ for a given function f?

For existence, the space X is generally a collection of points, sometimes with other special properties, but any set X can in theory be used. By Definitions I.2.1 and I.2.5, the **sigma algebra** $\sigma(X)$ is a collection of subsets of X with the following properties:

Definition 1.1 (Algebra of sets, sigma algebra) *A collection $\mathcal{A}$ of sets from a space X is called a **algebra of sets on** X:*

1. *If $A \in \mathcal{A}$ and $B \in \mathcal{A}$, then the **union** $A \bigcup B \in \mathcal{A}$ where:*
$$A \bigcup B \equiv \{x | x \in A \text{ or } x \in B\}.$$
2. *If $A \in \mathcal{A}$, then $\widetilde{A} \in \mathcal{A}$, where the **complement of** A is defined:*
$$\widetilde{A} \equiv \{x \in X | x \notin A\}.$$
The complement of A is also denoted A^c.

*A collection $\sigma(X)$ of sets from a space X is called a **sigma algebra on** X if $\sigma(X)$ is an algebra of sets, and:*

3. *If $\{A_i\}_{i=1}^{\infty} \subset \sigma(X)$, then $\bigcup_{i=1}^{\infty} A_i \in \sigma(X)$.*

Thus algebras are closed under finite unions and complementation, and also finite intersections by De Morgan's laws of Exercise I.2.2. A sigma algebra is closed under countable unions, and thus again countable intersections. In addition, algebras, and hence also sigma algebras, must contain X and the empty set $\emptyset$. For example, $A \in \mathcal{A}$ implies $\widetilde{A} \in \mathcal{A}$ and thus both $A \bigcup \widetilde{A} = X \in \mathcal{A}$ and $A \bigcap \widetilde{A} = \emptyset \in \mathcal{A}$.

The existence of algebras and sigma algebras can be demonstrated starting with any collection of sets. Such a collection then obtains both an algebra and a sigma algebra by defining $\mathcal{A}$ as the smallest algebra that contains this collection, and $\sigma(X)$ as the smallest such sigma algebra. This construction works because by Proposition I.2.8, the intersection of any collection of algebras is an algebra, and similarly for sigma algebras. To ensure that we do not have a vacuous intersection in this construction, note that the **power set** $\mathcal{P}(X)$, which contains all subsets of X, is one such algebra and sigma algebra.

An important example of such constructions are the Borel sigma algebras.

DOI: 10.1201/9781003264576-1

Example 1.2 (Borel sigma algebra) *By Definition I.2.13, the* ***Borel sigma algebra*** $\mathcal{B}(\mathbb{R}^n)$ *on* $\mathbb{R}^n$ *is the smallest sigma algebra that contains the open sets of* $\mathbb{R}^n$. *More generally, if* X *is a topological space (Definition 1.5), the Borel sigma algebra* $\mathcal{B}(X)$ *is defined as the smallest sigma algebra on* X *that contains all the open sets as defined by the topology* $\mathcal{T}$ *on* X.

By De Morgan's laws, Borel sigma algebras can also be defined as the smallest sigma algebras that contain the closed sets, but the above formulation is conventional.

Example 1.3 (Semi-algebra, algebra, and sigma algebra generated by E**)** *Generalizing Example 1.2, if* E *is any collection of subsets of a space* X, *then the smallest algebra* $\mathcal{A}(E)$ *and smallest sigma algebra* $\sigma(E)$ *that contain* E, *are well-defined. The same is true for the smallest semi-algebra* $\mathcal{A}'(E)$. *See Definition 1.14.*

It is also common to refer to these collections as the ***semi-algebra, algebra,*** *and* ***sigma algebra generated by*** E.

Recall the definition of open and closed sets. For more background on open and closed sets in various contexts, see Chapter 4 of **Reitano** (2010) or Section III.4.3.2, and also **Dugundji** (1970) or **Gemignani** (1967).

Definition 1.4 (Open sets in $\mathbb{R}$, $\mathbb{R}^n$, **and metric** X**)** *A set* $E \subset \mathbb{R}$ *is called* ***open*** *if for any* $x \in E$ *there is an open interval containing* x *that is also contained in* E. *In other words, there exists* $\epsilon_1, \epsilon_2 > 0$ *so that* $(x-\epsilon_1, x+\epsilon_2) \subset E$. *There is no loss of generality by requiring* $\epsilon_1 = \epsilon_2$.

A set $E \subset \mathbb{R}^n$ *is called* ***open*** *if for any* $x \in E$ *there is an open ball* $B_r(x)$ *about* x *of radius* $r > 0$:

$$B_r(x) \equiv \{y|\,|x-y| < r\},$$

so that $B_r(x) \subset E$. *Here* $|x-y|$ *denotes the standard metric on* $\mathbb{R}^n$:

$$|x-y| \equiv \left[\sum\nolimits_{i=1}^{n} (x_i - y_i)^2\right]^{1/2}.$$

More generally, if X *is a metric space with metric* d, *a set* $E \subset X$ *is called* ***open*** *if for any* $x \in E$ *there is an open ball* $B_r(x)$ *about* x *of radius* $r > 0$:

$$B_r(x) = \{y|d(x,y) < r\},$$

so that $B_r(x) \subset E$.

In all cases, a set F *is called* ***closed*** *if* $\widetilde{F}$, *the complement of* F, *is open.*

The above notions of an open set reflect the natural metric $|\cdot|$ on $\mathbb{R}$ and $\mathbb{R}^n$, where $d(x,y) \equiv |x-y|$, or more generally a metric d on a space X. For more on metrics, see the above references and Section III.4.3.1.

Open sets can also be defined without metrics, and this notion allows one to then define a continuous function. See Proposition 2.10.

Definition 1.5 (Topology) *A* ***topology*** *on a set* X *is a collection of subsets, denoted by* $\mathcal{T}$ *and called the "open sets," so that:*

1. $\emptyset, X \in \mathcal{T}$;
2. *If* $\{A_\alpha\}_{\alpha \in I} \subset \mathcal{T}$, *where the index set* I *is arbitrary (finite, countably infinite, uncountable), then* $\bigcup_{\alpha \in I} A_\alpha \in \mathcal{T}$;
3. *If* $\{A_i\}_{i=1}^n \subset \mathcal{T}$, *then* $\bigcap_{i=1}^n A_i \in \mathcal{T}$.

Exercise 1.6 (Natural topology on $\mathbb{R}^n$) *Prove that the collection of open sets on $\mathbb{R}$, as defined in Definition 1.4, is a topology on $\mathbb{R}$, and then generalize to $\mathbb{R}^n$. These topologies are often referred to as the* **natural topologies on** $\mathbb{R}$ *and* $\mathbb{R}^n$, *and also the* **Euclidean topologies.** *The same result is true for the metric space (X, d), and this is called the* **topology induced by the metric** *d. Hint: Recall Definition III.4.39 for definitional properties of metrics.*

Returning to the existence question for a measure space $(X, \sigma(X), \mu)$, since it is relatively easy to construct various $(X, \sigma(X))$ pairs, the real question of existence relates to the existence of a measure μ on $\sigma(X)$. Recalling Definition I.2.23:

Definition 1.7 (Measure space) *Let X be a set and $\sigma(X)$ a sigma algebra of subsets of X. A* **measure** μ **on** X *is a* **nonnegative set function** *defined on $\sigma(X)$, taking values in the nonnegative* **extended real numbers** $\overline{\mathbb{R}}^+ \equiv \mathbb{R}^+ \cup \{\infty\}$, *and which satisfies the following properties:*

1. $\mu(\emptyset) = 0$;

2. ***Countable Additivity:*** *If $\{A_i\}_{i=1}^{\infty}$ is a countable collection of* **pairwise disjoint** *sets in the sigma algebra $\sigma(X)$, meaning $A_j \bigcap A_k = \emptyset$ if $j \neq k$, then:*

$$\mu\left(\bigcup\nolimits_{i=1}^{\infty} A_i\right) = \sum\nolimits_{i=1}^{\infty} \mu(A_i). \tag{1.1}$$

In such a case, the sets in $\sigma(X)$ are said to be **measurable**, *and sometimes* μ-**measurable**, *and the triplet $(X, \sigma(X), \mu)$ is called a* **measure space.**

Definition 1.8 (Lebesgue, Borel measure spaces) *For these special measure spaces, a third requirement is added:*

- *For* **Lebesgue measure** $\mu \equiv m$ *and $X = \mathbb{R}$, Definition I.2.21 also adds the following requirement:*

3. *For any interval $I \equiv \langle a, b \rangle$, whether open, closed, or semi-closed,*

$$m(I) = |I| = b - a,$$

the length of I.

By the construction in Section I.7.6, this criterion generalizes for Lebesgue measure on $\mathbb{R}^n$, that for any measurable rectangle $R \equiv \prod_{i=1}^{n} \langle a_i, b_i \rangle$, whether such intervals are open, closed, or semi-closed:

$$m(R) = |R| = \prod\nolimits_{i=1}^{n} (b_i - a_i),$$

the volume of R.

- *For a* **Borel measure** μ, *$X = \mathbb{R}^n$ and $\sigma(X) = \mathcal{B}(\mathbb{R}^n)$, and Definitions I.5.1 and I.8.4 add the requirement:*

3'. *For any compact set $A \in \mathcal{B}(\mathbb{R}^n)$, $\mu(A) < \infty$.*

As noted above, while it is relatively easy to construct a sigma algebra on given set, it is by no means apparent how one then defines a measure on this sigma algebra.

Remark 1.9 (Additional properties of measures) *A few comments on the above definitions:*

- ***Lebesgue translation invariance:*** *For Lebesgue measure, it follows from item 3 that* m *is* ***translation invariant*** *on intervals:*

$$m(I) = m(I + x),$$

where $I + x \equiv \{x + y | y \in I\}$ *for any* x*. This property is then satisfied for all Lebesgue measurable sets on* $\mathbb{R}$ *by Propositions I.2.28 and I.2.35, and analogously generalizes to* $\mathbb{R}^n$ *by the construction of Chapter I.7.*

- ***Borel measures:*** *The notion of a Borel measure is not standardized, and it is not uncommon to see other restrictions added in place of item 3', such as the inner and outer regularity properties of Proposition I.5.29. Measure spaces that also satisfy item 3' are often called* ***Radon measures****, after* ***Johann Radon*** *(1887–1956).*

- ***Finite additivity:*** *Setting* $A_i = \emptyset$ *for* $i > n$ *in (1.1), it follows from this definition that all measures also satisfy* ***finite additivity*** *for pairwise disjoint sets:*

$$\mu\left(\bigcup\nolimits_{i=1}^{n} A_i\right) = \sum\nolimits_{i=1}^{n} \mu\left(A_i\right). \tag{1.2}$$

- ***Monotonicity:*** *Measures are* ***monotonic****, which means that if* $A \subset B$ *with both measurable, then:*

$$\mu\left(A\right) \leq \mu\left(B\right). \tag{1.3}$$

This follows from finite additivity since $B = A\bigcup(B - A)$ *as a disjoint union.*

- ***Subadditivity:*** *When* $\{A_i\}_{i=1}^{N}$ *is a finite* $(N < \infty)$ *or countable* $(N = \infty)$ *collection of not necessarily* ***pairwise disjoint*** *sets in the sigma algebra* $\sigma(X)$*, then a measure* μ *is in general* ***subadditive:***

$$\mu\left(\bigcup\nolimits_{i=1}^{N} A_i\right) \leq \sum\nolimits_{i=1}^{N} \mu\left(A_i\right). \tag{1.4}$$

This follows from Proposition I.2.20 that there exists pairwise disjoint $\{A_i'\}_{i=1}^{N}$ *with* $A_i' \subset A_i$ *and* $\bigcup_{i=1}^{m} A_i = \bigcup_{i=1}^{m} A_i'$ *for all* m*. Thus (1.4) follows from finite/countable additivity and monotonicity.*

The existence of measures is more readily justified when X has a finite or countable number of points.

Exercise 1.10 (Finite/countable X) *Verify that if* X *is a finite or countable set, then one can construct a measure space* $(X, \mathcal{P}(X), \mu)$*, where* $\mathcal{P}(X)$ *is the power set of* X*, defined to contain all subsets of* X*. Hint: If* $X = \{x_i\}_{i=1}^{n}$ *let* $\mu(x_i) = 1/n$*, if* $X = \{x_i\}_{i=1}^{\infty}$ *let* $\mu(x_i) = 2^{-n}$*, then* $\mu(X) = 1$*, and such measures are called* ***probability measures*** *and* $(X, \sigma(X), \mu)$ *is called a* ***probability space.***

For uncountable sets X, the construction is of necessity more subtle and sometimes obtains surprising results. In the next section, we review the special constructions of Lebesgue and Borel measure spaces of Chapters I.2 and I.5. We then summarize the general construction theory of Chapter I.6, and in the following section, recall Book I applications of this general framework to various special constructions.

1.1 Lebesgue and Borel Spaces on $\mathbb{R}$

Chapters I.2 and I.5 derived Lebesgue measure on $\mathbb{R}$ and then Borel measures on $\mathbb{R}$. Each of these derivations is summarized below, but suffice to say that they were fairly long, requiring a number of both detailed and subtle steps toward the final result. While somewhat different in their starting points and approaches, there was a certain redundancy in their development. And frankly, that was the point of dedicating two chapters to these important derivations, to illustrate that even with different starting points and objectives, many of the steps in the construction of a measure were very similar.

Being the first development of its kind, the Lebesgue construction is really the special case. Structurally, the Borel constructions introduced the framework that was generalized and abstracted in Chapter I.6.

1.1.1 Starting Points

Both constructions started with a special collection of sets and a rudimentary definition of the measure of such sets, and then an extension to the power sigma algebra $\mathcal{P}(\mathbb{R})$ of all subsets of $\mathbb{R}$ using the notion of an **outer measure.**

- **Lebesgue:** The collection of sets is the open intervals $\mathcal{G} \equiv \{(a,b)\}$, and the measure of $I = (a,b)$, denoted $|I|$, is defined as interval length:

$$|I| = b - a. \tag{1.5}$$

  If $A \in \mathcal{P}(\mathbb{R})$, the **Lebesgue outer measure of** A, denoted $m^*(A)$, is defined in I.(2.4):

$$m^*(A) = \inf\left\{\sum\nolimits_n |I_n| \;\middle|\; A \subset \bigcup\nolimits_n I_n,\right\}, \tag{1.6}$$

  where each I_n is an open interval, and $|I_n|$ denotes its interval length.

- **Borel:** The collection of sets is the right semi-closed intervals $\mathcal{A}' \equiv \{(a,b]\}$, and the measure of $I = (a,b]$, denoted $|I|_F$, is defined as the F-length of the interval:

$$|I|_F = F(b) - F(a), \tag{1.7}$$

  where $F(x)$ is an increasing, right continuous function.

  If $A \in \mathcal{P}(\mathbb{R})$, the **Borel outer measure of** A, denoted $\mu_F^*(A)$, is defined in I.(5.8):

$$\mu_{\mathcal{A}}^*(A) = \inf\left\{\sum\nolimits_n \mu_{\mathcal{A}}(A_n) \mid A \subset \bigcup\nolimits_n A_n\right\}, \tag{1.8}$$

  where $A_n \in \mathcal{A}$ and $\mu_{\mathcal{A}}(A_n)$ is F-length and defined in (1.7).

  Here, $\mathcal{A}$ is defined as the collection of all finite disjoint unions of sets from $\mathcal{A}'$, and thus $\mu_{\mathcal{A}}^*(A)$ can be equally well defined with all $A_n \in \mathcal{A}'$. In the next section, it will be seen that for general constructions, we use the collection $\mathcal{A}$.

The big question now becomes:

Can measures be created which extend the rudimentary notions of measure on $\mathcal{G}$-sets and $\mathcal{A}'$-sets to sigma algebras which contain these special sets, and if so, how?

1.1.2 Lebesgue Measure Space

Henri Lebesgue (1875–1941) wrote a collection of papers in the late 1800s and early 1900s that collectively created the foundational Lebesgue measure space $(\mathbb{R}, \mathcal{M}(\mathbb{R}), m)$ and an associated integration theory.

When studying the Lebesgue development for the first time, many would likely expect that the final result from the above starting point would be that m^* is a measure on $\mathcal{P}(\mathbb{R})$. After all, it seems like the perfect generalization of interval length to more general sets, defining $m^*(A)$ as the minimum of the total lengths of all "covers" of A by open intervals. This is reinforced by the results of Proposition I.2.28 that m^* is translation invariant, and for any interval I, whether open, closed, or semi-closed, that $m^*(I) = |I|$.

But this proves not to be. Remarkably, a construction was developed in 1905 by **Giuseppe Vitali** (1875–1932) using the **axiom of choice.** Background details on this axiom of set theory and Vitali's construction can be found in Section I.2.3. He demonstrated that the interval $[0,1]$ could be expressed as a countable **disjoint** union of translations of a given set A_0:

$$\bigcup\nolimits_{j=0}^{\infty} A_j = [0,1],$$

and then by translation invariance derived that all such sets must have the same outer measure.

This construction completely crushes any hope that m^* is countably additive. Indeed, it follows that depending on whether $m^*(A_0) = 0$ or $m^*(A_0) > 0$:

$$\sum\nolimits_{j=1}^{\infty} m^*(A_j) \in \{0, \infty\}.$$

But in no case can it be that $m^*([0,1]) = 1$.

As m^* is countably subadditive by Proposition I.2.29, it follows that $m^*(A_0) > 0$ and $\sum_{j=1}^{\infty} m^*(A_j) = \infty$. It then follows from this that m^* is not even finitely additive. So, the problem here is even more serious than it first appears.

We are left with two possibilities:

1. There is a Lebesgue measure on $\mathcal{P}(\mathbb{R})$, but it is not m^*;

2. The set function m^* is a Lebesgue measure, but on a smaller sigma algebra than $\mathcal{P}(\mathbb{R})$.

There is no hope for item 1, since any proposal will encounter the Vitali construction. To pursue item 2, there are two conventional approaches to identify the sets in $\mathcal{P}(\mathbb{R})$ on which m^* is indeed a measure, and these are discussed in Section I.2.4. The approach used there was one that generalizes well to other constructions.

In Definition I.2.33, the approach of **Constantin Carathéodory** (1873–1950) was used, an approach he developed for the general theory of outer measures.

Definition 1.11 (Lebesgue measurable set) *A set $A \subset \mathbb{R}$ is said to be **Lebesgue measurable** if it satisfies the **Carathéodory criterion**, that for any set $E \subset \mathbb{R}$:*

$$m^*(E) = m^*\left(A \bigcap E\right) + m^*\left(\widetilde{A} \bigcap E\right). \tag{1.9}$$

The collection of Lebesgue measurable sets is denoted $\mathcal{M}_L \equiv \mathcal{M}_L(\mathbb{R})$.

Thus by Carathéodory's criterion, a set A will be deemed to be Lebesgue measurable if A and its complement $\widetilde{A}$ can be used to split any set E in a way that preserves finite additivity of m^* into two subsets. By Proposition I.2.36, the satisfaction of this criterion is sufficient to ensure finite additivity generally, so m^* is thus finitely additive on $\mathcal{M}_L(\mathbb{R})$.

It was then proved in a series of results culminating in Proposition I.2.39 that $\mathcal{M}_L(\mathbb{R})$ is a sigma algebra that contains the intervals and hence the Borel sigma algebra $\mathcal{B}(\mathbb{R})$, and that m^* is a Lebesgue measure on this sigma algebra. Lebesgue measure m is then defined on $\mathcal{M}_L(\mathbb{R})$ by:

$$m \equiv m^*,$$

and $(\mathbb{R}, \mathcal{M}_L(\mathbb{R}), m)$ is a Lebesgue measure space.

It was also proved that $\mathcal{M}_L(\mathbb{R})$ is a **complete sigma algebra**, and thus $(\mathbb{R}, \mathcal{M}_L(\mathbb{R}), m)$ is a **complete measure space**. See Definition 1.16. As noted above, $\mathcal{M}_L(\mathbb{R})$ contains the intervals, and m agrees with the specification in (1.5) on this collection.

An important corollary of the definition of the outer measure m^* in (1.6) is that Lebesgue measurable sets can be approximated well with open sets, and various classes of sets defined using open sets. See Propositions I.2.42 for approximations, and I.2.43 for regularity of Lebesgue measure.

1.1.3 Borel Measure Spaces

Borel measures and the associated measure spaces are named for **Émile Borel** (1871–1956), an early pioneer in measure theory and probability theory. In many ways resembling the Lebesgue construction of Chapter I.2, Borel measures on $\mathbb{R}$ are constructed in Chapter I.5. Beginning with a study of such measures, it is proved in Proposition I.5.7 that every Borel measure μ on $\mathbb{R}$ induces an increasing, right continuous function $F_\mu(x)$ on $\mathbb{R}$ with the property that for all right semi-closed intervals $(a, b]$:

$$\mu\left[(a,b]\right] = F_\mu(b) - F_\mu(a). \tag{1.10}$$

To create a Borel measure space, it thus makes sense to investigate if **any** such increasing and right continuous function $F(x)$ can be used to induce a Borel measure μ_F. Indeed, the answer is in the affirmative. Initially defining the set function F-length on right semi-closed intervals $(a, b]$ by (1.7), which is now compelled by (1.10), a series of results extends this set function.

The first extension is to a set function $\mu_{\mathcal{A}}$ on the algebra $\mathcal{A}$, constructed as the collection of all finite disjoint unions of right semi-closed intervals. This extension is defined additively by disjointness. In Proposition I.5.13, $\mu_{\mathcal{A}}$ proves to be a measure on this algebra.

Definition 1.12 (Measure on an algebra) *A* ***measure on an algebra,*** *sometimes called a* ***pre-measure on an algebra,*** *is a nonnegative, extended real-valued set function μ defined on an algebra $\mathcal{A}$ with the properties that:*

1. $\mu(\emptyset) = 0,$

2. ***Countable Additivity:*** *If $\{A_j\}_{j=1}^\infty$ is a countable collection of* ***pairwise disjoint*** *sets in $\mathcal{A}$ with $\bigcup_{j=1}^\infty A_j \in \mathcal{A}$, then:*

$$\mu\left(\bigcup\nolimits_{j=1}^\infty A_j\right) = \sum\nolimits_{j=1}^\infty \mu\left(A_j\right). \tag{1.11}$$

An outer measure $\mu_{\mathcal{A}}^*$ is introduced as in (1.8), and as in the Lebesgue case, we utilize the **Carathéodory criterion** in Definition I.5.16:

Definition 1.13 (Carathéodory measurability w.r.t. $\mu_{\mathcal{A}}^*$) *Let $F(x)$ be an increasing, right continuous function and $\mu_{\mathcal{A}}^*$ the outer measure induced by $\mu_{\mathcal{A}}$ as in (1.8). A set $A \subset \mathbb{R}$*

is ***Carathéodory measurable with respect to*** $\mu_{\mathcal{A}}^*$ *, or simply* $\mu_{\mathcal{A}}^*$ ***measurable,*** *if for any set* $E \subset \mathbb{R}$ *:*

$$\mu_{\mathcal{A}}^*(E) = \mu_{\mathcal{A}}^*(A \cap E) + \mu_{\mathcal{A}}^*(\widetilde{A} \cap E). \tag{1.12}$$

The collection of $\mu_{\mathcal{A}}^*$ *measurable sets is denoted* $\mathcal{M}_{\mu_F}(\mathbb{R})$.

It was proved in a series of results culminating in Proposition I.5.23 that $\mathcal{M}_{\mu_F}(\mathbb{R})$ is a sigma algebra that contains the algebra $\mathcal{A}$ and hence the Borel sigma algebra $\mathcal{B}(\mathbb{R})$, and that $\mu_{\mathcal{A}}^*$ is a Borel measure on this sigma algebra. The Borel measure μ_F is then defined on $\mathcal{M}_{\mu_F}(\mathbb{R})$ by:

$$\mu_F \equiv \mu_{\mathcal{A}}^*,$$

and $(\mathbb{R}, \mathcal{M}_{\mu_F}(\mathbb{R}), \mu_F)$ is a Borel measure space.

It was also proved that $\mathcal{M}_{\mu_F}(\mathbb{R})$ is a **complete sigma algebra**, and thus $(\mathbb{R}, \mathcal{M}_{\mu_F}(\mathbb{R}), \mu_F)$ is a complete measure space. See Definition 1.16. Further, $\mathcal{A} \subset \mathcal{M}_{\mu_F}(\mathbb{R})$ as noted above, and μ_F agrees with the specification in (1.7) on $\mathcal{A}$.

An important corollary of the definition of the outer measure $\mu_{\mathcal{A}}^*$ in (1.6) is that sets in $\mathcal{M}_{\mu_F}(\mathbb{R})$ can be approximated well with various classes of sets defined using right semi-closed intervals. See Proposition I.5.26 and Remark I.5.27 for approximations and why these appear to differ structurally from the Lebesgue approximations, and Proposition I.5.29 for regularity of Borel measures.

1.2 General Extension Theory

While the Lebesgue and Borel constructions had a lot of similarities, the latter development better lends itself to generalization. The Lebesgue approach began with the class of open intervals, so any generalization to a general set X almost certainly requires X to at least have a topology so that the notion of "open" makes sense. The Borel approach starts with the collection of right semi-closed intervals $\{(a,b]\}$, which superficially looks quite similar to the collection of open intervals. However, this collection is different enough that one can easily create an algebra $\mathcal{A}$, and from this algebra, a sigma algebra is then "just" one step away.

Chapter I.6 exploits these structures within the Borel development, first identifying the essential property of the collection $\{(a,b]\}$, which allowed the simple step to an associated algebra $\mathcal{A}$.

Definition 1.14 (Semi-algebra of sets) *A collection* $\mathcal{A}'$ *of sets from a space* X *is called a* ***semi-algebra of sets on*** X*:*

1. *If* $A_1', A_2' \in \mathcal{A}'$, *then* $A_1' \bigcap A_2' \in \mathcal{A}'$, *and thus this holds by induction for all finite intersections.*

2. *If* $A' \in \mathcal{A}'$, *then there exists disjoint* $\{A_j'\}_{j=1}^n \subset \mathcal{A}'$ *so that* $\widetilde{A'} = \bigcup_{j=1}^n A_j'$.

   The collection $\mathcal{A}' = \mathcal{A}$, *an* ***algebra,*** *if in place of item 2 we have:*

2'. *If* $A \in \mathcal{A}$ *then* $\widetilde{A} \in \mathcal{A}$.

The collection $\{(a,b]\}$ is a semi-algebra, and by Exercise I.6.10, every semi-algebra $\mathcal{A}'$ generates an algebra $\mathcal{A}$ defined as the collection of all finite disjoint unions of $\mathcal{A}'$-sets. This implies that if we can define a set function on such a semi-algebra $\mathcal{A}'$, that the extension to a measure on $\mathcal{A}$ will follow additively as in the Borel case.

Next, the notion of an outer measure is introduced. This definition captured the key properties of the above special definitions that supported the conclusion that the collection of **Carathéodory measurable** sets is a sigma algebra, and that μ^* restricted to this collection is a measure.

Definition 1.15 (Outer measure) *Given a set X, a set function μ^* defined on the power sigma algebra $\sigma(P(X))$ of all subsets of X is an **outer measure** if:*

1. $\mu^*(\emptyset) = 0$.
2. ***Monotonicity:*** *For sets $A \subset B$:*
$$\mu^*(A) \leq \mu^*(B).$$
3. ***Countable Subadditivity:*** *Given a countable collection $\{A_j\}_{j=1}^{\infty}$:*
$$\mu^*\left(\bigcup\nolimits_{j=1}^{\infty} A_j\right) \leq \sum\nolimits_{j=1}^{\infty} \mu^*(A_j).$$

This definition looks nothing like the outer measure definitions in (1.6) and (1.8). But it can be noted that the first results developed after the Book I introductions of m^* and $\mu_{\mathcal{A}}^*$ were to establish that these outer measures had the above properties. Further, these properties were essential in the subsequent developments.

The results seen in the Borel development are then completely generalized in a series of Chapter I.6 results, which worked backward starting with the **Carathéodory extension theorem 1** of Proposition I.6.2 and named for **Constantin Carathéodory** (1873–1950). It asserts that outer measures always obtain complete measure spaces by the **Carathéodory criterion.**

Definition 1.16 (Complete measure space) *A measure space $(X, \sigma(X), \mu)$ is **complete** if for any $A \in \sigma(X)$ with $\mu(A) = 0$, then $B \in \sigma(X)$ for all $B \subset A$. It then follows by monotonicity of measures that $\mu(B) = 0$ for all such B.*

Equivalently, if $A, C \in \sigma(X)$ with $\mu(A) = \mu(C)$, then $B \in \sigma(X)$ for all $C \subset B \subset A$, and then $\mu(B) = \mu(A)$ for all such B.

Proposition 1.17 (Carathéodory extension theorem 1) *Let μ^* be an outer measure defined on a set X. Denote by $\mathcal{C}(X)$ the collection of all subsets of $\sigma(P(X))$ that are **Carathéodory measurable** with respect to μ^*. That is, $A \in \mathcal{C}(X)$ if for all $E \in \sigma(P(X))$:*

$$\mu^*(E) = \mu^*(A \cap E) + \mu^*(\widetilde{A} \cap E). \tag{1.13}$$

Then $\mathcal{C}(X)$ is a complete sigma algebra.

Further, if μ denotes the restriction of μ^* to $\mathcal{C}(X)$, then μ is a measure, and thus $(X, \mathcal{C}(X), \mu)$ is a complete measure space.

The **Hahn-Kolmogorov extension theorem** of Proposition I.6.4 is named for **Hans Hahn** (1879–1934) and **Andrey Kolmogorov** (1903–1987). It asserts that from a measure on an algebra, one can always induce an outer measure defined exactly as in (1.8), and thus a complete measure space by the above Carathéodory result. Specifically, given a measure $\mu_{\mathcal{A}}$ on an algebra $\mathcal{A}$, we define the associated outer measure $\mu_{\mathcal{A}}^*$ on $A \in \mathcal{P}(X)$ by:

$$\mu_{\mathcal{A}}^*(A) = \inf\left\{\sum\nolimits_n \mu_{\mathcal{A}}(A_n) \mid A \subset \bigcup\nolimits_n A_n,\ A_n \in \mathcal{A}\right\}. \tag{1.14}$$

Proposition 1.18 (Hahn–Kolmogorov extension theorem) *Let $\mathcal{A}$ be an algebra of sets on X and $\mu_{\mathcal{A}}$ a measure on $\mathcal{A}$ in the sense of Definition 1.12. Then, $\mu_{\mathcal{A}}$ gives rise to an outer measure $\mu^*_{\mathcal{A}}$ on $\sigma(P(X))$ such that $\mu^*_{\mathcal{A}}(A) = \mu_{\mathcal{A}}(A)$ for all $A \in \mathcal{A}$. In addition, there exists a complete sigma algebra $\mathcal{C}(X)$ with $\mathcal{A} \subset \mathcal{C}(X)$, and $\mu \equiv \mu^*_{\mathcal{A}}$ is a measure on $\mathcal{C}(X)$.*

Thus, $(X, \mathcal{C}(X), \mu)$ is a complete measure space and $\mu(A) = \mu_{\mathcal{A}}(A)$ for all $A \in \mathcal{A}$.

The Hahn-Kolmogorov extension theorem then obtains **approximations** of $\mathcal{C}(X)$-sets by various collections derived from $\mathcal{A}$-sets in Proposition I.6.5, exactly as in the Borel case.

The final step relates to the creation of a measure on an algebra. In the Borel case we began with the notion of F-length in (1.7) defined on $\mathcal{A}'$, and generalized this definition to $\mathcal{A}$ additively. Definition I.6.6 introduced the notion of a pre-measure on a collection of sets, which we restrict here to a pre-measure on a semi-algebra $\mathcal{A}'$. It is an exercise to check that F-length satisfies this definition.

Definition 1.19 (Pre-measure on a semi-algebra) *Given a semi-algebra of sets $\mathcal{A}'$, a set function $\mu_0 : \mathcal{A}' \to [0, \infty]$ is a* ***pre-measure*** *if:*

1. ***Either:*** *$\emptyset \in \mathcal{A}'$ and $\mu_0(\emptyset) = 0$,* ***or:***

1'. ***Finite additivity:*** *If $\{A_j\}_{j=1}^n \subset \mathcal{A}'$ is a disjoint finite collection of sets and $\bigcup_{j=1}^n A_j \in \mathcal{A}'$, then:*
$$\mu_0\left(\bigcup\nolimits_{j=1}^n A_j\right) = \sum\nolimits_{j=1}^n \mu_0(A_j);$$

and:

2. ***Countable additivity:*** *If $\{A_n\}_{j=1}^\infty \subset \mathcal{A}'$ is a disjoint countable collection of sets and $\bigcup_{j=1}^\infty A_n \in \mathcal{A}'$, then:*
$$\mu_0\left(\bigcup\nolimits_{j=1}^\infty A_j\right) = \sum\nolimits_{j=1}^\infty \mu_0(A_j).$$

The final existence result of Proposition I.6.13 is again attributable to **Constantin Carathéodory** (1873–1950).

Proposition 1.20 (Carathéodory extension theorem 2) *Let $\mathcal{A}'$ be a semi-algebra and μ_0 a pre-measure on $\mathcal{A}'$. Then, μ_0 can be extended to a measure $\mu_{\mathcal{A}}$ on the algebra $\mathcal{A}$, defined as the collection of all finite disjoint unions of sets in $\mathcal{A}'$, including $\emptyset$, if necessary.*

Beyond the existence theory summarized above, there is an associated uniqueness theory of Proposition I.6.14 for sigma-finite measure spaces. Recall Definition I.5.34:

Definition 1.21 (σ-finite measure space) *The measure space $(X, \sigma(X), \mu)$ is said to be* ***sigma finite,*** *or* ***σ-finite****, if there exists a countable collection $\{B_j\}_{j=1}^\infty \subset \sigma(X)$ with $\mu(B_j) < \infty$ for all j, and $X = \bigcup_{j=1}^\infty B_j$. In this case, it is also said that the measure μ is σ-finite.*

A measure $\mu_{\mathcal{A}}$ on an algebra $\mathcal{A}$ is sigma finite if such $\{B_j\}_{j=1}^\infty \subset \mathcal{A}$.

It should be noted that there is no loss of generality in assuming that $\{B_j\}_{j=1}^\infty$ is pairwise disjoint, recalling Proposition I.2.20 and monotonicity of measures.

The next result addresses the uniqueness of extensions from an algebra $\mathcal{A}$ is to $\sigma(\mathcal{A})$, the **smallest sigma algebra** that contains $\mathcal{A}$. For this result, note that both $\sigma(\mathcal{A}) \subset \mathcal{C}(X)$ and $\sigma(\mathcal{A}) \subset \sigma(X)$, and thus both μ and μ' are defined on $\sigma(\mathcal{A})$.

Originally proved as Proposition I.6.14, see Example 2.40 for an alternative proof using the monotone class theorem.

Proposition 1.22 (Uniqueness of Extensions to $\sigma(\mathcal{A})$) *Let $\mu_{\mathcal{A}}$ be a σ-finite measure on an algebra $\mathcal{A}$, and μ the extension of $\mu_{\mathcal{A}}$ to the sigma algebra $\mathcal{C}(X)$ induced by the outer measure $\mu_{\mathcal{A}}^*$. Let μ' be any other extension of $\mu_{\mathcal{A}}$ from $\mathcal{A}$ to a sigma algebra $\sigma(X)$.*

Then for all $B \in \sigma(\mathcal{A})$:

$$\mu(B) = \mu'(B).$$

More generally, if μ, μ' are σ-finite measures on a sigma algebra $\sigma(X)$, and $\mu(B) = \mu'(B)$ for all $B \in \mathcal{A}$ for some algebra $\mathcal{A} \subset \sigma(X)$, then $\mu = \mu'$ on $\sigma(\mathcal{A})$.

1.3 Measure Space Constructions

With the powerful tools of Chapter I.6, one can create a complete measure space without investigating all of the details seen in the Lebesgue and Borel developments.

The construction process is as follows:

1. **Required Step (Choose a or b):**

   (a) Define a set function μ_0 on a semi-algebra $\mathcal{A}'$ which can be proved to be a pre-measure.

   (b) Define a set function $\mu_{\mathcal{A}}$ on an algebra $\mathcal{A}$ which can be proved to be a measure.

   The algebra $\mathcal{A}$ in 1.b can be the algebra generated by a given semi-algebra $\mathcal{A}'$ or defined independently of a semi-algebra.

2. **"Free" Steps:**

   (a) From 1.a, if μ_0 is a pre-measure on a semi-algebra $\mathcal{A}'$, then μ_0 can be extended to a measure $\mu_{\mathcal{A}}$ on $\mathcal{A}$, the algebra generated by $\mathcal{A}'$, by the Carathéodory extension theorem 2. Alternatively from 1.b, we start with a measure $\mu_{\mathcal{A}}$ on an algebra $\mathcal{A}$.

   (b) In either case from step 2.a, $\mu_{\mathcal{A}}$ and $\mathcal{A}$ then generate an outer measure $\mu_{\mathcal{A}}^*$ on $\sigma(P(X))$, the power set sigma algebra on X. This follows from the Hahn-Kolmogorov extension theorem with $\mu_{\mathcal{A}}^*$ as defined in (1.14), and this outer measure then satisfies the conditions specified in the Carathéodory extension theorem 1.

   (c) The collection of sets that are Carathéodory measurable with respect to $\mu_{\mathcal{A}}^*$, as defined in (1.13), is then a complete sigma algebra $\mathcal{C}(X)$, and the restriction of $\mu_{\mathcal{A}}^*$ to $\mathcal{C}(X)$ is a measure μ. Hence, $(X, \mathcal{C}(X), \mu)$ is a complete measure space. Moreover, $\mathcal{A} \subset \mathcal{C}(X)$ and μ extends $\mu_{\mathcal{A}}$ in the sense that $\mu(A) = \mu_{\mathcal{A}}(A)$ for all $A \in \mathcal{A}$.

This framework was applied in Chapters I.7–I.9 to obtain measure spaces in three important contexts. While still requiring a certain amount of effort to obtain the required step, both the reader and the author were likely equally pleased to not have to then explicitly derive all the results obtained in the free steps.

1.3.1 Finite Products of Measure Spaces

Given measure spaces $\{(X_i, \sigma(X_i), \mu_i)\}_{i=1}^n$, the goal of Chapter I.7 was to create a product space and product measure. See Section 5.1 for more on product spaces.

Definition 1.23 (Product space and set function) *Given measure spaces* $\{(X_i, \sigma(X_i), \mu_i)\}_{i=1}^n$, *the* ***product space:***

$$X = \prod_{i=1}^n X_i,$$

is defined as:

$$X = \{x \equiv (x_1, x_2, ..., x_n) | x_i \in X_i\}. \tag{1.15}$$

A ***measurable rectangle*** *in* X *is a set* A*:*

$$A = \prod_{i=1}^n A_i = \{x \in X | x_i \in A_i\}, \tag{1.16}$$

where $A_i \in \sigma(X_i)$. *We denote the collection of measurable rectangles in* X *by* $\mathcal{A}'$.

The ***product set function*** μ_0 *is defined on* $A = \prod_{i=1}^n A_i \in \mathcal{A}'$ *by:*

$$\mu_0(A) = \prod_{i=1}^n \mu_i(A_i), \tag{1.17}$$

where we explicitly define $0 \cdot \infty = 0$.

Unsurprisingly given the notation, the collection $\mathcal{A}'$ proves to be a semi-algebra, and the set function μ_0 can be extended additively to a set function $\mu_{\mathcal{A}}$ on the associated algebra $\mathcal{A}$ of finite disjoint unions of $\mathcal{A}'$-sets. The derivation that $\mu_{\mathcal{A}}$ so defined is a measure on this algebra is subtle. This is in part due to the complexity of $\mathcal{A}$-sets, and also that the necessary results must be derived with (1.17) and properties of the measures $\{\mu_i\}_{i=1}^n$.

While generalized in Section 3.2.4, the Book I proof of countable additivity of $\mu_{\mathcal{A}}$ required that $\{\mu_i\}_{i=1}^n$ be σ-finite measures. This obtained σ-finiteness of $\mu_{\mathcal{A}}$ on $\mathcal{A}$ and of the resulting measure μ_X on the complete sigma algebra, there denoted $\sigma(X)$. The product measure space of Proposition I.7.20 is then denoted $(X, \sigma(X), \mu_X)$.

As noted in Notation I.7.21, it is common to express $\mu_X = \prod_{i=1}^n \mu_i$. This notation also reflects the fact that μ_X is an extension of $\mu_{\mathcal{A}}$ from $\mathcal{A}$ to $\sigma(X)$ and thus extends μ_0 from $\mathcal{A}'$ to $\sigma(X)$. So, for $A = \prod_{i=1}^n A_i \in \mathcal{A}'$, (1.17) can be expressed:

$$\mu_X(A) = \prod_{i=1}^n \mu_i(A_i). \tag{1.18}$$

When $\{(X_i, \sigma(X_i), \mu_i)\}_{i=1}^n = \{(\mathbb{R}, \mathcal{M}_{F_i}(\mathbb{R}), \mu_{F_i})\}_{i=1}^n$ are Borel measure spaces, which are sigma finite by item $3'$ of Definition 1.8, the final measure space $(\mathbb{R}^n, \mathcal{M}(\mathbb{R}^n), \prod_{i=1}^n \mu_{F_i})$ contains a Borel measure space $(\mathbb{R}^n, \mathcal{B}(\mathbb{R}^n), \prod_{i=1}^n \mu_{F_i})$ since $\mathcal{B}(\mathbb{R}^n) \subset \mathcal{M}(\mathbb{R}^n)$, and μ_X proves to be a Borel measure.

However, there are Borel measures on $\mathbb{R}^n$ other than these product measures, and this is the subject we discuss next.

1.3.2 Borel Measures on $\mathbb{R}^n$

Following the development of Chapter I.7, the next application of the Chapter I.6 extension theory is to general Borel measure spaces denoted $(\mathbb{R}^n, \mathcal{B}(\mathbb{R}^n), \mu)$. Generalizing the 1-dimensional case of Chapter I.5, any such Borel measure μ induces a multivariate function $F_\mu : \mathbb{R}^n \to \mathbb{R}$, which is continuous from above, and n-increasing. This is seen in Proposition I.8.10 for finite Borel measures, and Proposition I.8.12 in the general case, where:

Definition 1.24 (Continuous from above; n-increasing) *Given a function* $F_\mu : \mathbb{R}^n \to \mathbb{R}$*:*

1. F_μ ***is continuous from above at*** $x = (x_1, ..., x_n)$ *if given a sequence* $x^{(m)} = (x_1^{(m)}, ..., x_n^{(m)})$ *with* $x_i^{(m)} \geq x_i$ *for all* i *and* m, *and* $x^{(m)} \to x$ *as* $m \to \infty$, *then:*

$$F_\mu(x) = \lim_{m\to\infty} F_\mu(x^{(m)}). \tag{1.19}$$

We say that F_μ *is continuous from above if the above property is true for all* x.

2. F_μ ***satisfies the*** n***-increasing condition*** *if given any bounded right semi-closed rectangle* $A = \prod_{i=1}^n (a_i, b_i]$:

$$\sum_x sgn(x) F_\mu(x) \geq 0. \tag{1.20}$$

Each $x = (x_1, ..., x_n)$ *in the summation is one of the* 2^n *vertices of* A, *so* $x_i = a_i$ *or* $x_i = b_i$, *and* $sgn(x)$ *equals* -1 *if the number of* a_i*-components of* x *is odd, and equals* $+1$ *otherwise.*

It was derived in Propositions I.8.9 (finite Borel measures) and I.8.12 (general Borel measures) that the measure $\mu\left[\prod_{i=1}^n (a_i, b_i]\right]$ of a bounded right semi-closed rectangle can be expressed in terms of this induced function:

$$\mu\left[\prod_{i=1}^n (a_i, b_i]\right] = \sum_x sgn(x) F_\mu(x), \tag{1.21}$$

where this summation is defined above.

Exercise 1.25 (Product functions and measures) *If* $\{F_i(x_i)\}_{i=1}^n$ *are increasing and right continuous functions on* $\mathbb{R}$, *show that* $F(x) \equiv \prod_{i=1}^n F_i(x_i)$ *is continuous from above and* n*-increasing. Hint: Prove that the expression in (1.21) can be rewritten:*

$$\mu\left[\prod_{i=1}^n (a_i, b_i]\right] = \prod_{i=1}^n \left(F_i(b_i) - F_i(a_i)\right).$$

Note that when all $F_i(x_i) = x_i$, *that* $\mu\left[\prod_{i=1}^n (a_i, b_i]\right]$ *reduces to the Lebesgue measure of this rectangle:*

$$\mu\left[\prod_{i=1}^n (a_i, b_i]\right] = \prod_{i=1}^n (b_i - a_i).$$

Given the insights of this study of general Borel measures, the Chapter I.6 extension theory is then applied to investigate if a Borel measure space $(\mathbb{R}^n, \mathcal{B}(\mathbb{R}^n), \mu_F)$ can be constructed from a continuous from above and n-increasing function $F : \mathbb{R}^n \to \mathbb{R}$. Given such $F(x)$, we begin by defining the class of bounded right semi-closed rectangles:

$$\mathcal{A}'_B \equiv \left\{A \in \mathcal{B}(\mathbb{R}^n) | A = \prod_{i=1}^n (a_i, b_i], \text{ with } -\infty < a_i \leq b_i < \infty\right\},$$

and on $\mathcal{A}'_B$ define the set function μ_0 as in (1.21). It can be checked that $\mathcal{A}'_B$ is not a semi-algebra (Hint: consider $\widetilde{A}$), so this will need to be addressed later in the development.

It is then proved in Proposition I.8.13 that μ_0 is finitely additive and countably subadditive on $\mathcal{A}'_B$. Since $\mathcal{A}'_B$ is not a semi-algebra, Carathéodory's extension theorem 2 cannot be directly applied. Instead, the set function μ_F^* is defined on $A \subset \mathbb{R}^n$ by:

$$\mu_F^*(A) = \inf\left\{\sum_n \mu_0(A_n) \mid A \subset \bigcup A_n,\ A_n \in \mathcal{A}'_B\right\}, \tag{1.22}$$

and proved in Proposition I.8.14 to be an outer measure by Definition 1.15. Carathéodory's extension theorem 1 now assures the existence of a complete measure space $(\mathbb{R}^n, \mathrm{M}_F(\mathbb{R}^n), \mu_F)$, and in Proposition I.8.15, it is proved that μ_F extends the set function μ_0 defined on $\mathcal{A}'_B$ by showing that $\mathcal{A}'_B \subset \mathrm{M}_F(\mathbb{R}^n)$ and $\mu_F^*(A) = \mu_0(A)$ for all $A \in \mathcal{A}'_B$.

Example 1.26 (Borel measures as Borel/Lebesgue product measures) *When $F(x)$ is the product function of Exercise 1.25, then $(\mathbb{R}^n, M_F(\mathbb{R}^n), \mu_F)$ here is the product measure space obtained from $\{(\mathbb{R}, \mathcal{M}_{\mu_{F_i}}(\mathbb{R}), \mu_{F_i})\}_{i=1}^n$, while when all $F_i(x_i) = x_i$, then $(\mathbb{R}^n, M_F(\mathbb{R}^n), \mu_F)$ is the Lebesgue product space obtained from $\{(\mathbb{R}, \mathcal{M}_L(\mathbb{R}), m_i)\}_{i=1}^n$ where $m_i = m$ for all i. See Section I.7.6 for Lebesgue and Borel product spaces, and Proposition I.8.16, for uniqueness of extensions.*

1.3.3 Infinite Products of Probability Spaces

For infinite products of measure spaces, the theory essentially requires that these measure spaces be probability spaces. And while a somewhat more general derivation is possible, Chapter I.9 ultimately assumes that these probability spaces are defined on $\mathbb{R}$, the application of greatest interest.

Definition 1.27 (Infinite product space and set function) *Given probability spaces:*

$$\{(X_i, \sigma(X_i), \mu_i)\}_{i=1}^{\infty},$$

*define the **product space** $X = \prod_{i=1}^{\infty} X_i$ by:*

$$X = \{(x_1, x_2, ...)|x_i \in X_i\}. \tag{1.23}$$

*A **finite dimensional measurable rectangle** A **in** X, also called a **cylinder set**, is defined for any n and n-tuple of positive integers $J = (j(1), j(2), ..., j(n))$ by:*

$$A = \{x \in X|x_{j(i)} \in A_{j(i)}\}, \tag{1.24}$$

*where $A_{j(i)} \in \sigma(X_{j(i)})$. The cylinder set in (1.24) is said **to be defined by** J **and** $\prod_{i=1}^n A_{j(i)}$, and the collection of cylinder sets in X is denoted by $\mathcal{A}'$.*

*The **product set function** μ_0 is defined on $\mathcal{A}'$ as follows. If $A \in \mathcal{A}'$ is defined by J and $\prod_{i=1}^n A_{j(i)}$, then:*

$$\mu_0(A) = \prod\nolimits_{i=1}^{n} \mu_{j(i)}(A_{j(i)}). \tag{1.25}$$

The above restriction to probability spaces stems from the need to have $\mu_0(A)$ in (1.25) well defined. For example, if $J = (1, 2, ..., n)$ then:

$$A = \{x_i \in A_i\} = \{x_i \in A_i,\ x_{n+1} \in X_{n+1}\}.$$

For $\mu_0(A)$ to be well defined requires that:

$$\prod\nolimits_{i=1}^{n} \mu_i(A_i) = \mu_{n+1}(X_{n+1}) \prod\nolimits_{i=1}^{n} \mu_i(A_i),$$

and so $\mu_{n+1}(X_{n+1}) = 1$ is derived unless one of these sets has infinite or zero measure.

As the notation suggests, $\mathcal{A}'$ so defined is a semi-algebra. Further, μ_0 extends additively to $\mu_{\mathcal{A}}$ on the associated algebra $\mathcal{A}$, and $\mu_{\mathcal{A}}$ proves to be finitely additive and countably subadditive. For countable additivity, the algebra $\mathcal{A}$ needed to be enlarged to $\mathcal{A}^+$, and all X_i were then restricted to $\mathbb{R}$.

Definition 1.28 (Product space; general cylinder sets: $\mathcal{A}^+$) *Given probability spaces $\{(\mathbb{R}, \mathcal{B}(\mathbb{R}), \mu_i)\}_{i=1}^{\infty}$, where $\mathcal{B}(\mathbb{R})$ denotes the Borel sigma algebra, the **product space** $\mathbb{R}^{\mathbb{N}} = \prod_{i=1}^{\infty} \mathbb{R}_i$ is defined by:*

$$\mathbb{R}^{\mathbb{N}} = \{(x_1, x_2, ...)|x_i \in \mathbb{R}\}.$$

A ***general finite dimensional measurable rectangle*** *or* ***general cylinder set*** $A \subset \mathbb{R}^{\mathbb{N}}$ *is defined for any n-tuple of positive integers* $J = (j(1), j(2), ..., j(n))$ *and* $H \in \mathcal{B}\left(\prod_{i=1}^{n} \mathbb{R}_{j(i)}\right)$ *by:*

$$A = \{x \in \mathbb{R}^{\mathbb{N}} | (x_{j(1)}, x_{j(2)}, ...x_{j(n)}) \in H\}. \tag{1.26}$$

Here, $\mathcal{B}\left(\prod_{i=1}^{n} \mathbb{R}_{j(i)}\right) = \mathcal{B}(\mathbb{R}^n)$ *denotes the finite dimensional product space Borel sigma algebra associated with* $\{(\mathbb{R}, \mathcal{B}(\mathbb{R}), \mu_{j(i)})\}_{i=1}^{n}$, *a sigma subalgebra of the above denoted* $\sigma(X)$.

The cylinder set in (1.26) will be said ***to be defined by*** *H and J, and the collection of general cylinder sets in* $\mathbb{R}^{\mathbb{N}}$ *is denoted by* $\mathcal{A}^+$.

The cylinder set A can also be characterized in terms of the projection mapping:

$$\pi_J \equiv \prod_{i=1}^{n} \pi_{j(i)} : \mathbb{R}^{\mathbb{N}} \to \prod_{i=1}^{n} \mathbb{R}_{j(i)},$$

by:

$$A = \pi_J^{-1}(H). \tag{1.27}$$

For $A \in \mathcal{A}^+$ *defined by H and J, the* ***product set function*** μ_0 *is defined by:*

$$\mu_0(A) = \mu_J(H), \tag{1.28}$$

where μ_J *denotes the finite dimensional product space probability measure associated with* $\{(\mathbb{R}, \mathcal{B}(\mathbb{R}), \mu_{j(i)})\}_{i=1}^{n}$.

Then $\mathcal{A}^+$ again proves to be an algebra in Proposition I.9.17, and μ_0 is a measure on the algebra by this result and Proposition I.9.19. The Hahn-Kolmogorov extension theorem is applied to obtain the complete probability space $(\mathbb{R}^{\mathbb{N}}, \sigma(\mathbb{R}^{\mathbb{N}}), \mu_{\mathbb{N}})$. Further, $\mathcal{A}^+ \subset \sigma(\mathbb{R}^{\mathbb{N}})$ and $\mu_{\mathbb{N}}(A) = \mu_0(A)$ for all $A \in \mathcal{A}^+$.

1.4 Continuity of Measures

One of the most important properties of all measures used in various proofs is continuity, and specifically, **continuity from above** and **continuity from below**. These properties identify conclusions that can be drawn on the measures of a collection of **nested** measurable sets. By nested is meant that the collection $\{A_i\}_{i=1}^{\infty}$ satisfies:

$$A_i \subset A_{i+1}, \text{ for all } i,$$

or

$$A_{i+1} \subset A_i, \text{ for all } i.$$

In the former case, we are interested in the measure of the union, and in the latter, the measure of the intersection.

The properties identified in this proposition are often referred to in terms of the "continuity" of measures and understood in the following sense. Given a collection of measurable sets $\{B_j\}_{j=1}^{\infty}$, define A_i by:

1. $A_i = \bigcup_{j=1}^{i} B_j$, or,

2. $A_i = \bigcap_{j=1}^{i} B_j$, where it is assumed that $\mu(A_1) < \infty$.

Proposition 1.29 states that in both cases:

$$\mu\left(\lim_{i\to\infty} A_i\right) = \lim_{i\to\infty} \mu(A_i).$$

In the first case, $\{A_i\}_{i=1}^{\infty}$ is an **increasing sequence of sets** and the result is called **continuity from below**, whereas in the second case, the sequence is **decreasing sequence of sets** and the result is called **continuity from above**.

Proposition 1.29 (Continuity of all measures) *Given the measure space* $(X, \sigma(X), \mu)$ *and* $\{A_i\}_{i=1}^{\infty} \subset \sigma(X)$*:*

1. ***Continuity from Below:*** *If* $A_i \subset A_{i+1}$ *for all* i*:*

$$\mu\left(\bigcup\nolimits_{i=1}^{\infty} A_i\right) = \lim_{i\to\infty} \mu(A_i), \tag{1.29}$$

where the limit on the right may be finite or infinite.

2. ***Continuity from Above:*** *If* $A_{i+1} \subset A_i$ *for all* i *and* $\mu(A_1) < \infty$*:*

$$\mu\left(\bigcap\nolimits_{i=1}^{\infty} A_i\right) = \lim_{i\to\infty} \mu(A_i). \tag{1.30}$$

Proof. *To prove item 1, note that* $A_i \subset A_{i+1}$ *implies that* $\mu(A_i) \leq \mu(A_{i+1})$ *by monotonicity of* μ*. Define* $B_1 = A_1$*, and for* $i \geq 2$*, let* $B_i = A_i - A_{i-1}$*. Then* $\{B_i\}_{i=1}^{\infty} \subset \sigma(X)$ *are disjoint sets, and:*

$$\bigcup\nolimits_{i=1}^{\infty} A_i = \bigcup\nolimits_{i=1}^{\infty} B_i.$$

By countable additivity of μ*:*

$$\begin{aligned}\mu\left(\bigcup\nolimits_{i=1}^{\infty} A_i\right) &= \sum\nolimits_{i=1}^{\infty} \mu(B_i) \\ &= \mu(A_1) + \lim_{i\to\infty} \sum\nolimits_{j=2}^{i} \mu(A_j - A_{j-1}).\end{aligned}$$

Since A_{j-1} *and* $A_j - A_{j-1}$ *are disjoint with union* A_j*, finite additivity assures that:*

$$\mu(A_j - A_{j-1}) = \mu(A_j) - \mu(A_{j-1}).$$

Thus cancellation in this telescoping summation obtains (1.29):

$$\mu\left(\bigcup\nolimits_{i=1}^{\infty} A_i\right) = \lim_{i\to\infty} \mu(A_i).$$

For item 2, $\bigcap_{j=1}^{i} A_j = A_i$ *by the nesting property, while monotonicity and the assumption that* $\mu(A_1) < \infty$ *yields for all* i *:*

$$\mu\left(\bigcap\nolimits_{j=1}^{i} A_j\right) = \mu(A_i) < \mu(A_1).$$

Again by monotonicity, $\{\mu(A_i)\}_{i=1}^{\infty}$ *is a bounded, nonincreasing sequence, and thus has a well-defined limit as* $i \to \infty$*, which proves (1.30).* ■

2

Measurable Functions

We begin with a discussion of the definition and properties of a μ-measurable function defined on a measure space $(X, \sigma(X), \mu)$. For this definition, recall that for $f : X \to \overline{\mathbb{R}}$, with range in the extended real numbers of Definition I.3.1, the inverse f^{-1} is defined as a set function on any set $A \subset \overline{\mathbb{R}}$ by:

$$f^{-1}(A) \equiv \{x \in X | f(x) \in A\}. \tag{2.1}$$

While f^{-1} is always defined as a set function, it is only defined pointwise if f is one-to-one.

In Book I, Lebesgue measurability of a function defined on $(\mathbb{R}, \mathcal{M}_L(\mathbb{R}), m)$ could be characterized in Definition I.3.9 by various equivalent properties on the set function f^{-1}. The ultimate goal of such measurability was derived in Proposition I.3.26, and the following definition of a **measurable function** f on $(X, \sigma(X), \mu)$ focuses on this goal. Following this, we investigate equivalent formulations.

Definition 2.1 (Measurable function; transformation) *An **extended real-valued function** $f : X \to \overline{\mathbb{R}}$ defined on the measure space $(X, \sigma(X), \mu)$ is said to be **measurable,** or μ-**measurable,** or $\sigma(X)$-**measurable,** if for every Borel set $A \in \mathcal{B}(\mathbb{R})$:*

$$f^{-1}(A) \in \sigma(X).$$

*An **extended real-valued transformation** $f : X \to \overline{\mathbb{R}}^n$ defined on the measure space $(X, \sigma(X), \mu)$ is said to be **measurable,** etc., if $f^{-1}(A) \in \sigma(X)$ for every Borel set $A \in \mathcal{B}(\mathbb{R}^n)$.*

*More generally, a mapping $f(x)$ between measure spaces $(X, \sigma(X), \mu_X)$ and $(Y, \sigma(Y), \mu_Y)$ is **measurable** or $\sigma(X)/\sigma(Y)$-**measurable,** if for all $A \in \sigma(Y)$:*

$$f^{-1}(A) \in \sigma(X).$$

If $D \subset X$ and $f(x)$ is defined on D, then the criterion for measurability is the same as above, and thus of necessity, $D = f^{-1}(\mathbb{R}) \in \sigma(X)$.

Remark 2.2 (On $A \in \mathcal{B}(\mathbb{R})$ or $A \in \mathcal{B}(\mathbb{R}^n)$) *With $f(x)$ an extended real-valued function, $f : X \to \overline{\mathbb{R}}$ or $f : X \to \overline{\mathbb{R}}^n$, it may seem odd that the measurability criterion only addresses Borel sets in $\mathbb{R}$ and $\mathbb{R}^n$. But note that if f is measurable by the above definition, then:*

$$f^{-1}(\infty) = X - f^{-1}(\mathbb{R}/\mathbb{R}^n),$$

and so $f^{-1}(\infty) \in \sigma(X)$. Thus $f^{-1}(A \bigcup \{\infty\}) \in \sigma(X)$ for all such A.

Notation 2.3 ($\sigma(X)$-measurable) *Given the variety of labels used above to declare measurability, $\sigma(X)$-measurable is the most accurate for extended real-valued functions and transformations because the criterion $f^{-1}(A) \in \sigma(X)$ for all Borel A is a sigma algebra restriction. Measurability has nothing to do with the measure μ since there can be many*

DOI: 10.1201/9781003264576-2

measures defined on a sigma algebra, and these do not affect which functions are measurable and which are not. Nonetheless, the use of μ***-measurable*** *is fairly common and rarely causes confusion when there is one sigma algebra on the space.*

However, there will be instances in coming studies where we will encounter measure spaces $(X, \sigma_i(X), \mu)$ *with various sigma-algebras* $\sigma_i(X)$*. In other words, the space* X *is fixed as is the measure* μ*, but there can be various sigma algebras on which* μ *satisfies the definition of measure. A simple example but a common one is when* $(X, \sigma(X), \mu)$ *is a measure space and* $\sigma_i(X) \subset \sigma(X)$ *is a* ***sigma subalgebra,*** *then* $(X, \sigma_i(X), \mu)$ *is again a measure space. But there can also be multiple sigma algebras with no such inclusions.*

In these situations, the notion of a ***measurable function*** *can become ambiguous, as can the notion of a* μ***-measurable function.*** *Thus when there is more than one sigma algebra on* X*, it is necessary to say that* f *is* $\sigma(X)$***-measurable,*** *identifying the defining sigma algebra.*

When $f(x)$ *is a mapping between general measure spaces* $(X, \sigma(X), \mu_X)$ *and* $(Y, \sigma(Y), \mu_Y)$*, we will always say that* f ***is*** $\sigma(X)/\sigma(Y)$***-measurable*** *as noted above. Although a* $\sigma(X)$*-measurable function or transformation could be called* $\sigma(X)/\mathcal{B}(\mathbb{R})$***-measurable, or*** $\sigma(X)/\mathcal{B}(\mathbb{R}^n)$***-measurable,*** *this level of formality is rarely needed.*

Exercise 2.4 (Composition of measurable functions) *Show that if* $f(x)$ *is* $\sigma(X)/\sigma(Y)$*-measurable between measure spaces* $(X, \sigma(X), \mu_X)$ *and* $(Y, \sigma(Y), \mu_Y)$*, and* $g(y)$ *is* $\sigma(Y)/\sigma(Z)$*-measurable between measure spaces* $(Y, \sigma(Y), \mu_Y)$ *and* $(Z, \sigma(Z), \mu_Z)$*, then* $f(g(x))$ *is* $\sigma(X)/\sigma(Z)$*-measurable between measure spaces* $(X, \sigma(X), \mu_X)$ *and* $(Z, \sigma(Z), \mu_Z)$*.*

Although measurability of $f^{-1}(A)$ for all A in the range space sigma algebra is the requirement, it is not necessary to verify this condition for all such A to establish measurability. This was seen in Proposition I.3.4 for Lebesgue or Borel measurability, meaning where the respective domain space was $(\mathbb{R}, \mathcal{M}_L(\mathbb{R}), m)$ or $(\mathbb{R}, \mathcal{B}(\mathbb{R}), m)$. Then, measurability for all y of $f^{-1}((-\infty, y))$ is equivalent to this statement for all $f^{-1}([y, \infty))$, all $f^{-1}((y, \infty))$, or all $f^{-1}((-\infty, y])$. Using the same ideas, this is equivalent to measurability of all $f^{-1}((a, b))$. In any of these cases, Proposition I.3.26 obtains that this is equivalent to measurability of $f^{-1}(A)$ for all $A \in \mathcal{B}(\mathbb{R})$.

Exercise 2.5 *Generalize the prior paragraph to Lebesgue or Borel measurability of transformations defined on* $(\mathbb{R}^n, \mathcal{M}_L(\mathbb{R}^n), m)$ *or* $(\mathbb{R}^n, \mathcal{B}(\mathbb{R}^n), m)$*. Show that if* $f^{-1}\left(\prod_{i=1}^n(-\infty, y_i)\right)$ *is measurable for all* $y = (y_1, ..., y_n)$*, then this is equivalent to measurability of all* $f^{-1}(A)$ *for* A *defined as* $\prod_{i=1}^n[y_i, \infty)$*,* $\prod_{i=1}^n(y_i, \infty)$*,* $\prod_{i=1}^n(-\infty, y_i]$ *or* $\prod_{i=1}^n(a_i, b_i)$*.*

What is clear from the Book I development and the results of Exercise 2.5 is that these collections of sets, and there are many others, have the property that they generate the Borel sigma algebras $\mathcal{B}(\mathbb{R})$ or $\mathcal{B}(\mathbb{R}^n)$, the sigma algebras of the range spaces. This generalizes as might be expected.

In the following result, we specify this special collection as $\mathcal{A}'$, which is our standard notation for a semi-algebra. This result is true for any collection of sets that generate $\sigma(Y)$, not just semi-algebras. But we use this notation because it will often be the case that there is an apparent semi-algebra $\mathcal{A}'$ that generates an algebra $\mathcal{A}$, which in turn generates the range space sigma algebra $\sigma(Y)$. For example, this applies when $(Y, \sigma(Y), \mu_Y)$ is a measure space created by the extension theory of the prior chapter.

Example 2.6 (Sigma algebras generated by collections) *If* $\mathcal{A}'$ *is a semi-algebra and* μ_0 *a pre-measure on* $\mathcal{A}'$*, then* μ_0 *can be extended to a measure* $\mu_{\mathcal{A}}$ *on the associated algebra* $\mathcal{A}$ *of finite disjoint unions of* $\mathcal{A}'$*-sets by the second Carathéodory extension theorem of Proposition 1.20.*

Then, by the Hahn-Kolmogorov extension theorem of Proposition 1.18 (with notation changed), if $\mathcal{A}$ is an algebra of sets on Y and $\mu_{\mathcal{A}}$ a measure on $\mathcal{A}$, then $\mu_{\mathcal{A}}$ gives rise to an outer measure $\mu^*_{\mathcal{A}}$ on $\sigma(P(Y))$ such that $\mu^*_{\mathcal{A}}(A) = \mu_{\mathcal{A}}(A)$ for all $A \in \mathcal{A}$. In addition, there exists a complete sigma algebra $\mathcal{C}(Y)$ with $\mathcal{A} \subset \mathcal{C}(Y)$, and $\mu \equiv \mu^*_{\mathcal{A}}$ is a measure on $\mathcal{C}(Y)$.

Thus if $\sigma(\mathcal{A})$ denotes the smallest sigma algebra generated by $\mathcal{A}$, then $(Y, \sigma(\mathcal{A}), \mu)$ is an example of a measure space where the sigma algebra $\sigma(Y) \equiv \sigma(\mathcal{A})$ is generated by the semi-algebra $\mathcal{A}'$, or by the algebra $\mathcal{A}$.

Proposition 2.7 (Measurability test: Generating collections) *If $f(x)$ is a mapping between measure spaces $(X, \sigma(X), \mu_X)$ and $(Y, \sigma(Y), \mu_Y)$, and $\mathcal{A}'$ is a collection of sets that generates $\sigma(Y)$, then $f(x)$ is $\sigma(X)/\sigma(Y)$-measurable if and only if $f^{-1}(A) \in \sigma(X)$ for all $A \in \mathcal{A}'$.*

Proof. *"Only if" is true by definition. For the "if" direction, note that if $f^{-1}(\mathcal{A}') \subset \sigma(X)$, then $\sigma\left(f^{-1}(\mathcal{A}')\right) \subset \sigma(X)$, where $\sigma\left(f^{-1}(\mathcal{A}')\right)$ denotes the smallest sigma algebra generated by this collection of sets.*

It is an exercise to check that:

- $f^{-1}(\widetilde{A}) = \widetilde{f^{-1}(A)}$ *for all $A \in \mathcal{A}'$, where $\widetilde{A} \equiv Y - A$ and $\widetilde{f^{-1}(A)} \equiv X - f^{-1}(A)$ denote the complements of these sets.*
- $f^{-1}(\bigcup_{i=1}^n A_i) = \bigcup_{i=1}^n f^{-1}(A_i)$ *for all $\{A_i\}_{i=1}^n \subset \mathcal{A}'$.*

Thus if $A \in \sigma(\mathcal{A}')$, the smallest sigma algebra that contains $\mathcal{A}'$, then $f^{-1}(A) \in \sigma\left(f^{-1}(\mathcal{A}')\right)$, and the proof is complete. ■

The final definitional result relates the notions of measurable functions and measurable transformations. If $f(x) : X \to \overline{\mathbb{R}}^n$ is defined on the measure space $(X, \sigma(X), \mu)$, then $f(x) = (f_1(x), ..., f(x_n))$, and it is logical to inquire into the measurability of $f(x)$ vs. measurability of the component functions $\{f_i(x)\}_{i=1}^n$. The reader may recall a similar discussion related to random variables and random vectors that was summarized in Proposition II.3.32.

Proposition 2.8 (Measurability test: Transformations and functions) *If $f(x) : X \to \overline{\mathbb{R}}^n$ is defined on the measure space $(X, \sigma(X), \mu)$ with $f(x) = (f_1(x), ..., f(x_n))$, then $f(x)$ is a $\sigma(X)$-measurable transformation if and only if $f_i(x)$ is $\sigma(X)$-measurable for all i.*

Proof. *If $f(x)$ is $\sigma(X)$-measurable, then $f^{-1}(A) \in \sigma(X)$ for all $A \in \mathcal{B}(\mathbb{R}^n)$. Fixing i and taking $A = (a_i, b_i) \times \mathbb{R}^{n-1}$ with apparent notation obtains that $f^{-1}(A) = f_i^{-1}((a_i, b_i)) \in \sigma(X)$ for all (a_i, b_i). Since such sets generate $\mathcal{B}(\mathbb{R})$, $f_i(x)$ is $\sigma(X)$-measurable by Proposition 2.7.*

Conversely, $f_i^{-1}((a_i, b_i)) \in \sigma(X)$ for all i and (a_i, b_i) implies that:

$$f^{-1}\left(\prod\nolimits_{i=1}^{n}(a_i, b_i)\right) = \bigcap\nolimits_{i=1}^{n} f^{-1}\left((a_i, b_i)\right) \in \sigma(X),$$

and again Proposition 2.7 completes the proof. ■

2.1 Properties of Measurable Functions

Recall Example 1.2 that if the measure space $(X, \sigma(X), \mu)$ also has a topology $\mathcal{T}$, the **Borel sigma algebra on** X, denoted $\mathcal{B}(X)$, is the smallest sigma algebra that contains the open sets of X. A topology on X also allows one to define the notion of a continuous function.

To set the stage, we document the definition of continuous function in the more familiar settings.

Definition 2.9 (Continuous function on metric space) *The function $f : \mathbb{R} \to \mathbb{R}$ is* ***continuous at*** x_0 *if:*

$$\lim_{x \to x_0} f(x) = f(x_0). \tag{2.2}$$

That is, given $\epsilon > 0$ there is a $\delta \equiv \delta(x_0, \epsilon) > 0$ so that:

$$|f(x) - f(x_0)| < \epsilon \text{ if } |x - x_0| < \delta.$$

A function is said to be ***continuous on an interval*** $[a, b]$ *if it is continuous at each $x_0 \in (a, b)$, and also continuous at a and b where the limit in (2.2) is understood as one-sided, meaning for $x < b$ or $x > a$. A function is said to be* ***continuous*** *if it is continuous everywhere on its domain.*

The same definition in (2.2) applies to a function $f : \mathbb{R}^n \to \mathbb{R}$, but where $|x - x_0| \equiv d(x, x_0)$ is interpreted in terms of the standard metric on $\mathbb{R}^n$:

$$d(x, y) = \left[\sum\nolimits_{i=1}^{n} (x_i - y_i)^2\right]^{1/2}. \tag{2.3}$$

More generally, this definition applies to a function $f : (X_1, d_1) \to (X_2, d_2)$, where X_j is a metric space with metric d_j. The limit in (2.2) then means that given $\epsilon > 0$ there is a $\delta \equiv \delta(x_0, \epsilon) > 0$, so that:

$$d_2(f(x), f(x_0)) < \epsilon \text{ if } d_1(x, x_0) < \delta.$$

The following result is also true for $f : (X_1, d_1) \to (X_2, d_2)$ where, recalling Exercise 1.6, open sets in X_1 and X_2 are those induced by these metrics. Details are left as an exercise in changing notation.

Proposition 2.10 (Continuity and open sets) *Let $f : \mathbb{R}^n \to \mathbb{R}$ be a given function. Then, f is continuous if and only if for any open set $G \subset \mathbb{R}$, the set $f^{-1}(G)$ is open in $\mathbb{R}^n$ where:*

$$f^{-1}(G) \equiv \{x | f(x) \in G\}.$$

Proof. *Assume that f is continuous, that $G \subset \mathbb{R}$ is open, and that $x_0 \in f^{-1}(G)$. To prove that $f^{-1}(G)$ is open by Definition 1.4, it must be shown that there exists $r > 0$ and an open ball $B_r(x_0) \subset \mathbb{R}^n$ with $B_r(x_0) \subset f^{-1}(G)$. As G is open there exists $\epsilon > 0$ so that the open ball $B_\epsilon(y_0) \subset G$, where $y_0 = f(x_0)$. Thus $|y - y_0| < \epsilon$ for $y \in B_\epsilon(y_0)$. By definition of continuity, there exists δ so that $|f(x) - y_0| < \epsilon$ if $|x - x_0| < \delta$. Translating these statements obtains $f(B_\delta(x_0)) \subset B_\epsilon(y_0)$, and so with $r = \delta$:*

$$B_r(x_0) \subset f^{-1}(B_\epsilon(y_0)) \subset f^{-1}(G).$$

Conversely, assume $f^{-1}(G)$ is open for all open $G \subset \mathbb{R}$. Let $x_0 \in \mathbb{R}^n$ be given and $y_0 = f(x_0) \in \mathbb{R}$. Choose any open set $G \subset \mathbb{R}$ that contains y_0, for example, we could choose $G = \mathbb{R}$. By definition of open, there exists $\epsilon > 0$ so that $B_\epsilon(y_0) \subset G$. By assumption $f^{-1}(B_\epsilon(y_0))$ is open in $\mathbb{R}^n$ and contains x_0. Again by definition of open, there exists $B_\delta(x_0) \subset f^{-1}(B_\epsilon(y_0))$ and thus $f(B_\delta(x_0)) \subset B_\epsilon(y_0)$. This now translates to the $\epsilon - \delta$ definition for continuity and the proof is complete. ■

This result now provides an immediate extension of the notion of continuity to functions on topological spaces.

Definition 2.11 (Continuous function on a topological space) *Given a topological space X, a real-valued function $f : X \to \mathbb{R}$ is continuous if $f^{-1}(G)$ is open for any open $G \subset \mathbb{R}$.*

It now follows from this characterization of continuity that continuous functions are $\sigma(X)$-measurable.

Proposition 2.12 (Continuous $\Rightarrow$ μ-measurable) *Given a measure space $(X, \sigma(X), \mu)$, assume that X is also a topological space and that $\sigma(X)$ contains the open sets of X, and hence contains the Borel sigma algebra $\mathcal{B}(X)$. If $f : X \to \mathbb{R}$ is continuous, then f is $\sigma(X)$-measurable.*

Proof. *Consider the open set $G \equiv (a, b)$. Since continuous, $f^{-1}((a, b))$ is open in X, and thus by definition:*

$$f^{-1}((a, b)) \in \mathcal{B}(X) \subset \sigma(X).$$

The proof is complete by Proposition 2.7. ∎

The next result summarizes that measurability is preserved under simple arithmetic operations. We restrict to real-valued functions with range in $\mathbb{R}$. The reader is referred to Remark I.3.34 for a discussion on generalizing to extended real-valued functions with range in $\overline{\mathbb{R}}$. Furthermore, items 1 and 2 remain true for measurable real-valued transformations and this is left as an exercise.

Proposition 2.13 *Let $f(x)$ and $g(x)$ be real-valued $\sigma(X)$-measurable functions defined on a measure space $(X, \sigma(X), \mu)$, and let $a, b \in \mathbb{R}$. Then, the following are $\sigma(X)$-measurable:*

1. $af(x) + b$,
2. $f(x) \pm g(x)$,
3. $f(x)g(x)$,
4. $f(x)/g(x)$ *on* $\{x | g(x) \neq 0\}$.

Proof. *To simplify notation, the set $\{x | f(x) < r\}$ is denoted by $\{f(x) < r\}$, and so forth. Also, by Proposition 2.7, it is sufficient to prove that $h^{-1}(A) \in \sigma(X)$ for any collection of sets that generates $\mathcal{B}(\mathbb{R})$, where $h(x)$ denotes any function under consideration.*

1. *If $a = 0$, the function $g(x) = b$ is $\sigma(X)$-measurable since $g^{-1}(A) \in \{\emptyset, X\}$ for all A. For $a > 0$,*
$$\{af(x) + b < y\} = \{f(x) < (y - b)/a\},$$
which is measurable since $f(x)$ is $\sigma(X)$-measurable. A similar result applies to $a < 0$.

2. *Consider the sum since then by part 1, $-g(x)$ is measurable and this implies the result for $f(x) - g(x)$. For rational r, if $f(x) < r$ and $g(x) < y - r$, then $f(x) + g(x) < y$. Taking a union over all rational r:*
$$\bigcup\nolimits_r \left[\{f(x) < r\} \bigcap \{g(x) < y - r\}\right] \subset \{f(x) + g(x) < y\}.$$
On the other hand, if $f(x) + g(x) < y$ then $f(x) < y - g(x)$, and by density of the rationals, there exists rational r so that $f(x) < r < y - g(x)$. This implies $f(x) < r$ and $g(x) < y - r$.

   Hence,
$$\{f(x) + g(x) < y\} = \bigcup\nolimits_r \left[\{f(x) < r\} \bigcap \{g(x) < y - r\}\right],$$
and this set is measurable as a countable union of intersections of measurable sets.

3. First note that both $f^2(x)$ *and* $g^2(x)$ *are measurable. For* $f^2(x)$*, for example:*

$$\{f^2(x) < y\} = \begin{cases} \{f(x) < \sqrt{y}\} \bigcap \{f(x) > -\sqrt{y}\}, & y \geq 0, \\ \emptyset, & y < 0. \end{cases}$$

By parts 1 and 2, so too is $[f(x) + g(x)]^2$ *measurable, as is:*

$$f(x)g(x) = 0.5\left([f(x) + g(x)]^2 - f^2(x) - g^2(x)\right).$$

4. First, $D \equiv \{g(x) \neq 0\} \in \sigma(X)$*, since:*

$$D = g^{-1}(-\infty, 0) \bigcup g^{-1}(0, \infty),$$

and so $1/g(x)$ *is real-valued and well-defined on* D*. Measurability of* $1/g(x)$ *on* D *then follows since:*

- $y > 0$*:*
$$\{1/g(x) < y\} = \{g(x) > 1/y\} \bigcup \{g(x) < 0\}.$$
- $y = 0$*:*
$$\{1/g(x) < 0\} = \{g(x) < 0\}.$$
- $y < 0$*:*
$$\{1/g(x) < y\} = \{g(x) > 1/y\} \bigcap \{g(x) < 0\}.$$

Thus $1/g(x)$ *is measurable as is* $f(x)/g(x)$ *by item 3.*

■

The next result is a good example of when a measurability conclusion requires completeness of the measure space $(X, \sigma(X), \mu)$. Recall that when $f(x) = g(x)$, except on a set of μ-measure 0, this is often written as $f(x) = g(x)$ μ-a.e., and read, "μ almost everywhere."

Proposition 2.14 ($f(x) = g(x)$, μ-a.e.) *Let* $f(x)$ *be a* $\sigma(X)$*-measurable function on a* ***complete measure space*** $(X, \sigma(X), \mu)$*, and* $g(x)$ *a function with* $f(x) = g(x)$*,* μ*-a.e. Then,* $g(x)$ *is* $\sigma(X)$*-measurable.*
Proof. *If* $E \in \sigma(X)$ *is the set of* μ*-measure zero on which* $f(x) \neq g(x)$*, then:*

$$\begin{aligned} \{x | g(x) < y\} &= \{x \in E | g(x) < y\} \bigcup \{x \notin E | g(x) < y\} \\ &= \{x \in E | g(x) < y\} \bigcup \{x \notin E | f(x) < y\}. \end{aligned}$$

The first set is a subset of a set of μ*-measure zero and is hence* $\sigma(X)$*-measurable by completeness, wheras the second set is the intersection of measurable* $\widetilde{E}$*, the complement of* E*, and* $\sigma(X)$*-measurable* $\{x | f(x) < y\}$*.* ■

2.2 Limits of Measurable Functions

In this section, we investigate various limits of measurable functions and begin by recalling some definitions. The reader is referred to Section I.3.4.2 for a discussion of these limiting functions.

Definition 2.15 (Infimum/supremum) *Given a finite or countable sequence of functions $\{f_n(x)\}_{n=1}^{\infty}$, the **infimum** and **supremum** of the sequence are defined pointwise as follows.*

For each $x \in D \equiv \bigcap_{n=1}^{\infty} Dmn\{f_n\}$, where $Dmn\{f_n\}$ denotes the domain of the function f_n:

$$\inf_n f_n(x) = \begin{cases} -\infty, & \{f_n(x)\}_{n=1}^{\infty} \text{ unbounded below,} \\ \max\{y|y \leq f_n(x) \text{ all } n\}, & \{f_n(x)\}_{n=1}^{\infty} \text{ bounded below.} \end{cases} \tag{2.4}$$

$$\sup_n f_n(x) = \begin{cases} \infty, & \{f_n(x)\}_{n=1}^{\infty} \text{ unbounded above,} \\ \min\{y|y \geq f_n(x) \text{ all } n\}, & \{f_n(x)\}_{n=1}^{\infty} \text{ bounded above.} \end{cases} \tag{2.5}$$

*When $\{f_n(x)\}_{n=1}^{N}$ is a **finite collection**, $\inf_n f_n(x)$ is often denoted:*

$$\inf_n f_n(x) \equiv \min\{f_1(x), ..., f_N(x)\},$$

and $\sup_n f_n(x)$ is denoted:

$$\sup_n f_n(x) \equiv \max\{f_1(x), ..., f_N(x)\}.$$

Definition 2.16 (Limits inferior/superior) *Given a sequence of functions $\{f_n(x)\}_{n=1}^{\infty}$, the **limit inferior** and **limit superior** of the sequence are defined pointwise as follows.*

For each $x \in D \equiv \bigcap_{n=1}^{\infty} Dmn\{f_n\}$:

$$\liminf_{n\to\infty} f_n(x) = \sup_n \inf_{k\geq n} f_k(x), \tag{2.6}$$

$$\limsup_{n\to\infty} f_n(x) = \inf_n \sup_{k\geq n} f_k(x). \tag{2.7}$$

When clear from the context, the subscript $n \to \infty$ is often dropped from the lim inf *and* lim sup *notation.*

Notation 2.17 *The limit superior of a function sequence is alternatively denoted $\overline{\lim} f_n(x)$, and the limit inferior denoted $\underline{\lim} f_n(x)$, but we will use the above notation throughout these books.*

Exercise 2.18 (The lim in lim inf and lim sup) *From Definition 2.16, it may not be apparent where the notion of limit appears. Prove that:*

$$\begin{aligned} \liminf_{n\to\infty} f_n(x) &= \lim_{n\to\infty} \inf_{k\geq n} f_k(x), \\ \limsup_{n\to\infty} f_n(x) &= \lim_{n\to\infty} \sup_{k\geq n} f_k(x). \end{aligned}$$

In other words, the limit inferior is the limit of infima, wheras the limit superior is the limit of suprema. Hint: Consider how $\inf_{k\geq n} f_k(x)$ and $\sup_{k\geq n} f_k(x)$ vary with n.

As anticipated, these limiting functions are $\sigma(X)$-measurable when the functions in the sequence are $\sigma(X)$-measurable. For item 7, we recall Corollary I.3.46 that $\lim f_n(x)$ exists at x if and only if:

$$-\infty < \liminf f_n(x) = \limsup f_n(x) < \infty. \tag{2.8}$$

Proposition 2.19 (Measurability of functions derived from $\{f_n(x)\}_{n=1}^{\infty}$) *Given a sequence of $\sigma(X)$-measurable functions $\{f_n(x)\}_{n=1}^{\infty}$ defined on measurable domains $\{D_n\}_{n=1}^{\infty}$ of the measure space $(X, \sigma(X), \mu)$, the following functions are also $\sigma(X)$-measurable as defined on $D \equiv \bigcap_{n=1}^{\infty} D_n$:*

1. $\min_{n\leq N}\{f_n(x)\}$, *for all* N;
2. $\max_{n\leq N}\{f_n(x)\}$, *for all* N;
3. $\inf f_n(x)$;
4. $\sup f_n(x)$;
5. $\liminf f_n(x)$;
6. $\limsup f_n(x)$;
7. *If* $h(x) \equiv \lim f_n(x)$ *exists on* $D' \subset D$, *then* $h(x)$ *is* $\sigma(X)$*-measurable on* D'.

Proof. *By Proposition 2.7, it is sufficient to prove that* $h^{-1}(A) \in \sigma(X)$ *for any collection of sets that generate* $\mathcal{B}(\mathbb{R})$ *where* $h(x)$ *denotes any function under consideration.*

Item 1 *follows from item* 3, *and* 2 *from* 4, *by defining* $f_n(x) = f_N(x)$ *for* $n \geq N$.

For item 3, *if* $h(x)$ *is defined by* $h(x) = \inf f_n(x)$, *then by (2.4):*

$$h(x) > y \iff f_n(x) > y \text{ for all } n.$$

Thus:

$$\{x|h(x) > y\} = \bigcap\nolimits_{n=1}^{\infty}\{x|f_n(x) > y\},$$

and is measurable as the intersection of measurable sets.

Similarly, with $g(x) = \sup f_n(x)$*:*

$$\{x|g(x) < y\} = \bigcap\nolimits_{n=1}^{\infty}\{x|f_n(x) < y\},$$

and this set is again measurable as the intersection of measurable sets.

Now let $h(x) = \liminf f_n(x)$, *which by (2.6) means* $h(x) = \sup_n \inf_{k\geq n} f_k(x)$. *Then for each* n, $F_n(x) \equiv \inf_{k\geq n} f_k(x)$ *is measurable by* 3, *and hence* $h(x) = \sup F_n(x)$ *is measurable by* 4. *The same approach proves that* $\limsup f_n(x)$ *is measurable using (2.7).*

If $\lim f_n(x)$ *exists everywhere, then* $\sigma(X)$*-measurability follows from (2.8). If* D' *is the set on which* $\lim f_n(x)$ *exists, then* $D' = k^{-1}(-\infty, 0] \bigcap k^{-1}[0, \infty)$ *where* $k(x) = \limsup f_n(x) - \liminf f_n(x)$, *and so* $D' \in \sigma(X)$. *Thus* $\lim f_n(x) = \chi_{D'}(x) \limsup f_n(x)$ *is* $\sigma(X)$*-measurable on* D' *by item* 6 *and Definition 2.1, where* $\sigma(X)$*-measurable* $\chi_{D'}(x)$ *equals* 1 *on* D' *and* 0 *elsewhere.* ■

Corollary 2.20 (Measurability on complete $(X, \sigma(X), \mu)$**)** *Given a sequence of* $\sigma(X)$*-measurable functions* $\{f_n(x)\}_{n=1}^{\infty}$ *defined on* $\sigma(X)$*-measurable domains* $\{D_n\}_{n=1}^{\infty}$ *of the complete measure space* $(X, \sigma(X), \mu)$, *if* $h(x)$ *denotes any of the functions identified in Proposition 2.19, and* $g(x) = h(x)$ μ*-a.e., then* $g(x)$ *is* $\sigma(X)$*-measurable.*

Proof. *This is Proposition 2.14.* ■

2.3 Results on Function Sequences

In this section, we investigate a few implications of the convergence of $\sigma(X)$-measurable functions. The first states that pointwise convergence of measurable functions assures something more outside arbitrarily small sets. This "something more" initially resembles a uniform convergence result. However, as discussed below, this result does not assure uniform convergence outside a set of measure 0, nor even outside an arbitrarily small set.

Proposition 2.21 *Given $(X, \sigma(X), \mu)$, let $\{f_n(x)\}_{n=1}^{\infty}$ be a sequence of real-valued $\sigma(X)$-measurable functions defined on a measurable set D with $\mu(D) < \infty$, and let $f(x)$ be a real-valued function so that $f_n(x) \to f(x)$ pointwise for $x \in D$. Then given $\epsilon > 0$ and $\delta > 0$, there is a measurable set $A \subset D$ and an integer N, so that $\mu(A) < \delta$, and for all $x \in D - A$ and all $n \geq N$:*

$$|f_n(x) - f(x)| < \epsilon.$$

Proof. *Given $\epsilon > 0$, define:*

$$G_n = \{x| \, |f_n(x) - f(x)| \geq \epsilon\},$$

and $D_N = \bigcup_{n=N}^{\infty} G_n$:

$$D_N = \{x| \, |f_n(x) - f(x)| \geq \epsilon \text{ for some } n \geq N\}.$$

Then, $\{D_N\}_{N=1}^{\infty}$ is a nested sequence, $D_{N+1} \subset D_N \subset D$.

Since $f_n(x) \to f(x)$ for each $x \in D$, it follows that for every such x there is a D_N with $x \notin D_N$. Hence, $\bigcap_{N=1}^{\infty} D_N = \emptyset$, and since $\mu(D) < \infty$, it follows from continuity from above of μ by Proposition 1.29 that $\lim_{N\to\infty} \mu[D_N] \to 0$. Thus given $\delta > 0$, there is an N with $\mu[D_N] < \delta$.

Defining $A \equiv D_N$, then $\mu(A) < \delta$ and if $x \notin A$, then $|f_n(x) - f(x)| < \epsilon$ for all $n \geq N$. ■

Corollary 2.22 *If $(X, \sigma(X), \mu)$ is complete, the conclusion of the above proposition remains valid if $f_n(x) \to f(x)$ for each $x \in D$ outside a set of μ-measure 0.*

Proof. *Everything in the above proof remains the same, except that we can now only conclude that for every $x \in D$ outside an exceptional set of measure 0, that there exists D_N with $x \notin D_N$, and hence $\bigcap_N D_N$ equals this set of measure 0. But then, $\lim_{N\to\infty} \mu[\,D_N] \to 0$ again by Proposition 1.29, and the proof follows as above, with a final application of Corollary 2.20.* ■

This proposition does not imply that $f_n(x)$ converges uniformly to $f(x)$ on $D - A$ because the set A depends on the given ϵ and δ. This result is close to but not equivalent to **Littlewood's third principle** of Chapter I.4, named for **J. E. Littlewood** (1885–1977).

To improve this result to Littlewood's conclusion of "*nearly uniform convergence*," it must be shown that A can be chosen so that $f_n(x) \to f(x)$ uniformly on $D - A$. That is, we need to find a fixed set A with $\mu(A) < \delta$, so that for any $\epsilon > 0$, there is an N, such that $|f_n(x) - f(x)| < \epsilon$ for all $x \in D - A$ and all $n \geq N$. See the introduction to Chapter I.4 for more on Littlewood's principles.

This next result formalizes Littlewood's third principle and is known as **Egorov's theorem,** named for **Dmitri Fyodorovich Egorov** (1869–1931), and sometimes phonetically translated to **Egoroff**. It is also known as the **Severini–Egorov theorem** in recognition of the somewhat earlier and independent proof by **Carlo Severini** (1872–1951).

Proposition 2.23 (Severini-Egorov theorem) *Given $(X, \sigma(X), \mu)$, let $\{f_n(x)\}_{n=1}^{\infty}$ be a sequence of $\sigma(X)$-measurable functions defined on a measurable set D with $\mu(D) < \infty$, and let $f(x)$ be a $\sigma(X)$-measurable function so that $f_n(x) \to f(x)$ pointwise for $x \in D$. Then given $\delta > 0$, there is a measurable set $A \subset D$ with $\mu(A) < \delta$, so that $f_n(x) \to f(x)$ uniformly on $D - A$.*

That is, for $\epsilon > 0$, there is an N, so that $|f_n(x) - f(x)| < \epsilon$ for all $x \in D - A$ and $n \geq N$.

Proof. *Given* $\delta > 0$, *for each* m *define* $\epsilon_m = 1/m$ *and* $\delta_m = \delta/2^m$ *and apply Proposition 2.21. The result is a set* A_m *with* $\mu(A_m) < \delta_m$, *and an integer* N_m, *so that* $|f_n(x) - f(x)| < \epsilon_m$ *for* $n \geq N_m$ *and all* $x \in D - A_m$. *With* $A \equiv \bigcup_{m=1}^{\infty} A_m$, *countable subadditivity obtains* $\mu(A) \leq \sum_{m=1}^{\infty} \mu(A_m) = \delta$. *We now claim that* $f_n(x) \to f(x)$ *uniformly on* $D - A$.

Given ϵ *there is an* m *so that* $\epsilon_m < \epsilon$, *and hence an* N_m *so that* $|f_n(x) - f(x)| < \epsilon_m < \epsilon$ *for* $n \geq N_m$ *and all* $x \in D - A_m$. *But then this statement is also true for* $x \in D - A$ *since* $A_m \subset A$. ■

Corollary 2.24 (Severini–Egorov theorem) *If* $(X, \sigma(X), \mu)$ *is complete, the above result remains valid if* $f_n(x) \to f(x)$ μ*-a.e. for* $x \in D$.

Proof. *Left as an exercise.* ■

Remark 2.25 (On $\mu(D) < \infty$) *To perhaps state the obvious, the above results apply without the explicit need for the restriction of* $\mu(D) < \infty$ *in finite measure spaces, and in particular, in probability spaces. In such a space, we can conclude that pointwise convergence on any measurable set assures nearly uniform convergence.*

2.4 Approximating $\sigma(X)$-Measurable Functions

In this section, we investigate various approximations of measurable functions with simple functions. We begin by re-introducing the notion of a simple function, as originally seen in Books I and III, but generalized somewhat and framed in the current context. The reader should confirm that simple functions are $\sigma(X)$-measurable.

Definition 2.26 (Simple function) *A* ***simple function*** $\varphi : X \to \mathbb{R}$ *on a measure space* $(X, \sigma(X), \mu)$ *is defined by:*

$$\varphi(x) = \sum\nolimits_{i=1}^{n} a_i \chi_{A_i}(x), \tag{2.9}$$

where:

1. $\{A_i\}_{i=1}^{n} \subset \sigma(X)$ *are disjoint* ***measurable sets****;*
2. $\chi_{A_i}(x)$ *is the* ***characteristic function*** *or* ***indicator function*** *for* A_i, *defined as* $\chi_{A_i}(x) = 1$ *for* $x \in A_i$ *and* $\chi_{A_i}(x) = 0$ *otherwise;*
3. $\{a_i\}_{i=1}^{n} \subset \mathbb{R}$ *and* $a_i \geq 0$ *for all* i.

A ***vector-valued simple function*** $\varphi : X \to \mathbb{R}^m$ *on a measure space* $(X, \sigma(X), \mu)$ *is defined by:*

$$\varphi(x) = (\varphi_1(x), ..., \varphi_m(x)),$$

where $\{\varphi_i\}_{i=1}^{m}$ *are given as in (2.9). Equivalently (see the proof of Corollary 2.30):*

$$\varphi(x) = \sum\nolimits_{i=1}^{n} \overline{a}_i \chi_{A_i}(x),$$

where items 1 *and* 2 *apply, with* $\overline{a}_i = (a_{i,1}, ..., a_{i,m})$*:*

3′. $\{\overline{a}_i\}_{i=1}^{n} \subset \mathbb{R}^m$ *and* $a_{i,j} \geq 0$ *for all* i *and* j.

Remark 2.27 (On $a_i \geq 0$; disjoint $\{A_i\}_{i=1}^n$) *We do not restrict the definition of a simple function to require that $\mu(\bigcup_{i=1}^n A_i) < \infty$ as in the Lebesgue case of Definition III.2.2. Each definition reflects the approach taken in the development of the associated integration theory. For the forthcoming definition of the μ-integral of such $\varphi(x)$ in (3.1), eliminating the restriction that $\mu(\bigcup_{i=1}^n A_i) < \infty$ will now require that $a_i \geq 0$ for all i. This assumption avoids the potential problem of having a definition, which in effect contains terms of $\pm\infty$, a problem avoided in the Lebesgue development by requiring all $\mu(A_i) < \infty$.*

Such nonnegative simple functions will initially support the integration theory of nonnegative functions, but this is not a constraint. As seen in Book III, general $\sigma(X)$-measurable functions can be expressed as a difference of such nonnegative functions. See Definition 3.37 for this decomposition, and Remark 2.29 for an illustration of its use.

Similar to the Lebesgue case, while it is not strictly necessary to assume that $\{A_i\}_{i=1}^n$ are disjoint, a more generally defined simple function $\varphi(x)$ remains measurable and can always be expressed with disjoint sets. This follows because the range of such a simple function is still finite, say $\{b_j\}_{j=1}^m$. Then, measurability of $\varphi(x)$ obtains that $B_j \equiv \varphi^{-1}(b_j)$ is μ-measurable, $\{B_j\}_{j=1}^m$ are disjoint, and:

$$\varphi(x) = \sum\nolimits_{j=1}^{m} b_j \chi_{B_j}(x).$$

That simple functions will be useful in the general development of an integration theory is predicted by the following proposition. Note that this result does not require that D have finite measure.

Proposition 2.28 (Approximating nonnegative, $\sigma(X)$-measurable $f : X \to \mathbb{R}$) *Let $f(x)$ be a nonnegative $\sigma(X)$-measurable function defined on a measurable set D of a measure space $(X, \sigma(X), \mu)$. Then there is an increasing sequence of simple functions $\{\varphi_n(x)\}_{n=1}^\infty$, so that $\varphi_n(x) \to f(x)$ for all $x \in D$.*

Proof. *Given n, define $N \equiv n2^n + 1$ disjoint measurable sets $\{A_j^{(n)}\}_{j=1}^N$ by:*

$$A_j^{(n)} = \begin{cases} \{x \in D | (j-1)2^{-n} \leq f(x) < j2^{-n}\}, & 1 \leq j \leq N-1, \\ \{x \in D | n \leq f(x)\}, & j = N, \end{cases}$$

and the simple function $\varphi_n(x)$ by:

$$\varphi_n(x) = \left\{ \ (j-1)2^{-n}, \quad x \in A_j^{(n)}, \ 1 \leq j \leq N. \right.$$

Then, $\{\varphi_n(x)\}_{j=1}^\infty$ is an increasing sequence of simple functions with $\varphi_n(x) \to f(x)$ for all $x \in D$. In particular, on $\{x \in D | f(x) < n\} \equiv D - A_N^{(n)}$:

$$|f(x) - \varphi_n(x)| \leq 2^{-n},$$

while $\varphi_n(x) \to \infty$ on $\bigcap_{n=1}^\infty A_N^{(n)} = \{f(x) = \infty\}$. ■

Remark 2.29 (On general $f(x)$) *The above proposition and results below can be applied more generally than only to nonnegative functions.*

For example, if $f(x)$ is a $\sigma(X)$-measurable function defined on a measurable set D of a measure space $(X, \sigma(X), \mu)$, express $f(x) = f^+(x) - f^-(x)$ where $f^+(x)$ and $f^-(x)$ are nonnegative functions in (3.20) and (3.21) of Definition 3.37. Then there are increasing sequences of simple functions $\{\varphi_n^+(x)\}_{n=1}^\infty$ and $\{\varphi_n^-(x)\}_{n=1}^\infty$ so that $\varphi_n^+(x) \to f^+(x)$ and $\varphi_n^-(x) \to f^-(x)$ for all $x \in D$. Defining $\varphi_n(x) = \varphi_n^+(x) - \varphi_n^-(x)$, then $\varphi_n(x) \to f(x)$ for all $x \in D$, though this convergence is in general not monotonic. However, $\varphi_n(x)$ is nonnegative and increasing if $f(x) \geq 0$, and negative and decreasing if $f(x) \leq 0$.

In addition, defining the simple function sequence, $\{|\varphi_n(x)|\}_{j=1}^\infty \equiv \{\varphi_n^+(x) + \varphi_n^-(x)\}_{j=1}^\infty$, this sequence is increasing and $|\varphi_n(x)| \to |f(x)|$ for all $x \in D$.

The above result and those below can generally be applied to derive similar approximations to $\sigma(X)$-measurable transformations $f : X \to \mathbb{R}^m$ by generalizing the notion of a simple function, so that all a_i terms are now m-vectors. We illustrate this with the above result.

Corollary 2.30 (Approximating nonnegative, $\sigma(X)$-measurable $f : X \to \mathbb{R}^m$) *Let $f(x)$ be a nonnegative $\sigma(X)$-measurable transformation $f : X \to \mathbb{R}^m$, defined on a measurable set D of a measure space $(X, \sigma(X), \mu)$. Then there is an increasing sequence of vector-valued simple functions $\{\varphi_n(x)\}_{n=1}^{\infty}$, so that $\varphi_n(x) \to f(x)$ for all $x \in D$.*

Proof. *Expressing $f(x) \equiv (f_1(x), ..., f_m(x))$, each $f_k(x)$ is a nonnegative $\sigma(X)$-measurable function by Proposition 2.8. Thus for each n, there exists m simple functions $\{\varphi_n^{(k)}(x)\}_{k=1}^{m}$, such that for each k, $\{\varphi_n^{(k)}(x)\}_{n=1}^{\infty}$ is an increasing sequence with $\varphi_n^{(k)}(x) \to f_k(x)$ as $n \to \infty$ for all $x \in D$. Further, $\left|f_k(x) - \varphi_n^{(k)}(x)\right| \leq 2^{-n}$ on $\{x \in D | f_k(x) < n\}$.*

Express:

$$\varphi_n^{(k)}(x) = \sum\nolimits_{j=1}^{N} a_{j,n}^{(k)} \chi_{A_{j,n}^{(k)}}(x),$$

where in the notation of Proposition 2.28, $A_{j,n}^{(k)} = A_j^{(n)}$ defined in terms of $f_k(x)$, and then $a_{j,n}^{(k)} = (j-1)2^{-n}$ for all k.

Then:

$$\begin{aligned}\left(\varphi_n^{(1)}(x), ..., \varphi_n^{(m)}(x)\right) &= \left(\sum\nolimits_{j=1}^{N} a_{j,n}^{(1)} \chi_{A_{j,n}^{(1)}}(x), ..., \sum\nolimits_{j=1}^{N} a_{j,n}^{(m)} \chi_{A_{j,n}^{(m)}}(x)\right) \\ &= \sum\nolimits_{I} \left(a_{j_1,n}^{(1)}, ..., a_{j_m,n}^{(m)}\right) \prod\nolimits_{k=1}^{m} \chi_{A_{j_k,n}^{(k)}}(x),\end{aligned}$$

where the indexing set $I = \{J = (j_1, ..., j_m) | j_k \in \{1, ..., N\}\}$ has N^m vectors.

Defining this last expression as $\varphi_n(x)$, this is a vector-valued simple function since:

$$\prod\nolimits_{k=1}^{m} \chi_{A_{j_k,n}^{(k)}}(x) = \chi_{A_{J,n}}(x),$$

with:

$$A_{J,n} = \bigcap\nolimits_{k=1}^{m} A_{j_k,n}^{(k)}.$$

Finally by (2.3), for $x \in \bigcap_{k=1}^{m} \{x \in D | f_k(x) < n\}$:

$$|f(x) - \varphi_n(x)|^2 \equiv \sum\nolimits_{k=1}^{m} \left|f_k(x) - \varphi_n^{(k)}(x)\right|^2 \leq m2^{-2n},$$

and thus:

$$|f(x) - \varphi_n(x)| \leq \sqrt{m}2^{-n}.$$

It is left as an exercise to check that on the various subsets of D for which one or more $f_k(x)$ is infinite, that again $\varphi_n(x) \to f(x)$ in the sense that $\varphi_n^{(k)}(x) \to f_k(x)$ for bounded $f_k(x)$, and $\varphi_n^{(k)}(x) \to \infty$ otherwise. ■

Exercise 2.31 *Confirm the last statement of the prior proof. Hint: For each k:*

$$D = \{f_k < \infty\} \bigcup \{f_k = \infty\},$$

and so:

$$D = \bigcap\nolimits_{k=1}^{m} \left(\{f_k < \infty\} \bigcup \{f_k = \infty\}\right).$$

This is a union of 2^m measurable intersection sets, one of which is the set where all f_k are bounded and the convergence rate in the proof can be applied to subsets defined by $\bigcap_{k=1}^{m} \{x \in D | f_k(x) < n\}$.

The question addressed next is: when does each $\varphi_n(x)$ of Proposition 2.28 have **finite support?** The support of a function $\varphi_n(x)$ is the set $\{x|\varphi_n(x) \neq 0\}$, and thus we seek conditions that assure that $\mu\{\varphi_n(x) \neq 0\} < \infty$. For this result, recall the notion of a sigma finite measure space of Definition 1.21.

The third condition below is apparently out of place here, since we have not yet even defined μ-integrability. But this result belongs here and so we formally prove it. For now, the reader can exploit the intuitive framework of the Lebesgue integral of Book III and later generalize once μ-integrals have been developed.

Corollary 2.32 (When $\mu\{x|\varphi_n(x) \neq 0\} < \infty$) *If $f(x)$ is a nonnegative $\sigma(X)$-measurable function defined on a measurable set D of a measure space $(X, \sigma(X), \mu)$, then $\{\varphi_n(x)\}_{j=1}^{\infty}$ defined in Proposition 2.28 can be constructed so that each $\varphi_n(x)$ is zero outside a set of finite measure in the following cases:*

1. *$\mu(D) < \infty$;*
2. *$(X, \sigma(X), \mu)$ is σ-finite;*
3. *$f(x)$ is μ-integrable on D.*

Proof. *Item 1 needs no proof. For item 2, Definition 1.21 obtains a countable collection of measurable sets $\{B_j\}_{j=1}^{\infty}$ so that $X = \bigcup_{j=1}^{\infty} B_j$ and $\mu(B_j) < \infty$ for all j. Without loss of generality, we can assume that this collection is nested, $B_j \subset B_{j+1}$, since given a general collection $\{B_i'\}$, we define $B_j = \bigcup_{i \leq j} B_i'$.*

Now redefine each $\varphi_n(x)$ by:

$$\varphi_n(x) = \left\{ \; (j-1)2^{-n}, \quad x \in A_j^{(n)} \cap B_n, \; 1 \leq j \leq N. \right.$$

For the last result, if $f(x)$ is nonnegative and integrable, then $0 \leq \varphi_n(x) \leq f(x)$ will assure that each $\varphi_n(x)$ is integrable. But a simple function can be integrable by Definition 3.2 if and only if it is zero outside a set of finite measure. See Proposition 3.15 to formalize this observation. ■

The final result generalizes Proposition 2.28 to allow more control over the choice of the $A_j^{(n)}$-sets in the case where the measure space is constructed using the Hahn-Kolmogorov extension theorem of Proposition 1.18. Recall that in (1.8), an outer measure $\mu_{\mathcal{A}}^*$ is defined relative to an algebra $\mathcal{A}$ and a measure $\mu_{\mathcal{A}}$ on this algebra, that $\sigma(X) \equiv \mathcal{C}(X)$ is the complete sigma algebra of **Carathéodory-measurable sets** defined relative to $\mu_{\mathcal{A}}^*$, and that the measure μ is defined to equal $\mu_{\mathcal{A}}^*$ restricted to $\sigma(X)$.

Here, we assume as is often the case that the algebra $\mathcal{A}$ is generated by a semi-algebra $\mathcal{A}'$. Recalling Exercise I.6.10, $\mathcal{A}$ equals the collection of all finite disjoint unions of $\mathcal{A}'$-sets.

The result below then states that the simple functions used to approximate $f(x)$ can be constructed with characteristic functions of $\mathcal{A}'$-sets. The utility of being able to construct simple functions this way is that such $\mathcal{A}'$-sets are typically very simple and relatively easy to work with. See Example 2.34. The price of this utility is that the conclusion is now that $\psi_n \to f$, μ-a.e.

Proposition 2.33 (Simple functions with $\mathcal{A}'$-sets.) *Let $f(x)$ be a nonnegative $\sigma(X)$-measurable function defined on a measurable set D of a complete measure space $(X, \sigma(X), \mu)$, where $\sigma(X)$ is the complete sigma algebra of Carathéodory-measurable sets defined with respect to an outer measure $\mu_{\mathcal{A}}^*$ that reflects an algebra of sets $\mathcal{A}$ and a measure $\mu_{\mathcal{A}}$ on this algebra.*

If $\mathcal{A}$ is generated by a semi-algebra $\mathcal{A}'$, and either:

1. $\mu(D) < \infty$, *or,*
2. $(X, \sigma(X), \mu)$ *is* σ*-finite, or,*
3. $f(x)$ *is* μ*-integrable on* D,

then there is a sequence $\{\psi_n(x)\}_{j=1}^{\infty}$ *of simple functions so that:*

- *Each* $\psi_n(x)$ *is defined by characteristic functions of sets in* $\mathcal{A}'$;
- $\mu\{\psi_n(x) \neq 0\} < \infty$ *for all* n;
- $\psi_n(x) \to f(x)$ *for almost all* $x \in D$.

Proof. *If* $\{\varphi_n(x)\}_{j=1}^{\infty}$ *is the sequence constructed in Corollary 2.32 in these three cases, then each* $\varphi_n(x)$ *is defined on* $N = n2^n + 1$ *disjoint measurable* $A_j^{(n)}$*-sets of finite total measure. By Proposition I.6.5 (note notational change), for each* n *and* j*, there exists* $B_j^{(n)} \in \mathcal{A}_\sigma$*, the collection of countable unions of sets in the algebra* $\mathcal{A}$*, so that* $A_j^{(n)} \subset B_j^{(n)}$ *and*

$$\mu(B_j^{(n)} - A_j^{(n)}) < 1/2N^2, \tag{1}$$

with N *as above. Thus* $\mu(B_j^{(n)}) < \infty$ *by finite additivity since* $\mu(A_j^{(n)}) < \infty$.

Now $B_j^{(n)} \in \mathcal{A}_\sigma$ *implies that* $B_j^{(n)} = \bigcup_{k=1}^{\infty} B_{jk}^{(n)}$ *with* $B_{jk}^{(n)} \in \mathcal{A}$*, and hence:*

$$\bigcap\nolimits_M \left(B_j^{(n)} - \bigcup\nolimits_{k=1}^M B_{jk}^{(n)}\right) = \emptyset.$$

Since $\mu(B_j^{(n)}) < \infty$*, continuity from above of* μ *of Proposition 1.29 obtains that there is an* $M_j(n)$ *so that:*

$$\mu\left(B_j^{(n)} - \bigcup_{k=1}^{M_j(n)} B_{jk}^{(n)}\right) < 1/2N^2. \tag{2}$$

Subadditivity and monotonicity of μ*, and (1) and (2) obtain:*

$$\begin{aligned} &\mu\left[\left(\bigcup_{k=1}^{M_j(n)} B_{jk}^{(n)} - A_j^{(n)}\right) \bigcup \left(A_j^{(n)} - \bigcup_{k=1}^{M_j(n)} B_{jk}^{(n)}\right)\right] \\ \leq\ &\mu\left(B_j^{(n)} - A_j^{(n)}\right) + \mu\left(B_j^{(n)} - \bigcup_{k=1}^{M_j(n)} B_{jk}^{(n)}\right) \\ <\ &1/N^2. \end{aligned}$$

Thus for each j,

$$\mu\left(A_j^{(n)} \triangle \bigcup_{k=1}^{M_j(n)} B_{jk}^{(n)}\right) < 1/N^2, \tag{3}$$

where $A \triangle B$ *is the* ***symmetric set difference*** *of Definition I.4.1, and* $B_{jk}^{(n)} \in \mathcal{A}$ *for all* k*. Since the algebra* $\mathcal{A}$ *is closed under finite unions,* $\bigcup_{k=1}^{M_j(n)} B_{jk}^{(n)} \equiv C_j^{(n)} \in \mathcal{A}$ *for all* j *and* n*, and the estimate in (3) obtains for each* n:

$$\sum\nolimits_{j=1}^{N} \mu\left[A_j^{(n)} \triangle C_j^{(n)}\right] < 1/N. \tag{4}$$

Define $\psi_n(x)$ by:

$$\psi_n(x) = \left\{ \ (j-1)2^{-n}, \quad x \in C_j^{(n)}, \ 1 \leq j \leq N. \right.$$

Then:

$$|f(x) - \psi_n(x)| \leq |f(x) - \varphi_n(x)| + |\varphi_n(x) - \psi_n(x)|.$$

Now, $|f(x) - \varphi_n(x)| \to 0$ for all x by Corollary 2.32. For the second term:

$$\begin{aligned} \varphi_n(x) &= \sum\nolimits_{j=1}^{N} (j-1)2^{-n} \chi_{A_j^{(n)}}(x), \\ \psi_n(x) &= \sum\nolimits_{j=1}^{N} (j-1)2^{-n} \chi_{C_j^{(n)}}(x), \end{aligned} \tag{5}$$

and it follows by the estimate in (4):

$$\mu\{|\varphi_n(x) - \psi_n(x)| \neq 0\} \leq \sum\nolimits_{j=1}^{N} \mu\left[A_j^{(n)} \triangle C_j^{(n)}\right] < 1/N.$$

Hence, $|\varphi_n(x) - \psi_n(x)| \to 0$ μ-a.e.

In addition, $\mu\left(C_j^{(n)}\right) < \infty$ by (2) since $\mu(B_j^{(n)}) < \infty$, and thus by finite subadditivity $\mu\left(\bigcup_{j=1}^{N} C_j^{(n)}\right) < \infty$. To complete the proof, recall that sets in $\mathcal{A}$ are finite disjoint unions of $\mathcal{A}'$-sets, so $C_j^{(n)} = \bigcup_{k=1}^{N_j(n)} C_{jk}^{(n)}$ with $C_{jk}^{(n)} \in \mathcal{A}'$. Thus:

$$\chi_{C_j^{(n)}}(x) = \sum\nolimits_{k=1}^{N_j(n)} \chi_{C_{jk}^{(n)}}(x),$$

and $\psi_n(x)$ in (5) can be rewritten as a simple function with $\mathcal{A}'$-sets. ■

Example 2.34 *If $(X, \sigma(X), \mu) = (\mathbb{R}, \mathcal{M}_L, m)$, Lebesgue measure space, or $(X, \sigma(X), \mu) = (\mathbb{R}, \mathcal{M}_{\mu_F}(\mathbb{R}), \mu_F)$, a Borel measure space, then X is σ-finite in either case. The above proposition applies with $\mathcal{A}'$ defined as the semi-algebra of right semi-closed intervals.*

Thus if $f(x)$ is a nonnegative measurable function defined on a measurable set D, there is a sequence $\{\psi_n(x)\}_{j=1}^{\infty}$ of simple functions defined in terms of characteristic functions of right semi-closed intervals $C_{jk}^{(n)} \equiv (a_{jk}^{(n)}, b_{jk}^{(n)}] \in \mathcal{A}'$, so that $\psi_n(x) \to f(x)$ for almost all $x \in D$, and each $\psi_n(x)$ is zero outside a set of finite measure. This last observation assures that all such $(a_{jk}^{(n)}, b_{jk}^{(n)}]$-intervals are bounded in the Lebesgue and general Borel measure applications.

2.5 Monotone Class Theorems

In this section, we study the **monotone class theorem**, a characterizing result on sigma algebras attributed to **Paul Halmos** (1916–2006). This result then gives rise to the **functional monotone class theorem**, which proves that if a generally defined class of functions satisfies a few relatively easily verifiable properties, then it contains all bounded, measurable functions.

Though generally applicable, these results will be very important in the study of stochastic processes and the associated integration theory in Books VII–VIII.

2.5.1 Monotone Class Theorem

We begin with a definition of a **monotone class of sets.**

Definition 2.35 (Monotone class of sets) *A finite or countable collection $\{A_j\}$ is* ***monotone*** *if either:*

1. ***Monotone increasing:*** $A_j \subset A_{j+1}$ *for all j, or,*

2. ***Monotone decreasing:*** $A_{j+1} \subset A_j$ *for all j.*

A nonempty class of sets M is a ***monotone class*** *if given any monotone collection* $\{A_j\} \subset M$*:*

1. ***Monotone increasing:*** $\lim A_j \equiv \bigcup_j A_j \in M$,

2. ***Monotone decreasing:*** $\lim A_j \equiv \bigcap_j A_j \in M$.

Example 2.36 *1. Every sigma algebra is a monotone class since it is closed under all countable unions and intersections, not just unions and intersections of monotone collections of sets.*

2. *An algebra $\mathcal{A}$ that is a monotone class is in fact a sigma algebra, since given $\{A_j\}_{j=1}^{\infty} \subset \mathcal{A}$, the collection $\left\{\bigcup_{k=1}^{j} A_k\right\}_{j=1}^{\infty}$ is monotone increasing. Hence, if $\mathcal{A}$ is a monotone class:*
$$\bigcup_j \left[\bigcup_{k=1}^{j} A_k\right] = \bigcup_{j=1}^{\infty} A_j \in \mathcal{A},$$
and so $\mathcal{A}$ is closed under countable unions and is thus a sigma algebra.

3. *A monotone class need not be a sigma algebra. For example, let $\mathcal{C} = \{A \subset \mathbb{R} | A$ is countable$\}$. Then $\mathcal{C}$ is a monotone class since any countable union or intersection of countable sets is countable. But $\mathcal{C}$ is not a sigma algebra since, for example, $\mathcal{C}$ is not closed under complements. In fact, $\mathcal{C}$ is not even a semi-algebra.*

Exercise 2.37 (Intersection of monotone classes) *If $\{M_i\}$ is a finite or countable collection of monotone classes, define:*
$$M \equiv \bigcap_i M_i = \{A | A \in M_i \text{ for all } i\}.$$
Prove that M is a monotone class unless $M = \emptyset$.

The following definition reflects another construction of the type illustrated in Example 1.3. This notion is again well-defined because $\mathcal{P}(X)$, the collection of all subsets of X, is a monotone class that contains E, Then, the intersection of all such monotone classes is a monotone class by Exercise 2.37, since this intersection cannot be empty.

Definition 2.38 (Monotone class generated by E) *Given a collection E of subsets of a space X, define $M(E)$* ***as the smallest monotone class that contains*** *E. The monotone class $M(E)$ is also called* ***the monotone class generated by*** *E.*

The monotone class theorem was originally stated and proved in the context of rings and σ-rings of sets, an alternative approach to measure theory from the algebras and σ-algebras of sets used in these books. We state and prove this result from the latter perspective.

Proposition 2.39 (Monotone class theorem) *If $\mathcal{A}$ is an algebra of sets, then $M(\mathcal{A}) = \sigma(\mathcal{A})$, the smallest sigma algebra that contains $\mathcal{A}$. Hence, any monotone class that contains $\mathcal{A}$ contains $\sigma(\mathcal{A})$.*

Proof. *By item 1 of Example 2.36, $M(\mathcal{A}) \subset \sigma(\mathcal{A})$ since $\sigma(\mathcal{A})$ is a monotone class. To complete the proof, we will show that $M(\mathcal{A})$ is an algebra, and hence by item 2 of this example, $M(\mathcal{A})$ is a sigma algebra and so $\sigma(\mathcal{A}) \subset M(\mathcal{A})$.*

To show that $M(\mathcal{A})$ is an algebra, we introduce new collections of subsets. For any set A, define $C(A) \subset M(\mathcal{A})$ by:

$$C(A) = \left\{B \in M(\mathcal{A}) | A - B,\ B - A,\ \text{and } A \bigcup B \text{ are in } M(\mathcal{A})\right\}. \tag{1}$$

We first show that for any A, $C(A)$ is a monotone class if it is not empty.

If $\{B_j\}_{j=1}^{\infty} \subset C(A)$ is a monotone sequence, then since $\{A \bigcup B_j\}_{j=1}^{\infty} \subset M(\mathcal{A})$ is also monotone it follows that $\lim (A \bigcup B_j) \in M(\mathcal{A})$, where this limit is given as in Definition 2.35. In either case 1 or 2 of this definition:

$$\lim \left(A \bigcup B_j\right) = A \bigcup \lim B_j,$$

and so $A \bigcup \lim B_j \in M(\mathcal{A})$. A similar argument shows that $A - \lim B_j$ and $\lim B_j - A$ are also in $M(\mathcal{A})$, and thus by definition $\lim B_j \in C(A)$. So $C(A)$ is a monotone class if it is not empty, and this conclusion is true for any set A.

Given $A \in \mathcal{A}$, then $B \in C(A)$ for all $B \in \mathcal{A}$ since algebras are closed under the finite operations in (1). Thus $\mathcal{A} \subset \mathcal{C}(A)$ for any such $A \in \mathcal{A}$, and since $C(A)$ is a monotone class and $M(\mathcal{A})$ is the smallest monotone class containing $\mathcal{A}$, this obtains that $M(\mathcal{A}) \subset C(A)$ for any $A \in \mathcal{A}$. It then follows that if $A \in M(\mathcal{A})$ and $B \in \mathcal{A}$, then $A \in C(B)$ and by symmetry in (1), $B \in C(A)$. Thus $\mathcal{A} \subset \mathcal{C}(A)$, and again as $C(A)$ is a monotone class, $M(\mathcal{A}) \subset C(A)$ for any $A \in M(\mathcal{A})$.

We now can show that $M(\mathcal{A})$ is an algebra. If A, $B \in M(\mathcal{A})$, then $A \in C(B)$ implies that $A \bigcup B \in M(\mathcal{A})$. Similarly, given $A \in M(\mathcal{A})$, then $A \bigcup \widetilde{A} \in \mathcal{A} \subset M(\mathcal{A})$, and it follows from above that $A \in C(A \bigcup \widetilde{A})$. Thus $A \bigcup \widetilde{A} - A = \widetilde{A} \in M(\mathcal{A})$, and so $M(\mathcal{A})$ is an algebra.
■

Example 2.40 (Uniqueness of extensions of measures) *Proposition 1.22 stated that the Proposition 1.18 extension of a sigma finite measure from an algebra to a sigma algebra is effectively unique. Originally proved as Proposition I.6.14, we provide another proof using the monotone class theorem.*

Proposition 1.22: *Let $\mu_{\mathcal{A}}$ be a sigma finite measure on an algebra $\mathcal{A}$, and μ the Proposition 1.18 extension of $\mu_{\mathcal{A}}$ to $\mathcal{C}(X)$ induced by the outer measure $\mu_{\mathcal{A}}^*$. By extension is meant that $\mu(A) = \mu_{\mathcal{A}}(A)$ for all $A \in \mathcal{A}$.*

If μ' is any other extension of $\mu_{\mathcal{A}}$, then $\mu(B) = \mu'(B)$ for all $B \in \sigma(\mathcal{A})$, where $\sigma(\mathcal{A}) \subset \mathcal{C}(X)$ denotes the smallest sigma algebra that contains $\mathcal{A}$.

Proof. *To apply the monotone class theorem, we prove this result by showing that the class of sets on which $\mu = \mu'$ is a monotone class, and since this class contains the algebra $\mathcal{A}$ by assumption, it must contain $\sigma(\mathcal{A})$.*

Since $\mu_{\mathcal{A}}$ is sigma finite, Definition 1.21 obtains that the measure space X can be expressed as a countable union of $\mathcal{A}$-sets of finite measure:

$$X = \bigcup\nolimits_{j=1}^{\infty} X_j, \quad \mu_{\mathcal{A}}(X_j) < \infty.$$

We can assume that such sets are nested by replacing these with finite unions, $X'_k \equiv \bigcup_{j=1}^{k} X_j$, so that $X'_k \subset X'_{k+1}$.

Returning to the original notation obtains that $\mu(X_j) = \mu'(X_j)$ *for all* j. *Assume that it can be shown that for every* j *and all* $A \in \sigma(\mathcal{A})$:

$$\mu\left(X_j \bigcap A\right) = \mu'\left(X_j \bigcap A\right). \tag{1}$$

Then by nesting and continuity from below of measures, (1) *will obtain that for all* $A \in \sigma(\mathcal{A})$:

$$\mu(A) = \lim_{j\to\infty} \mu\left(X_j \bigcap A\right) = \lim_{j\to\infty} \mu'\left(X_j \bigcap A\right) = \mu'(A).$$

To prove (1), *fix* j *and define:*

$$\mathcal{C} = \left\{A \in \sigma(\mathcal{A}) | \mu\left(X_j \bigcap A\right) = \mu'\left(X_j \bigcap A\right)\right\}.$$

Then $\mathcal{A} \subset \mathcal{C}$ *by assumption, and to show that* $\mathcal{C}$ *is a monotone class, let* $\{A_k\} \subset \mathcal{C}$ *be a monotone increasing sequence. By continuity from below and the definition of* $\mathcal{C}$:

$$\mu\left(X_j \bigcap \lim_k A_k\right) = \lim_k \mu\left(X_j \bigcap A_k\right) = \mu'\left(X_j \bigcap \lim_k A_k\right),$$

and so $\lim A_k \in \mathcal{C}$. *If this sequence is monotone decreasing, continuity from above is applicable because* μ *and* μ' *are finite measures restricted to* X_j, *and thus again* $\lim A_k \in \mathcal{C}$.

Hence, $\mathcal{C}$ *is a monotone class that contains* $\mathcal{A}$, *and it now follows from Proposition 2.39 that* $\sigma(\mathcal{A}) \subset \mathcal{C}$. ■

2.5.2 Functional Monotone Class Theorem

As noted in the introduction, there is a functional counterpart to the monotone class theorem known as the **functional monotone class theorem.** This result allows one to conclude that if a given class of functions satisfies a few, often easily verified properties, then it contains all **bounded measurable functions**. This result is useful and will be applied in forthcoming books in the study of Markov processes in Book VII, and the development of stochastic integrals in Book VIII.

For this result, the sigma algebra $\sigma(\mathcal{A}')$ identified in the statement is the same as the sigma algebra $\sigma(\mathcal{A})$ used above when the algebra $\mathcal{A}$ is generated by $\mathcal{A}'$. This follows because $\mathcal{A}$ is the collection of finite unions of $\mathcal{A}'$-sets, and thus $\mathcal{A} \subset \sigma(\mathcal{A}')$ and then $\sigma(\mathcal{A}') = \sigma(\mathcal{A})$ by definition.

The sigma algebra $\sigma(\mathcal{A}')$ is identified in item 1, and in applications this specification simplifies verification by requiring only $\mathcal{A}'$-sets be considered, rather than all $\mathcal{A}$-sets. Since a semi-algebra need not include the space X, item 1 requires separately that $\chi_X \in \mathcal{L}$.

Proposition 2.41 (Functional monotone class theorem) *Let* $(X, \sigma(X), \mu)$ *be a measure space and* $\mathcal{A}'$ *a semi-algebra that generates* $\sigma(X)$, *meaning that* $\sigma(X) = \sigma(\mathcal{A}')$, *the smallest sigma algebra that contains* $\mathcal{A}'$.

Let $\mathcal{L}$ *denote a class of functions with the following properties:*

1. $\chi_A \in \mathcal{L}$ *for all* $A \in \mathcal{A}'$, *and* $\chi_X \in \mathcal{L}$.
2. $\mathcal{L}$ *is a vector space: If* $f, g \in \mathcal{L}$ *then* $af + bg \in \mathcal{L}$ *for all* $a, b \in \mathbb{R}$.
3. *If* $f : X \to \mathbb{R}^+$ *is bounded and the pointwise limit of* $\{f_n\}_{n=1}^{\infty} \subset \mathcal{L}$, *then* $f \in \mathcal{L}$.

Then $\mathcal{L}$ contains all bounded $\sigma(X)$-measurable functions defined on X.

Proof. *We first show that $\chi_A \in \mathcal{L}$ for all $A \in \mathcal{A}$, the algebra generated by $\mathcal{A}'$. Let $\mathcal{K}$ denote the class of all $A \in \sigma(X)$ such that $\chi_A \in \mathcal{L}$. Then by item 1, $\mathcal{A}' \subset \mathcal{K}$ and $X \in \mathcal{K}$. If $A = \bigcup_{k=1}^{n} A_k \in \mathcal{A}$, a disjoint union of $\{A_k\}_{k=1}^{n} \subset \mathcal{A}'$, then $\chi_A \equiv \sum_{k=1}^{j} \chi_{A_k} \in \mathcal{L}$ by items 1 and 2, and hence $A \in \mathcal{K}$. Letting $a = b = 0$ in 2, it follows that $0 = \chi_\emptyset \in \mathcal{L}$ and so $\emptyset \in \mathcal{K}$. This proves that $\mathcal{A} \subset \mathcal{K}$.*

We next show that $\chi_A \in \mathcal{L}$ for all $A \in \sigma(X)$, and thus $\sigma(X) \subset \mathcal{K}$. Let $\{A_k\}_{k=1}^{\infty} \subset \mathcal{K}$ be a monotone sequence. Then $A = \lim A_j \in \mathcal{K}$ by item 3 because $\chi_A = \lim_{j\to\infty} \chi_{A_j} \in \mathcal{L}$. Thus $\mathcal{K}$ is a monotone class that contains the algebra $\mathcal{A}$, and by Proposition 2.39, $\mathcal{K}$ contains the sigma algebra $\sigma(\mathcal{A}) = \sigma(X)$, as noted above. Hence, $\mathcal{L}$ contains the characteristic functions of all sets in $\sigma(X)$.

If f is a bounded $\sigma(X)$-measurable function, write $f = f^+ - f^-$, with f^+ and f^- nonnegative and bounded and defined in (3.20) and (3.21). By item 2, $f \in \mathcal{L}$ if f^+, $f^- \in \mathcal{L}$. To simplify notation, assume that f is a nonnegative, bounded measurable function. If f has finite range $\{y_j\}_{j=1}^{n}$, then with $A_j = f^{-1}(y_j)$,

$$f(x) = \sum\nolimits_{j=1}^{n} y_j \chi_{A_j}(x),$$

and so $f \in \mathcal{L}$ by item 2 because $A_j \in \sigma(X)$.

More generally, for a nonnegative, bounded $\sigma(X)$-measurable function f, define:

$$f_n(x) = 2^{-n} \lfloor 2^n f(x) \rfloor .$$

*Here, $\lfloor 2^n f(x) \rfloor$ denotes the **floor function** or **greatest integer function,** defined as the greatest integer less than or equal to $2^n f(x)$. Since f is nonnegative and bounded, $f_n(x)$ is finite valued and so $f_n \in \mathcal{L}$ for all n. But $\lfloor 2^n f(x) \rfloor = 2^n f(x) - \varepsilon_n(x)$ with $0 \leq \varepsilon_n(x) < 1$, and so $f_n(x) = f(x) - 2^{-n}\varepsilon_n(x)$. Consequently, $f_n(x) \to f(x)$ pointwise, and by item 3, $f \in \mathcal{L}$.* ■

Exercise 2.42 ($\chi_A \in \mathcal{L}$ **for all** $A \in \mathcal{A}$) *Provide another proof that $\chi_A \in \mathcal{L}$ for all $A \in \mathcal{A}$. Hint: For $\{A_j\}_{j=1}^{n} \subset \mathcal{A}'$ and $A = \bigcup_{j=1}^{n} A_j$, prove that:*

$$\chi_A = \chi_X - \prod\nolimits_{j=1}^{n} \left(1 - \chi_{A_j}\right).$$

The right-hand expression can be expanded as a linear combination of χ_{B_k} for $B_k \in \mathcal{A}'$, since:

$$\chi_{B_k} = \prod\nolimits_{k=1}^{n_k} \chi_{A_{j_k}},$$

for $B_k \equiv \bigcap_{k=1}^{n_k} A_{j_k}$.

Exercise 2.43 (Generalize to 3′: Increasing $\{f_n\}_{n=1}^{\infty} \subset \mathcal{L}$**)** *Revise the above proof if item 3 above is changed to:*

*3′. If $f : X \to \mathbb{R}^+$ is bounded and the pointwise limit of **increasing** $\{f_n\}_{n=1}^{\infty} \subset \mathcal{L}$, then $f \in \mathcal{L}$. Hint: Recall Proposition 2.28, on approximations by simple functions.*

3

General Integration Theory

In this chapter, we generalize the integration theory of the Lebesgue measure space $(\mathbb{R}^n, \mathcal{M}_L(\mathbb{R}^n), m)$ of Book III to Borel measure spaces $(\mathbb{R}^n, \mathcal{M}_F(\mathbb{R}^n), \mu_F)$, in which case these integrals are known as **Lebesgue-Stieltjes integrals**, and to more general measure spaces $(X, \sigma(X), \mu)$. This latter collection of spaces includes finite products of measure spaces $\{(X_i, \sigma(X_i), \mu_i)\}_{i=1}^n$ constructed in Chapter I.7 :

$$(X, \sigma(X), \mu) \equiv \left(\prod_{i=1}^n X_i, \sigma\left(\prod_{i=1}^n X_i\right), \prod_{i=1}^n \mu_i\right),$$

in which case the integrals are known as **product measure integrals** or **product space integrals**.

The development of the **Lebesgue integral** of Book III followed the sequential steps:

1. Define the Lebesgue integral of simple functions that equal zero outside sets of finite measure.

2. Extend this definition to bounded measurable functions that again equal zero outside sets of finite measure, using "limits" of integrals of subordinate and dominant simple functions defined in step 1. It was then seen that a bounded function on a set E with $m(E) < \infty$ was Lebesgue integrable if and only if it was Lebesgue measurable.

3. Extend the definition to nonnegative Lebesgue measurable functions based on the integrals of bounded functions from step 2.

4. Extend the definition of Lebesgue integral to general measurable functions, splitting such functions into positive and negative parts and applying the results of step 3.

Along the way, important "integration to the limit" results for the Lebesgue integral were developed. Given a sequence of functions $\{f_j(x)\}_{j=1}^\infty$ that converge pointwise to a function $f(x)$, such results relate integrability of this limit function, and the value of this integral, to limits of the integrals of the function sequence. Such results are fundamental to the applications of the theory.

The above development reflected the conventional approach to the Lebesgue theory. However, we could well have jumped from step 1 to step 3, defining the integral of nonnegative functions directly in terms of the integrals of simple functions and done this without the restriction that simple functions are zero outside sets of finite measure. In a general measure space, the Lebesgue approach must in fact be modified in this way, because any restriction to functions that are zero outside a set of finite measure may create a counter intuitive result.

Example 3.1 *Consider a measure space $(X, \sigma(X), \mu)$ with the trivial sigma algebra $\sigma(X) = \{\emptyset, X\}$ and μ defined by $\mu(\emptyset) = 0$, $\mu(X) = \infty$. An example of this is Lebesgue measure $\mu = m$ defined on the trivial sigma algebra on $\mathbb{R}$. Logically one expects that any integration theory will obtain:*

$$\int_X 1 d\mu = \mu(X) = \infty,$$

DOI: 10.1201/9781003264576-3

but we cannot derive this conclusion from the sequence of steps used above. The problem is that there are no simple functions defined to be zero outside a set of finite measure.

This is indeed an extreme example, and one that cannot occur on a σ-finite measure space. In such a space, simple functions can be defined on sets of finite measure by definition.

Rather than restrict the current development to σ-finite measure spaces, we use a general measure space $(X, \sigma(X), \mu)$ and adapt the above approach. In this general setting, we abandon the notion that a general domain $D \subset X$ with $\mu(D) = \infty$ contains measurable domains with finite measure or that a given function $f(x)$ defined on X can be approximated with simple functions defined on such domains.

Consequently, the approach taken in this chapter will be to:

1. Define the μ-integrals of simple functions without restrictions on the μ-measure of their domains.

2. Extend this definition to nonnegative $\sigma(X)$-measurable functions, using the μ-integrals of simple functions of step 1.

3. Extend the definition of μ-integral to general $\sigma(X)$-measurable functions, splitting such functions into positive and negative parts and applying the results of step 2.

Along the way, we will develop various important integration to the limit theorems that will be fundamental to the applications of this theory.

3.1 Integrating Simple Functions

Simple functions were characterized in (2.9) of Definition 2.26:

$$\varphi(x) = \sum\nolimits_{i=1}^{n} a_i \chi_{A_i}(x),$$

there $\{A_i\}_{i=1}^n \subset \sigma(X)$ are disjoint μ-**measurable sets**, and $a_i \geq 0$ for all i.

Next, we **define** the μ-integral of a simple function, which of necessity will require an investigation into well-definedness. While $a_i \geq 0$ for all i assures that $\int \varphi(x)d\mu \geq 0$, the integral as defined need not be finite since there is no definitional restriction on the μ-measures of the disjoint sets $\{A_i\}_{i=1}^n$.

Definition 3.2 ($\int \varphi(x)d\mu$) *Given a simple function $\varphi(x)$ in (2.9),* ***the μ-integral of*** *$\varphi(x)$ is defined as:*

$$\int \varphi(x)d\mu \equiv \sum\nolimits_{i=1}^{n} a_i \mu(A_i). \tag{3.1}$$

Before proceeding further, we must show that this definition is well-defined since a simple function can be represented with infinitely many choices of coefficients $\{a_i\}_{i=1}^n$ and sets $\{A_i\}_{i=1}^n$. In addition, we show that the value of this integral will not be changed by redefinitions of $\varphi(x)$ on sets of μ-measure 0.

The following result echoes Proposition III.2.7 in the Lebesgue development but must address the generalization that the A_i-sets need not have finite measure.

Proposition 3.3 (Well-definedness of $\int \varphi(x)d\mu$) *With the simple function $\varphi(x)$ defined above, assume that $\varphi(x) = \psi(x)$ μ-a.e., where the simple function $\psi(x)$ is defined with disjoint $\{A'_j\}_{j=1}^m$ and $a'_j \geq 0$ for all j by:*

$$\psi(x) = \sum\nolimits_{j=1}^{m} a'_j \chi_{A'_j}(x).$$

Then:

$$\int \varphi(x)d\mu = \int \psi(x)d\mu,$$

meaning:

$$\sum\nolimits_{j=1}^{m} a'_j \mu(A'_j) = \sum\nolimits_{i=1}^{n} a_i \mu(A_i). \tag{3.2}$$

Further, if $a_i > 0$ and $a'_j > 0$ for all i, j :

$$\mu\left(\bigcup\nolimits_{j=1}^{m} A'_j - \bigcup\nolimits_{i=1}^{n} A_i\right) = \mu\left(\bigcup\nolimits_{i=1}^{n} A_i - \bigcup\nolimits_{j=1}^{m} A'_j\right) = 0.$$

Proof. *For any i, j define $B_{i,j} \equiv A_i \bigcap A'_j$, and also define $B_{i,0} \equiv A_i \bigcap \left(\bigcup\nolimits_{j=1}^{m} A'_j\right)^c$ and $B_{0,j} \equiv \left(\bigcup\nolimits_{i=1}^{n} A_i\right)^c \bigcap A'_j$ where $A^c \equiv \widetilde{A}$. Then the full collection $\{B_{i,j}\}_{i,j}$ are disjoint, and:*

$$A_i = \bigcup\nolimits_{j=0}^{m} B_{i,j}, \quad A'_j = \bigcup\nolimits_{i=0}^{n} B_{i,j}.$$

By finite additivity,

$$\mu(A_i) = \sum\nolimits_{i=0}^{m} \mu\left[B_{i,j}\right], \quad \mu\left(A'_j\right) = \sum\nolimits_{i=0}^{n} \mu\left[B_{i,j}\right],$$

and since $\varphi(x) = \psi(x)$ μ-a.e., it follows that for every i either $\mu\left[B_{i,0}\right] = 0$ or $a_i = 0$, and similarly, for every j either $\mu\left[B_{0,j}\right] = 0$ or $a'_j = 0$. Thus:

$$\begin{aligned}
\sum\nolimits_{i=1}^{n} a_i \mu(A_i) &= \sum\nolimits_{i=1}^{n} \sum\nolimits_{j=0}^{m} \mu\left[B_{i,j}\right] a_i = \sum\nolimits_{i=1}^{n} \sum\nolimits_{j=1}^{m} \mu\left[B_{i,j}\right] a_i, \\
\sum\nolimits_{j=1}^{m} a'_j \mu(A'_j) &= \sum\nolimits_{j=1}^{m} \sum\nolimits_{i=0}^{n} \mu\left[B_{i,j}\right] a'_j = \sum\nolimits_{j=1}^{m} \sum\nolimits_{i=1}^{n} \mu\left[B_{i,j}\right] a'_j.
\end{aligned}$$

The identity in (3.2) now follows. First, both sums are finite or infinite together. For example, the first sum is infinite if and only if at least one $\mu(A_i) = \infty$, and this occurs if and only if $\mu\left[B_{i,j}\right] = \infty$ for at least one j, which implies that $\mu(A'_j) = \infty$ and the second sum is infinite. When both sums are finite, they must agree because for every i, j, either $\mu\left[B_{i,j}\right] = 0$, or, $\mu\left[B_{i,j}\right] > 0$ and then of necessity $a_i = a'_j$.

If $a_i > 0$ and $a'_j > 0$ for all i, j, then as noted above, $\mu\left[B_{i,0}\right] = 0$ all i and $\mu\left[B_{0,j}\right] = 0$ all j. Thus:

$$0 = \mu\left[\bigcup\nolimits_{i=1}^{n} B_{i,0}\right] = \mu\left[\left(\bigcup\nolimits_{i=1}^{n} A_i\right) \bigcap \left(\bigcup\nolimits_{j=1}^{m} A'_j\right)^c\right],$$

and this last expression is $\mu\left(\bigcup\nolimits_{j=1}^{m} A'_j - \bigcup\nolimits_{i=1}^{n} A_i\right)$. The same conclusion applies to $\mu\left(\bigcup\nolimits_{i=1}^{n} A_i - \bigcup\nolimits_{j=1}^{m} A'_j\right)$. ■

Example 3.4 (Lebesgue/Borel measures) *We exemplify (3.1) in the context of the Lebesgue and Borel measure spaces discussed in Chapter 1.*

1. *If* $(X,\sigma(X),\mu) = (\mathbb{R},\mathcal{M}_{\mu_F}(\mathbb{R}),\mu_F)$*, a Borel measure space on* $\mathbb{R}$ *as in Section 1.1.3, let* $\{(a_j,b_j]\}_{j=1}^m \subset \mathcal{M}_{\mu_F}(\mathbb{R})$ *be any disjoint collection of right semi-closed intervals. If* $\varphi(x) = \sum_{j=1}^m c_j\chi_{(a_j,b_j]}(x)$*, then since* $\mu_F[(a_j,b_j]] = F(b_j) - F(a_j)$ *:*

$$\int \varphi(x)d\mu_F = \sum\nolimits_{j=1}^m c_j\left[F(b_j) - F(a_j)\right].$$

If $F(x) = x$*, then this obtains the Lebesgue integral* $(\mathcal{L})\int\varphi(x)dx$ *of Definition III.2.4 since* $m[(a_j,b_j]] = b_j - a_j$*.*

2. *If* $(X,\sigma(X),\mu) = (\mathbb{R}^n,\mathcal{M}(\mathbb{R}^n),\mu_F)$ *with* $\mu_F \equiv \prod_{i=1}^n \mu_{F_i}$*, a product measure from the Borel measure spaces* $\{(\mathbb{R},\mathcal{M}_{\mu_{F_i}}(\mathbb{R}),\mu_{F_i})\}_{i=1}^n$ *as in Section 1.3.1, let* $\{A_j\}_{j=1}^m \subset \mathcal{M}(\mathbb{R}^n)$ *be any disjoint collection of right semi-closed rectangles with* $A_j \equiv \prod_{i=1}^n (a_{ji},b_{ji}]$*. If* $\psi(x) = \sum_{j=1}^m c_j\chi_{A_j}(x)$*, then since* $\mu_F\left[\prod_{i=1}^n (a_{ji},b_{ji}]\right] = \prod_{i=1}^n\left[F_i(b_{ji}) - F_i(a_{ji})\right]$ *:*

$$\int \psi(x)d\mu_F = \sum\nolimits_{j=1}^m c_j\prod\nolimits_{i=1}^n\left[F_i(b_{ji}) - F_i(a_{ji})\right].$$

Again, if $F_i(x) = x$ *for all* i*, then this obtains the Lebesgue integral* $(\mathcal{L})\int\varphi(x)dx$ *of Definition III.2.4 since* $m[A_j] = \prod_{i=1}^n (b_{ji} - a_{ji})$*.*

3. *If* $(X,\sigma(X),\mu) = (\mathbb{R}^n,\mathcal{M}_F(\mathbb{R}^n),\mu_F)$*, a general Borel measure space of Section 1.3.2 induced by a continuous from above and* n*-increasing function* $F(x_1,...,x_n)$*, let* $\{A_j\}_{j=1}^m \subset \mathcal{M}_F(\mathbb{R}^n)$ *be any disjoint collection of right semi-closed rectangles as in item 2. If* $\psi(x) = \sum_{j=1}^m c_j\chi_{A_j}(x)$*, then since:*

$$\mu_F\left[\prod\nolimits_{i=1}^n (a_{ji},b_{ji}]\right] = \sum\nolimits_{x_{j_k}} sgn(x_{j_k})F(x_{j_k}),$$

it follows that:

$$\int \psi(x)d\mu_F = \sum\nolimits_{j=1}^m c_j\sum\nolimits_{x_{j_k}} sgn(x_{j_k})F(x_{j_k}).$$

Here, for each j*, the summation over* x_{j_k} *denotes the sum over the* 2^n *vertices of* A_j*, with* $sgn(x_{j_k}) \equiv -1$ *if the number of components of* x_{j_k} *equal to* a_{ji} *is odd, and* $sgn(x_{j_k}) = 1$ *otherwise.*

By Exercise 1.25, if $F(x) = \prod_{i=1}^n F_i(x_i)$*, this obtains the product space integral of item 2, whereas if* $F(x) = \prod_{i=1}^n x_i$*, this obtains the Lebesgue integral* $(\mathcal{L})\int\varphi(x)dx$ *of Definition III.2.4.*

We now establish a couple of properties of integrals of simple functions. For the second part of item 1, this upper bound is only useful when $\mu(\bigcup_{i=1}^n A_i) < \infty$, which is not required by the definition of simple function.

Proposition 3.5 (Properties of integrals of simple functions) *Let* $\varphi(x)$ *and* $\psi(x)$ *be simple functions defined on the measure space* $(X,\sigma(X),\mu)$*. Then:*

1. *If* $\varphi(x) \le \psi(x)$ *except on a set of* μ*-measure* 0*:*

$$\int \varphi(x)d\mu \le \int \psi(x)d\mu. \tag{3.3}$$

Thus with $a \equiv \max a_i$ *and* $A \equiv \bigcup_{i=1}^n A_i$*:*

$$\int \varphi(x)d\mu \le a\mu(A).$$

2. For any nonnegative constants a and b :

$$\int [a\varphi(x) + b\psi(x)]d\mu = a\int \varphi(x)d\mu + b\int \psi(x)d\mu. \tag{3.4}$$

Proof. *Left as an exercise. For part 2, decompose $a\varphi(x) + b\psi(x)$ as in the proof of Proposition 3.3.* ■

Corollary 3.6 (Nondisjoint $\{A_i\}_{i=1}^n$) *The integral of a simple function is well-defined by (3.1) even if the measurable collection $\{A_i\}_{i=1}^n$ is not disjoint.*

Proof. *This follows from part 2 of the prior result since then $\varphi(x) = \sum_{j=1}^m a_j\varphi_j(x)$ with $\varphi_j(x) \equiv \chi_{A_j}(x)$.* ■

We end this section by introducing the definition of the integral of a simple function over a μ-measurable set E. This definition will be applicable in the general setting below, so it is stated here in that general notational context. The open question of existence of such integrals will be addressed below. See Definition 3.11 and item 3 of Remark 3.12 for an example of how this definition will be utilized in a more general context.

Given the current definitional vagueness, it may seem odd that we are restricting this definition to μ-measurable $f(x)$ rather than to a more general function. However, the reader can verify that even in the case of a simple function, if defined with nonmeasurable $\{A_i\}_{i=1}^n$, the integral in (3.1) would never be well-defined. The same goes for the required measurability of E.

Definition 3.7 ($\int_E f(x)d\mu$) *Given the measure space $(X, \sigma(X), \mu)$, let E be a measurable set and $\chi_E(x)$ the* ***characteristic function*** *of E. If $f(x)$ a $\sigma(X)$-measurable function, define :*

$$\int_E f(x)d\mu \equiv \int \chi_E(x)f(x)d\mu, \tag{3.5}$$

when the integral on the right exists.

Extending the convention of Definition III.2.9, when $(X, \sigma(X), \mu) = (\mathbb{R}, \sigma(\mathbb{R}), \mu)$ and $E = [a, b]$, it is customary to write:

$$\int_{[a,b]} f(x)d\mu = \int_a^b f(x)d\mu.$$

Remark 3.8 (On $\int_b^a f(x)d\mu$, $a < b$) *Recalling Remark III.1.35 on the Riemann integral, which could well have been repeated after Proposition III.3.62 for the Lebesgue integral, when $a < b$ one defines for either integral:*

$$\int_b^a f(x)dx = -\int_a^b f(x)dx.$$

When $f(x) = g'(x)$ for integrable $g'(x)$, this definition is compelled by the respective fundamental theorems of Propositions III.1.30 and III.3.62.

Similarly, such integrals can be encountered on $\mathbb{R}$ when the interval $E = [a, b]$ is transformed by a decreasing function g, and one seeks to evaluate $\int_{g^{-1}(E)} f(x)d\mu$. Thus we continue this convention in this general case and define:

$$\int_b^a f(x)d\mu = -\int_a^b f(x)d\mu. \tag{3.6}$$

Proposition 3.9 **($\int_E \varphi(x)d\mu$ is well-defined)** *Let $\varphi(x)$ be a simple function as in (2.9). Then (3.5) is well-defined for all $E \in \sigma(X)$.*

Further, if $\mu(E) = 0$:

$$\int_E \varphi(x)d\mu = 0. \tag{3.7}$$

Proof. *By (3.5) and (3.1):*

$$\int_E \varphi(x)d\mu \equiv \int \chi_E(x)\varphi(x)d\mu = \sum\nolimits_{i=1}^{n} a_i\mu\left(A_i \bigcap E\right). \tag{3.8}$$

Since $\{A_i \bigcap E\}_{i=1}^n$ are disjoint and $A_i \bigcap E \in \sigma(X)$ for all i, this integral is well-defined and (3.7) follows. ∎

Exercise 3.10 **($E \equiv E_1 \bigcup E_2$)** *Show that if E_1 and E_2 are disjoint and measurable and $E \equiv E_1 \bigcup E_2$, then for any simple function $\varphi(x)$:*

$$\int_E \varphi(x)d\mu = \int_{E_1} \varphi(x)d\mu + \int_{E_2} \varphi(x)d\mu.$$

3.2 Integrating Nonnegative Measurable Functions

Continuing with the program outlined in the introduction, we start with the definition:

Definition 3.11 **($\int_E f(x)d\mu$, nonnegative $f(x)$)** *Let $f(x)$ be a nonnegative, $\sigma(X)$-measurable, extended real-valued function defined on a measurable set E of a measure space $(X, \sigma(X), \mu)$. Define the **the μ-integral of $f(x)$ over E** by:*

$$\int_E f(x)d\mu = \sup_{\varphi \leq f} \int_E \varphi(x)d\mu, \tag{3.9}$$

*where $\varphi(x)$ denotes a **simple function**, and the integrals on the right are given in Definitions 3.2 and 3.7.*

*When this supremum is finite, $f(x)$ is said to be **μ-integrable.** When the supremum is infinite, so $\int_E f(x)d\mu = \infty$, then $f(x)$ is said to be **not μ-integrable.***

Remark 3.12 *Several comments on the above definition are warranted.*

1. ***On $\varphi \leq f$:*** *The observant reader may recognize the Darboux approach in (3.9) as discussed in Remark III.1.8 and utilized extensively in Book III. Named for **Jean-Gaston Darboux** (1842–1917), this approach typically requires that the supremum of the subordinate simple (or step) function integrals agrees with the infimum of the superior simple (or step) function integrals. But logically, this can only work if the function to be integrated is bounded.*

   Here, we use only the subordinate functions, because the nonnegative function $f(x)$ to be integrated is, in general, an extended real-valued function, and thus unbounded. Hence, superior simple functions need not exist, recalling that $\{a_i\}_{i=1}^n \subset \mathbb{R}$ and all $a_i \geq 0$ for simple functions by Definition 2.26. See Remark 3.13 for more on this.

Three observations on this approach:

(a) ***"Best" simple functions:*** *Given disjoint* $\{A_i\}_{i=1}^n \subset \sigma(X)$ *and* $\varphi(x) = \sum_{i=1}^n a_i \chi_{A_i}(x)$ *with* $\varphi(x) \leq f(x)$*, for the purposes of the above definition, we are almost always better off using* $\widetilde{\varphi}(x) = \sum_{i=1}^n a_i' \chi_{A_i}(x)$ *where:*

$$a_i' \equiv \inf\{f(x)|x \in A_i\}.$$

By "better off" is meant that $\int_E \varphi(x)d\mu \leq \int_E \widetilde{\varphi}(x)d\mu$*, and thus this modification is beneficial to the above supremum goal. Further, given* $\{A_i\}_{i=1}^n$ *we can do no better than* $\widetilde{\varphi}(x)$ *and preserve* $\widetilde{\varphi}(x) \leq f(x)$*. Thus given* $\{A_i\}_{i=1}^n$*, such simple functions are the "best" to use in the above definition, though not required.*

(b) ***Refinements of best simple functions:*** *Given disjoint* $\{A_i\}_{i=1}^n \subset \sigma(X)$*, we call disjoint* $\{A_j'\}_{j=1}^m \subset \sigma(X)$ *a* ***refinement of*** $\{A_i\}_{i=1}^n$ *if for every i there exists* $\{A_{j_i}'\}$ *so that* $A_i = \bigcup A_{j_i}'$*. This is consistent with refinements of partitions of Book III.*

Then if $\varphi_1(x)$ *is defined in terms of* $\{a_i'\}_{i=1}^n$ *and* $\{A_i\}_{i=1}^n$ *as in item 1.a., and* $\varphi_2(x)$ *is similarly defined in terms of* $\{a_j'\}_{j=1}^m$ *and* $\{A_j'\}_{j=1}^m$*, then:*

$$\int_E \varphi_1(x)d\mu \leq \int_E \varphi_2(x)d\mu.$$

In other words, refinements of "best" simple functions always improve the result relative to the supremum goal because $\mu(A_i) = \sum \mu\left(A_{j_i}'\right)$ *but* $a_i' = \min\{a_{j_i}'\}$*.*

(c) ***Redefinition of*** $f(x)$ ***on sets of*** μ***-measure*** 0*: A consequence of item 2.b. is that the above definition of* $\int_E f(x)d\mu$ *is independent of redefinitions of* $f(x)$ *on sets of* μ*-measure* 0*.*

If $f(x) = g(x)$ μ*-a.e., let B denote the set on which* $f(x) \neq g(x)$ *and thus* $\mu(B) = 0$*. Given* $\{A_i\}_{i=1}^n$ *underlying a simple function* φ*, refine this collection by splitting each* $A_i = (A_i - B) \bigcup (A_i \bigcap B) \equiv A_i' \bigcup A_i''$*. Then for all i:*

$$\inf\{f(x)|x \in A_i'\} = \inf\{g(x)|x \in A_i'\},$$

while on A_i''*, these infima need not agree but then* $\mu\left(A_i''\right) = 0$*. Thus* $\int_E \varphi_f(x)d\mu = \int_E \varphi_g(x)d\mu$*, where these are best simple functions defined relative to the refined set collection. As this can be done for any initial collection* $\{A_i\}_{i=1}^n$*, it follows that:*

$$f(x) = g(x), \ \mu\text{-a.e.} \quad \Rightarrow \quad \int_E f(x)d\mu = \int_E g(x)d\mu. \tag{3.10}$$

In other words, both $f(x)$ *and* $g(x)$ *are* μ*-integrable and the integrals agree or both are not* μ*-integrable.*

2. ***On*** μ***-integrability:*** *While Definition 3.11 allows* $f(x)$ *to be a nonnegative,* $\sigma(X)$*-measurable, and extended real-valued function, a* μ***-integrable*** *function is real-valued almost everywhere:*

If $f(x)$ *is* μ*-integrable, then:*

$$\mu(\{f(x) = \infty\}) = 0.$$

Letting $E' \equiv \{f(x) = \infty\}$*, define the sequence of simple functions by* $\varphi_n(x) = n\chi_{E'}(x)$*. Then by Definition 3.11 and (3.1):*

$$\int_E f(x)dx \geq \int_{E'} f(x)dx \geq \int_{E'} \varphi_n(x)dx = nm(E').$$

Hence, if $m(E') > 0$*, then* $f(x)$ *cannot be* μ*-integrable.*

3. ***On $\int_E f(x)d\mu$ and Definition 3.7:*** *The definition of $\int_E f(x)d\mu$ in (3.9) is consistent with that in Definition 3.7.*

   First, with $E = X$, the above definition obtains:

$$\int f(x)d\mu = \sup_{\varphi \leq f} \int \varphi(x)d\mu, \tag{1}$$

   and then by (3.5) of Definition 3.7:

$$\begin{aligned} \int_E f(x)d\mu &\equiv \int \chi_E(x)f(x)d\mu \\ &= \sup_{\chi_E\varphi \leq \chi_E f} \int \varphi(x)d\mu. \end{aligned}$$

   The last step is (1)*, but noting that if $\varphi \leq \chi_E f$, then such φ is 0 outside E and so $\varphi = \chi_E \varphi$.*

   By the same logic and (3.5):

$$\begin{aligned} \sup_{\chi_E\varphi \leq \chi_E f} \int \varphi(x)d\mu &= \sup_{\chi_E\varphi \leq \chi_E f} \int \chi_E(x)\varphi(x)d\mu \\ &= \sup_{\varphi \leq f} \int \chi_E(x)\varphi(x)d\mu \\ &\equiv \sup_{\varphi \leq f} \int_E \varphi(x)d\mu. \end{aligned}$$

   Combining obtains (3.9).

Remark 3.13 (The Lebesgue integral) *The above step from simple to nonnegative measurable functions differs from that initiated as the second step for the Lebesgue integral. In Book III, the second step generalized the integral from simple functions to bounded measurable functions. Specifically, Proposition III.2.15 obtained in the current notation that for a bounded f and $E \in \sigma(X)$ with $\mu(E) < \infty$:*

$$\inf_{\psi \geq f} \int_E \psi(x)d\mu = \sup_{\varphi \leq f} \int_E \varphi(x)d\mu, \tag{3.11}$$

if and only if f is Lebesgue measurable. The integral $\int_E f(x)d\mu$ was then defined as this common value in Definition III.2.16.

Although we do not do this here, the Lebesgue approach for bounded measurable functions could also be applied in the current context. This first of all requires that such simple functions have both positive and negative real coefficients, and then as noted in Remark 2.27 that the defining sets of such simple functions satisfy $\mu\left(\bigcup_{i=1}^n A_i\right) < \infty$.

Then we can prove the following generalization of Proposition III.2.15. This statement contains the explicit requirement of completeness for $(X, \sigma(X), \mu)$, but recall that Lebesgue measure space is complete.

Proposition 3.14 ($\int_E f(x)d\mu$ for $|f(x)| \leq M$, $\mu(E) < \infty$: Integrable vs. measurable) *Let $f(x)$ be defined and bounded on a measurable set E in a measure space $(X, \sigma(X), \mu)$ with $\mu(E) < \infty$. If $f(x)$ is μ-measurable, then (3.11) is satisfied. Conversely, if (3.11) is satisfied and $(X, \sigma(X), \mu)$ is complete, then $f(x)$ is μ-measurable.*

Proof. *Left as an exercise in rethinking the earlier proof in this general context. Hint: Note that the validity of (3.11) will imply only that $f(x)$ is equal to a μ-measurable function μ-a.e., and hence the need for the completeness of $(X, \sigma(X), \mu)$ to justify an application of Proposition 2.14.* ■

Returning to the general approach of Definition 3.11, we first prove item 3 of Corollary 2.32.

Proposition 3.15 (Corollary 2.32, item 3) *Let $f(x)$ be a nonnegative, μ-measurable function defined on a μ-measurable set E of a measure space $(X, \sigma(X), \mu)$, which is μ-integrable by Definition 3.11. Then each $\varphi_n(x)$ of Proposition 2.28 equals zero outside a set of finite measure.*

Proof. *Because $\{\varphi_n(x)\}_{j=1}^{\infty}$ is an increasing sequence and $\varphi_n(x) \leq f(x)$, it follows from Definition 3.11 and (3.3) that for every n:*

$$\int_E \varphi_n(x)d\mu = \sup_{\varphi \leq \varphi_n} \int_E \varphi(x)d\mu \leq \sup_{\varphi \leq f} \int_E \varphi(x)d\mu = \int_E f(x)d\mu.$$

Hence, μ-integrability of $f(x)$ assures that $\int_E \varphi_n(x)d\mu < \infty$ for all n.

By (3.1), this assures that $\mu\left(A_j^{(n)}\right) < \infty$ for all j, and thus also that $\mu\left(\bigcup_{j=1}^{N} A_j^{(n)}\right) < \infty$. ■

The integrability criterion in Definition 3.11 is not yet very useful for developing deeper properties of this integral because there is no apparent way to organize and evaluate the typically uncountably many simple functions this definition reflects. For some basic properties, such as those summarized next, the above definition is sufficient.

For completeness, we repeat (3.10) here.

Proposition 3.16 (Simpler properties of integrals of nonnegative functions) *Given nonnegative μ-measurable functions $f(x)$ and $g(x)$ defined on $(X, \sigma(X), \mu)$ and $E \in \sigma(X)$:*

1. ***μ-Equivalence:*** *If $f(x) = g(x)$ μ-a.e., then:*

$$\int_E f(x)d\mu = \int_E g(x)d\mu.$$

2. ***Monotonicity:*** *If $f(x) \leq g(x)$ μ-a.e., then:*

$$\int_E f(x)d\mu \leq \int_E g(x)d\mu.$$

3. ***Constant multipliers:*** *If $a > 0$,*

$$\int_E af(x)d\mu = a \int_E f(x)d\mu.$$

4. ***Domain Decomposition:*** *If E_1 and E_2 are disjoint and measurable and $E \equiv E_1 \bigcup E_2$, then:*

$$\int_E f(x)d\mu = \int_{E_1} f(x)d\mu + \int_{E_2} f(x)d\mu.$$

Proof. *The proof of item 1 was seen in item 1.c. of Remark 3.12. The proof of item 2 is left as an exercise and uses (3.3) as in the proof of Proposition 3.15.*

For item 3, given a simple function φ, let $\varphi_a \equiv \varphi/a$, then:

$$\begin{aligned} \int_E af(x)d\mu &\equiv \sup_{\varphi \leq af} \int \chi_E(x)\varphi(x)d\mu \\ &= \sup_{\varphi_a \leq f} \int \chi_E(x)a\varphi_a(x)d\mu, \end{aligned}$$

and the result follows from (3.4)

Since $\chi_E(x) = \chi_{E_1}(x) + \chi_{E_2}(x)$ *and* $\chi_{E_j}(x)\varphi(x)$ *are simple functions, (3.4) obtains:*

$$\begin{aligned}\int_E f(x)d\mu &= \sup_{\varphi\leq f}\int \chi_E(x)\varphi(x)d\mu \\ &\leq \sup_{\varphi\leq f}\int \chi_{E_1}(x)\varphi(x)d\mu + \sup_{\varphi\leq f}\int \chi_{E_2}(x)\varphi(x)d\mu \\ &= \int_{E_1} f(x)d\mu + \int_{E_2} f(x)d\mu.\end{aligned}$$

On the other hand, given $\varphi_1, \varphi_2 \leq f$*, define the simple function* $\widetilde{\varphi}(x) = \chi_{E_1}(x)\varphi_1(x) + \chi_{E_2}(x)\varphi_2(x)$*. Then* $\widetilde{\varphi} \leq f$ *and by (3.4):*

$$\int \chi_E(x)\widetilde{\varphi}(x)d\mu = \int \chi_{E_1}(x)\varphi(x)d\mu + \int \chi_{E_2}(x)\varphi(x)d\mu.$$

Taking supremums:

$$\begin{aligned}&\int_{E_1} f(x)d\mu + \int_{E_2} f(x)d\mu. \\ =\ & \sup_{\varphi_1\leq f}\int \chi_{E_1}(x)\varphi_1(x)d\mu + \sup_{\varphi_2\leq f}\int \chi_{E_2}(x)\varphi_2(x)d\mu \\ =\ & \sup_{\widetilde{\varphi}\leq f}\int \chi_E(x)\widetilde{\varphi}(x)d\mu. \\ \leq\ & \int_E f(x)d\mu.\end{aligned}$$

This last inequality follows since $\{\,\widetilde{\varphi}|\widetilde{\varphi} \leq f\} \subset \{\,\varphi|\varphi \leq f\}$*, and the proof is complete.*

■

While item 3 is part of the definition of linearity for the integral, we would also like to establish the other part related to additivity. The following discussion suggests the need for additional tools beyond that provided by the supremum definition.

Claim 3.17 (Linearity of μ-integration) *Given nonnegative μ-measurable functions* $f(x)$ *and* $g(x)$ *defined on* $(X, \sigma(X), \mu)$ *and* $E \in \sigma(X)$*:*

$$\int_E [f(x) + g(x)]\,d\mu = \int_E f(x)d\mu + \int_E g(x)d\mu.$$

Discussion. *If* $\varphi_1 \leq f$ *and* $\varphi_2 \leq g$*, then repeating the second half of the proof of item 3 with* $\widetilde{\varphi}(x) = \varphi_1(x) + \varphi_2(x)$ *obtains:*

$$\begin{aligned}&\int_E f(x)d\mu + \int_E g(x)d\mu. \\ =\ & \sup_{\varphi_1\leq f}\int \chi_E(x)\varphi_1(x)d\mu + \sup_{\varphi_2\leq g}\int \chi_{E_2}(x)\varphi_2(x)d\mu \\ =\ & \sup_{\widetilde{\varphi}\leq f+g}\int \chi_E(x)\widetilde{\varphi}(x)d\mu. \\ \leq\ & \int_E f(x)d\mu.\end{aligned}$$

But for the first half of this proof, the assumption that $\varphi \leq f + g$ *does not readily reveal simple functions* φ_1 *and* φ_2 *dominated by* f *and* g*, separately.* ■

In the next two sections, we develop useful tools that will simplify the proof of μ-integrability and the derivation of the value of a μ-integral.

3.2.1 Fatou's Lemma

What will make derivations of various properties of integrals easier are limits theorems, collectively referenced as **integration to the limit** results. For example, if $\{f_n(x)\}_{n=1}^{\infty}$ is a sequence of functions with $f_n(x) \to f(x)$ in some manner, when does it follow that:

$$\int_E f(x)d\mu = \lim_{n\to\infty} \int_E f_n(x)d\mu.$$

Any such result will be useful because by Section 2.4, measurable functions can be approximated by simple functions which are easy to integrate by (3.1).

In the Lebesgue development of Book III, the **bounded convergence theorem, Lebesgue's monotone convergence theorem**, and **Lebesgue's dominated convergence theorem** were seen. Each provided exactly this kind of conclusion for different categories of measurable functions $f(x)$, based on different criteria on the given sequences.

As for the Book III Lebesgue analysis for nonnegative measurable functions, we first state and prove **Fatou's lemma,** named after its discoverer **Pierre Fatou** (1878–1929), and then turn to **Lebesgue's monotone convergence theorem,** named for **Henri Léon Lebesgue** (1875–1941).

Remark 3.18 (Fatou's lemma and convergence) *Fatou's lemma does not require the sequence $\{f_n(x)\}_{n=1}^{\infty}$ to converge pointwise to a function $f(x)$. But from the existence of a finite limit inferior of the sequence $\left\{\int_E f_n(x)d\mu\right\}_{n=1}^{\infty}$, this result assures integrability of the function:*

$$f(x) \equiv \liminf_{n\to\infty} f_n(x).$$

Recall Definition 2.16 that $\liminf_{n\to\infty} f_n(x)$ exists pointwise for all x, but need not be finite:

$$\liminf_{n\to\infty} f_n(x) \equiv \sup_n \inf_{k\geq n} f_k(x).$$

The limit inferior of the integral sequence is defined similarly, and again need not be finite.

If this function sequence actually converges pointwise, then by (2.8):

$$\liminf_{n\to\infty} f_n(x) = \lim_{n\to\infty} f_n(x),$$

and thus Fatou's result proves integrability of this limit function based on the same criterion.

Fatou's lemma is sometimes stated, dropping the $n \to \infty$:

$$\int_E [\liminf f_n(x)]\, d\mu \leq \liminf \int_E f_n(x)d\mu. \tag{3.12}$$

This representation emphasizes that Fatou's lemma is a result on interchanging two limiting processes, the limit inferior of a sequence, and the integral of a function, which is defined in terms of the supremum of the integrals of subordinate functions.

Proposition 3.19 (Fatou's lemma) *If $\{f_n(x)\}_{n=1}^{\infty}$ is a sequence of nonnegative μ-measurable functions on $(X, \sigma(X), \mu)$, and $f(x) \equiv \liminf f_n(x)$ on a μ-measurable set E, then:*

$$\int_E f(x)d\mu \leq \liminf \int_E f_n(x)d\mu. \tag{3.13}$$

Proof. *First, $f(x)$ is μ-measurable by Proposition 2.19 and nonnegative by definition of limit inferior. To prove (3.13), we show that if $\varphi(x)$ is a simple function with $\varphi(x) \leq f(x)$, then:*

$$\int_E \varphi(x)d\mu \leq \liminf \int_E f_n(x)d\mu. \tag{1}$$

This then assures the result for the supremum of all such φ-integrals with $\varphi \leq f$, which is the result in (3.13).

To this end, first assume that for some $\varphi(x) \leq f(x)$ that $\int_E \varphi(x)d\mu = \infty$. Then by (3.8), there is at least one μ-measurable set $A \subset E$ with $\mu(A) = \infty$, and for which $\varphi(x) > a > 0$ for $x \in A$. Since by definition $f(x) > a > 0$ for $x \in A$, define $B_n = \bigcap_{k=n}^{\infty}\{x \in E | f_k(x) > a\}$. Each B_n is μ-measurable, and $\{B_n\}_{n=1}^{\infty}$ is a nested sequence, $B_n \subset B_{n+1}$.

For each $x \in A$, $f(x) = \liminf f_n(x) > a$ implies by (2.6) that $f_k(x) > a$ for $k \geq N$ for some N, and so $A \subset \bigcup_{n=1}^{\infty} B_n$. Hence, $\mu\left(\bigcup_{n=1}^{\infty} B_n\right) = \infty$ by monotonicity, and as this is a nested increasing sequence, $\lim \mu(B_n) = \infty$ by continuity from below. By nonnegativity:

$$\int_E f_n(x)d\mu > \int_{B_n} f_n(x)d\mu > a\mu(B_n),$$

it thus follows that $\liminf \int_E f_n(x)d\mu = \infty$ and (1) is satisfied in this case.

Next, assume that for all $\varphi(x) \leq f(x)$ that $\int_E \varphi(x)d\mu < \infty$. Then for any such φ, $\mu(A_i) < \infty$ for each of the n defining sets in (2.9), and we can assume without loss of generality that all $a_i > 0$. Defining $A = \bigcup_{i \leq n} A_i$, then $A \subset E$, $\mu(A) < \infty$, and $\varphi(x) = 0$ on $E - A$. Also, $\varphi(x) \leq a \equiv \max_{i \leq n} a_i$ for $x \in A$.

Given $\epsilon > 0$ define:

$$B_n = \{x \in E | \inf_{k \geq n} f_k(x) > (1-\epsilon)\varphi(x)\}.$$

Since $(1-\epsilon)\varphi(x) < f(x) \equiv \sup_n \inf_{k \geq n} f_k(x)$ and $\{\inf_{k \geq n} f_k(x)\}_{n=1}^{\infty}$ is increasing for all x, it follows that $B_n \neq \emptyset$ for any $\epsilon > 0$. Further, $\{B_n\}_{n=1}^{\infty}$ is an increasing nested sequence of μ-measurable sets, $B_n \subset B_{n+1}$.

Now $A \subset \bigcup_{n=1}^{\infty} B_n$, because for $x \in A$:

$$0 < \varphi(x) \leq f(x) = \sup_n \inf_{k \geq n} f_k(x),$$

so $x \in B_n$ for all $n \geq N(\epsilon)$. Thus $\{A - B_n\}_{n=1}^{\infty}$ is a decreasing sequence with limit $\emptyset$, and since $\mu(A) < \infty$, continuity from above of μ from Proposition 1.29 obtains that $\lim \mu(A - B_n) = 0$.

For given $\epsilon > 0$, choose $N = N(\epsilon)$ so that $\mu(A - B_n) < \epsilon$ for $n \geq N$. Then for $n \geq N$, recalling Proposition 3.5:

$$\begin{aligned}
\int_E f_n(x)d\mu &\geq \int_{B_n} f_n(x)d\mu \\
&\geq (1-\epsilon)\int_{B_n} \varphi(x)d\mu \\
&= (1-\epsilon)\int_E \varphi(x)d\mu - (1-\epsilon)\int_{E-B_n} \varphi(x)d\mu \\
&\geq (1-\epsilon)\int_E \varphi(x)d\mu - \int_{E-B_n} \varphi(x)d\mu.
\end{aligned}$$

Since $\varphi = 0$ *on* $E - A$, $\mu(A - B_n) < \epsilon$, *and* $\varphi(x) \leq a$ *for* $x \in A$:

$$\begin{aligned}\int_E f_n(x)d\mu &\geq (1-\epsilon)\int_E \varphi(x)d\mu - \int_{A-B_n}\varphi(x)d\mu \\ &\geq \int_E \varphi(x)d\mu - \epsilon\left[\int_E \varphi(x)d\mu + a\right]. \end{aligned} \tag{2}$$

It now follows from (2) *that for* $n \geq N$:

$$\int_E \varphi(x)d\mu \leq \int_E f_n(x)d\mu,$$

and thus (1) *is satisfied in this case, completing the proof.* ■

When $f(x) \equiv \liminf_{n\to\infty} f_n(x)$ μ-a.e., Fatou's lemma can still apply, but measurability of such $f(x)$ becomes the main concern.

Remark 3.20 (On nonnegativity requirement for $f(x)$) *Certainly* $f(x) \geq 0$ μ*-a.e. by definition, but below we specify that such* $f(x)$ *is a nonnegative function. This is done simply because we have not yet defined* $\int_E f(x)d\mu$ *if* $f < 0$ *even on a set of measure zero. But once such integrals are defined in Section 3.3.1, it will be clear that* $f < 0$ *on a set of measure zero does not change the value of this integral, and this "nonnegative" qualifier can be omitted.*

Corollary 3.21 (Fatou's lemma, μ-a.e.) *Let* $\{f_n(x)\}_{n=1}^{\infty}$ *be a sequence of nonnegative* μ*-measurable functions defined on a measure space* $(X, \sigma(X), \mu)$, *and* $f(x)$ *a nonnegative function with* $f(x) \equiv \liminf_{n\to\infty} f_n(x)$ μ*-a.e. on a* μ*-measurable set* E. *Then (3.13) holds if either:*

1. $f(x)$ *is* μ*-measurable, or,*
2. $(X, \sigma(X), \mu)$ *is complete.*

Proof. *Let* $\tilde{f}(x) \equiv \liminf_{n\to\infty} f_n(x)$, *which is* μ*-measurable by Proposition 2.19. Then by the above result:*

$$\int_E \tilde{f}(x)d\mu \leq \liminf \int_E f_n(x)d\mu.$$

1. *If* $f(x)$ *is* μ*-measurable, then by* $f(x) = \tilde{f}(x)$ μ*-a.e and item 1 of Proposition 3.16:*

$$\int_E \tilde{f}(x)d\mu = \int_E f(x)d\mu. \tag{1}$$

2. *If* $(X, \sigma(X), \mu)$ *is complete, then* $f(x) = \tilde{f}(x)$ μ*-a.e. assures that* $f(x)$ *is* μ*-measurable by Proposition 2.14, and* (1) *again follows.*

In either case, the proof is complete. ■

As a final result, we consider a variation of Fatou's lemma that reflects the limit superior. If the sequence of functions $\{f_n(x)\}_{n=1}^{\infty}$ is dominated by an integrable function, which is the assumption seen below in Lebesgue's dominated convergence theorem, then Fatou's lemma has a limit superior version. The proof requires the linearity property of integrals in item 1 of Proposition 3.29, which was initially discussed in Claim 3.17.

Proposition 3.22 (Fatou's lemma and limit superior) *If* $\{f_n(x)\}_{n=1}^{\infty}$ *is a sequence of nonnegative* μ*-measurable functions defined on* $(X, \sigma(X), \mu)$*, and there exists* μ*-measurable* $g(x)$ *with* $f_n(x) \leq g(x)$ *on* E *for all* n *and* $\int_E g(x)d\mu < \infty$*, then:*

$$\int_E [\limsup f_n(x)]\, d\mu \geq \limsup \int_E f_n(x) d\mu. \tag{3.14}$$

Proof. *Letting* $h_n = g - f_n$*, then* $\{h_n(x)\}_{n=1}^{\infty}$ *is a nonnegative sequence of* μ*-measurable functions, so by (3.12) and linearity of the integral:*

$$\begin{aligned} \int_E [\liminf (g(x) - f_n(x))]\, d\mu &\leq \liminf \int_E (g(x) - f_n(x))\, d\mu \qquad (1) \\ &= \liminf \left(\int_E g(x)d\mu - \int_E f_n(x)d\mu \right). \end{aligned}$$

Given a numerical sequence $c_n \equiv a - b_n$*, then by Definition 2.16:*

$$\begin{aligned} \liminf c_n &\equiv \sup_n \inf_{k \geq n} (a - b_k) \\ &= \sup_n \left[a - \sup_{k \geq n} b_k \right] \\ &= a - \inf_n \sup_{k \geq n} b_k \\ &= a - \limsup b_n. \end{aligned}$$

Using this in (1)*, and then linearity of the integral on the left:*

$$\int_E g(x)d\mu - \int_E \limsup f_n(x)d\mu \leq \int_E g(x)d\mu - \limsup \int_E f_n(x)d\mu.$$

Since $\displaystyle\int_E g(x)d\mu < \infty$*, this can be subtracted and the proof is complete.* ■

3.2.2 Lebesgue's Monotone Convergence Theorem

Fatou's lemma provides "only" an upper bound to the value of the μ-integral of $f(x) \equiv \liminf f_n(x)$ over E vis-a-vis the μ-integrals of the given sequence of nonnegative functions $\{f_n(x)\}_{n=1}^{\infty}$. It therefore also provides an upper bound to this integral when $f(x)$ exists as a limit. But this result is the key ingredient for a short proof of the final integration to the limit result for nonnegative measurable functions.

Lebesgue's monotone convergence theorem, named for **Henri Léon Lebesgue** (1875–1941), replaces Fatou's inequality with equality subject to the additional constraint that the given sequence $\{f_n(x)\}_{n=1}^{\infty}$ is increasing. As was the case for Fatou's lemma, the next result does not require that the functions in the sequence be μ-integrable, which is to say, have finite integrals.

Lebesgue's result will be generalized below to apply to an increasing sequence of measurable, but not necessarily nonnegative, functions. It is then known as **Beppo Levi's theorem,** named for **Beppo Levi** (1875–1961). For this more general result, it must then be assumed that the functions in the sequence are in fact integrable and that the associated integral values have a finite supremum.

As in (3.12), Lebesgue's result is sometimes expressed as:

$$\int_E \lim_{n\to\infty} f_n(x)d\mu = \lim_{n\to\infty} \int_E f_n(x)d\mu, \tag{3.15}$$

to emphasize the interchanging of two limiting processes.

Proposition 3.23 (Lebesgue's monotone convergence theorem) *If $\{f_n(x)\}_{n=1}^{\infty}$ is an increasing sequence of nonnegative μ-measurable functions defined on $(X, \sigma(X), \mu)$ which converge on a μ-measurable set E to a function $f(x)$, then:*

$$\int_E f(x)d\mu = \lim_{n\to\infty} \int_E f_n(x)d\mu. \tag{3.16}$$

Proof. *As a limit of nonnegative μ-measurable functions, $f(x)$ is nonnegative by definition and μ-measurable by Proposition 2.19. By Fatou's lemma:*

$$\int_E f(x)d\mu \leq \liminf_{n\to\infty} \int_E f_n(x)d\mu. \tag{1}$$

As $\{f_n(x)\}_{n=1}^{\infty}$ is an increasing sequence, item 2 of Proposition 3.16 obtains for all n:

$$\int_E f_n(x)d\mu \leq \int_E f(x)d\mu.$$

Hence:

$$\limsup_{n\to\infty} \int_E f_n(x)d\mu \leq \int_E f(x)d\mu. \tag{2}$$

As the limit superior cannot be smaller than the limit inferior, the inequalities in (1) *and* (2) *imply that these limits are equal, and (3.16) is proved by Corollary I.3.46 restated in (2.8).* ■

Remark 3.24 (On Lebesgue's monotone convergence theorem) *Two clarifying comments on Lebesgue's theorem:*

1. ***On integrability:*** *It is not assumed that the functions in the given sequence $\{f_n(x)\}_{n=1}^{\infty}$ are integrable, nor is it concluded that $f(x)$ is integrable. Since $\{f_n(x)\}_{n=1}^{\infty}$ is an increasing sequence, so too is $\left\{\int_E f_n(x)d\mu\right\}_{n=1}^{\infty}$ by item 2 of Proposition 3.16. Specifically:*

   - $\int_E f_n(x)d\mu \leq \int_E f_{n+1}(x)d\mu$ *when both are finite;*
   - *If* $\int_E f_n(x)d\mu = \infty$, *then* $\int_E f_m(x)d\mu = \infty$ *for* $m > n$.

   Thus the proof confirms that $f(x)$ is integrable if and only if $\int_E f_n(x)d\mu \leq K < \infty$ for all n.

2. ***On decreasing*** $\{f_n(x)\}_{n=1}^{\infty}$: *Perhaps surprisingly, Lebesgue's monotone convergence theorem does not apply to decreasing function sequences.*

   A simple example from Lebesgue integration noted in Example III.2.38 is to define $f_n(x) = \chi_{[n,\infty)}(x)$, so $(\mathcal{L})\int f_n(x)dx = \infty$ for all n, while $f_n(x) \to 0$ pointwise and so $\int f(x)dx = 0$. But note that Fatou's lemma still applies.

The next result generalizes to almost everywhere convergence, and as for Fatou's generalization, the issue is μ-measurability of the given $f(x)$. For the nonnegativity requirement on $f(x)$, recall Remark 3.20.

Corollary 3.25 (Lebesgue's monotone convergence theorem, μ-a.e.) *Let $\{f_n(x)\}_{n=1}^{\infty}$ be an increasing sequence of nonnegative μ-measurable functions defined on a measure space $(X, \sigma(X), \mu)$, which converge almost everywhere on a μ-measurable set E to a nonnegative function $f(x)$. Then (3.16) holds if either:*

1. *$f(x)$ is μ-measurable, or,*

2. *$(X, \sigma(X), \mu)$ is complete.*

Proof. *Let $\widetilde{f}(x) \equiv \lim_{n\to\infty} f_n(x)$, which is μ-measurable by Proposition 2.19. Then by (3.16):*

$$\int_E \widetilde{f}(x)d\mu = \lim_{n\to\infty} \int_E f_n(x)d\mu.$$

1. *If $f(x)$ is μ-measurable, then by $f(x) = \widetilde{f}(x)$ μ-a.e and item 1 of Proposition 3.16:*

$$\int_E \widetilde{f}(x)d\mu = \int_E f(x)d\mu. \tag{1}$$

2. *If $(X, \sigma(X), \mu)$ is complete, then $f(x) = \widetilde{f}(x)$ μ-a.e. assures that $f(x)$ is μ-measurable by Proposition 2.14, and* (1) *again follows.*

In either case, the proof is complete. ■

Remark 3.26 (Evaluating integrals of nonnegative functions) *While the definition of the integral $\int_E f(x)d\mu$ in (3.9) potentially contemplates uncountably many simple functions in the supremum calculation, Lebesgue's monotone convergence theorem gives a more practical and useful way to evaluate the μ-integral of a nonnegative function.*

*Specifically, if $\{f_n(x)\}_{n=1}^{\infty}$ is **any** sequence of **increasing, μ-measurable simple functions** defined on a μ-measurable set E with $f_n(x) \to f(x)$ for all $x \in E$, then:*

$$\int_E f(x)d\mu = \lim_{n\to\infty} \int_E f_n(x)d\mu.$$

In theory, the μ-integral can be evaluated using any such increasing sequence of functions, but by using simple functions from Proposition 2.28, say, the integrals of the function sequence are simply evaluated by (3.1).

The same conclusion follows if only $f_n(x) \to f(x)$ for μ-a.e. on E if $f(x)$ is known to be μ-measurable, or in the general case if $(X, \sigma(X), \mu)$ is complete.

Example 3.27 ($\int_0^2 x^2 d\mu_F$) *If $(X, \sigma(X), \mu) = ([0, \infty), \mathcal{M}_{\mu_F}([0, \infty)), \mu_F)$, the Borel measure space on $[0, \infty)$ defined with $F(x) = x^2$, we evaluate $\int_E f(x)d\mu_F$ for $E = [0, 2]$ and $f(x) = x^2$.*

Because E is an interval, it is natural to utilize a sequence of simple functions defined as step functions. If we can then show that $f(x)$ is the pointwise limit of an increasing sequence of such functions, the value of this integral can be obtained by Lebesgue's monotone convergence theorem.

Given n, let $\{A_i\}_{i=1}^{n}$ be defined by $A_i = [2(i-1)/n, 2i/n]$ and define the simple function $\varphi_n(x) \le f(x)$ by:

$$\varphi_n(x) = \sum\nolimits_{i=1}^{n} \left[\frac{2(i-1)}{n}\right]^2 \chi_{A_i}(x).$$

To ensure that $\{\varphi_n(x)\}_{n=1}^{\infty}$ *is an increasing sequence, we use* ***interval bisection****, choosing* $n = 2m$ *for positive integers* m*.*

Now since $\mu_F[(a,b]] = F(b) - F(a)$ *by (1.10):*

$$\int_E \chi_{A_i}(x)d\mu_F = \left(\frac{2i}{n}\right)^2 - \left(\frac{2(i-1)}{n}\right)^2 = \frac{4[2i-1]}{n^2}.$$

Then by (3.1):

$$\int_E \varphi_n(x)d\mu_F = 16\sum\nolimits_{i=1}^{n} \frac{(i-1)^2(2i-1)}{n^4} = 16\sum\nolimits_{i=1}^{n} \frac{2i^3 - 5i^2 + 4i - 1}{n^4}.$$

Since $\sum\nolimits_{i=1}^{n} i^k = O(n^{k+1})$*:*

$$\int_E \varphi_n(x)d\mu_F = 32\sum\nolimits_{i=1}^{n} \frac{i^3}{n^4} + O(\frac{1}{n}).$$

Now $\sum\nolimits_{i=1}^{n} i^3 = \frac{1}{4}n^2(n+1)^2$*, and thus:*

$$\int_0^2 x^2 d\mu_F = \lim_{n\to\infty} \int_E \varphi_n(x)d\mu_F = 8.$$

3.2.3 Properties of Integrals

With the help of Lebesgue's monotone convergence theorem, we now establish various properties of the μ-integral of nonnegative measurable functions, restating the results from Proposition 3.16 for completeness. These properties will be generalized below to integrals of general measurable functions.

Remark 3.28 (On μ-integrability) *Recall that* $\int_E f(x)d\mu$ *and other such expressions below are defined in (3.9) in terms of a supremum of the integrals of subordinated simple functions, which need not be finite. The following proposition makes statements about such suprema. However, this result also has implications for* μ*-integrability, which requires that such suprema be finite.*

For example, in item 1, if both $f(x)$ *and* $g(x)$ *are* μ*-integrable, then so too is* $af(x) + bg(x)$*, while if either is not, then neither is this linear combination. Similarly for item 4,* μ*-integrability of* $g(x)$ *implies such integrability for* $f(x)$*, while if* $f(x)$ *is not integrable, then neither is* $g(x)$*.*

The reader should spend a moment investigating how the following results translate to statements on μ*-integrability.*

Proposition 3.29 (Properties of integrals of nonnegative functions) *If* $f(x)$ *and* $g(x)$ *are nonnegative* μ*-measurable functions defined on a measure space* $(X, \sigma(X), \mu)$ *and* $E \in \sigma(X)$*, then:*

1. *For any* $a, b > 0$*:*

$$\int_E [af(x) + bg(x)]d\mu = a\int_E f(x)d\mu + b\int_E g(x)d\mu.$$

2. *If $\mu(E) > 0$:*

$$\int_E f(x)d\mu = 0 \Leftrightarrow f(x) = 0, \ \mu\text{-a.e. on } E.$$

3. *If $\mu(E) = 0$, then:*

$$\int_E f(x)d\mu = 0.$$

4. *If $f(x) = g(x)$ μ-a.e.:*

$$\int_E f(x)d\mu = \int_E g(x)d\mu.$$

5. *If $f(x) \leq g(x)$ μ-a.e.:*

$$\int_E f(x)d\mu \leq \int_E g(x)d\mu.$$

6. *If $C_1 \leq f(x) \leq C_2$ μ-a.e., and $\mu(E) < \infty$:*

$$C_1\mu(E) \leq \int_E f(x)d\mu \leq C_2\mu(E).$$

7. *If $E' \subset E$ is μ-measurable:*

$$\int_{E'} f(x)d\mu \leq \int_E f(x)d\mu.$$

8. *If $E = E_1 \bigcup E_2$, a disjoint union of μ-measurable sets:*

$$\int_E f(x)d\mu = \int_{E_1} f(x)d\mu + \int_{E_2} f(x)d\mu.$$

Proof. *Taking these results in turn:*

1. *If $\{\varphi_n(x)\}_{n=1}^{\infty}$ is an increasing sequence of simple functions converging to $f(x)$ by Proposition 2.28, and $\{\psi_n(x)\}_{n=1}^{\infty}$ is increasing and converging to $g(x)$, then $\{a\varphi_n(x) + b\psi_n(x)\}_{n=1}^{\infty}$ is increasing and converges to $af(x) + bg(x)$. So by Lebesgue's monotone convergence theorem and Proposition 3.5:*

$$\begin{aligned}\int_E [af(x) + bg(x)]d\mu &= \lim_{n\to\infty} \int_E [a\varphi_n(x) + b\psi_n(x)]d\mu \\ &= \lim_{n\to\infty} \left[a\int_E \varphi_n(x)d\mu + b\int_E \psi_n(x)d\mu\right] \\ &= a\int_E f(x)d\mu + b\int_E g(x)d\mu.\end{aligned}$$

2. *That $\int_E 0d\mu = 0$ is apparent by definition in (3.9) and (3.1), and similarly $\int_E f(x)d\mu = 0$ if $f(x) = 0$ μ-a.e. by item 1.c. of Remark 3.12. Conversely, if $\int_E f(x)d\mu = 0$, let $A_n = \{x \in E | f(x) \geq 1/n\}$. Then $f(x) \geq \chi_{A_n}(x)/n$ obtains that $\mu(A_n) \leq n\int_E f(x)d\mu = 0$ by (3.1). But $\{x \in E | f(x) > 0\} = \bigcup_{n=1}^{\infty} A_n$, and hence this set has μ-measure 0 by countable subadditivity of μ.*

3. *By (3.5):*
$$\int_E f(x)d\mu = \int_E \chi_E(x)f(x)d\mu,$$
and $\mu(E) = 0$ *obtains* $\chi_E(x)f(x) = 0$ μ*-a.e. If a simple function satisfies* $\varphi(x) \leq \chi_E(x)f(x)$, *then* $\varphi(x) = 0$ μ*-a.e. and* $\int_E \varphi(x)d\mu = 0$ *by (3.1). Now apply (3.9).*

4. *Here,* $f(x) - g(x) = 0$ μ*-a.e. on* E, *and this result follows from items 2 then 1.*

5. *Now,* $f(x)-g(x) \geq 0$ μ*-a.e on* E, *so* $\int_E [f(x) - g(x)]\, d\mu \geq 0$ *by definition of this integral. Now apply item 1.*

6. *Given* $C_1\chi_E(x) \leq f(x) \leq C_2\chi_E(x)$ μ*-a.e., this result follows from item 5 and (3.1).*

7. *Since* $f(x)\chi_{E'}(x) \leq f(x)$ *on* E, *the result is an application of item 5 and (3.5).*

8. *By definition,* $f(x) = f(x)\chi_{E_1}(x) + f(x)\chi_{E_2}(x)$ *for* $x \in E$, *and the result reflects item 1 and (3.5),*

■

Part 4 of this proposition provides a simple corollary that states that every nonnegative μ-integrable function can be used as a **test function** to identify other μ-integrable functions.

Corollary 3.30 (Integrability comparison test) *If* $g(x)$ *is a nonnegative* μ*-measurable function that is* μ*-integrable on a* μ*-measurable set* E, *then for any nonnegative* μ*-measurable function* $f(x)$ *with* $f(x) \leq g(x)$ μ*-a.e., it follows that* $f(x)$ *is also* μ*-integrable.*
Proof. *Immediate from item 5 above.* ■

We next record two important corollaries to Lebesgue's monotone convergence theorem. The first applies to the integral of a function series, providing a condition that allows the reversal of the two limiting processes of summation and integration. The second allows the decomposition of an integral into a countable number of disjoint domains. As noted in Remark 3.28, these again provide statements on μ-integrability.

Corollary 3.31 (μ-integral of a series) *If* $\{f_n(x)\}_{n=1}^{\infty}$ *is a sequence of nonnegative* μ*-measurable functions defined on measure space* $(X, \sigma(X), \mu)$, *and* $f(x) = \sum_{n=1}^{\infty} f_n(x)$ *on* $E \in \sigma(X)$, *then:*
$$\int_E f(x)d\mu = \sum_{n=1}^{\infty} \int_E f_n(x)d\mu. \tag{3.17}$$

Proof. *Define* $g_m(x) = \sum_{n=1}^{m} f_n(x)$. *Then for all* m, $g_m(x)$ *is nonnegative,* μ*-measurable by Proposition 2.13, and the sequence* $\{g_m(x)\}_{m=1}^{\infty}$ *is increasing. Further,* $f(x) = \lim_{m\to\infty} g_m(x)$ *on* E, *so by Lebesgue's monotone convergence theorem:*
$$\int_E f(x)dx = \lim_{m\to\infty} \int_E g_m(x)dx.$$

By item 1 of Proposition 3.29:
$$\int_E g_m(x)dx = \sum_{n=1}^{m} \int_E f_n(x)dx,$$

and the result follows. ■

Corollary 3.32 (μ-integral over a countable union) *If $f(x)$ is a nonnegative μ-measurable function defined on a measure space $(X, \sigma(X), \mu)$, and $E = \bigcup_{n=1}^{\infty} E_n$, a disjoint union of μ-measurable sets, then:*

$$\int_E f(x)d\mu = \sum_{n=1}^{\infty} \int_{E_n} f(x)d\mu. \tag{3.18}$$

Proof. *Left as an exercise. Hint: $1 = \sum_{n=1}^{\infty} \chi_{E_n}(x)$ on E.* ■

3.2.4 Product Space Measures Revisited

The reader may recall Chapter I.7, where it was a mighty challenge to prove that the product set function μ_0, defined on the semi-algebra $\mathcal{A}'$ of measurable rectangles in (1.17), was a pre-measure (Definition 1.19) on this semi-algebra. Indeed, with the tools then at hand, we could only prove in Proposition I.7.15 that μ_0 was finitely additive on $\mathcal{A}'$.

For the countable additivity proof in Proposition I.7.18, we needed to expand $\mathcal{A}'$ to the algebra $\mathcal{A}$ of finite disjoint unions of $\mathcal{A}'$-rectangles, thereby proving μ_0 could be extended to a measure on $\mathcal{A}$. And even then, it was necessary to assume that the component measure spaces were σ-finite, an extraneous assumption, but one needed to justify the continuity from above argument that was utilized.

Of course, that μ_0 was ultimately proved to be a measure on $\mathcal{A}$ logically implied that it was also a pre-measure on $\mathcal{A}'$. But a direct proof of this latter result is still desirable, especially one that omits the σ-finite assumption.

Recalling Definition 1.23, given measure spaces $\{(X_i, \sigma(X_i), \mu_i)\}_{i=1}^n$, a measurable rectangle $A = \prod_{i=1}^n A_i$ in $X = \prod_{i=1}^n X_i$ is a set:

$$A = \{x \in X | x_i \in A_i\},$$

where $A_i \in \sigma(X_i)$. The collection of measurable rectangles in X is denoted $\mathcal{A}'$, contains $\emptyset$ by definition, and was proved to be a semi-algebra in Proposition I.7.2. The product set function μ_0 is defined on $\mathcal{A}'$ in (1.17) by:

$$\mu_0(A) = \prod_{i=1}^n \mu_i(A_i).$$

We now prove that μ_0 is a pre-measure on $\mathcal{A}'$ with the aid of Lebesgue's monotone convergence theorem.

Remark 3.33 (On σ-finiteness of $\{(X_i, \sigma(X_i), \mu_i)\}_{i=1}^n$) *The following result does not require that the component measure spaces be σ-finite, and thus the existence result of Proposition I.7.20 is generalized to arbitrary measure spaces $\{(X_i, \sigma(X_i), \mu_i)\}_{i=1}^n$.*

However, σ-finiteness is still needed to assure the uniqueness of this extension from a measure $\mu_{\mathcal{A}}$ defined on the algebra $\mathcal{A}$ to a measure μ defined on $\sigma(\mathcal{A})$, the smallest sigma algebra that contains $\mathcal{A}$. This follows because the uniqueness statement of Proposition 1.22 restated Proposition I.6.14, the proof of which required this assumption.

Proposition 3.34 (μ_0 is a pre-measure on $\mathcal{A}'$) *The product set function μ_0 is a pre-measure on $\mathcal{A}'$.*

Proof. *We prove countable additivity, that if $\{B_j\}_{j=1}^{\infty} \subset \mathcal{A}'$ is a disjoint collection of measurable rectangles with $\bigcup_{j=1}^{\infty} B_j \in \mathcal{A}'$, then:*

$$\mu_0\left(\bigcup_{j=1}^{\infty} B_j\right) = \sum_{j=1}^{\infty} \mu_0(B_j). \tag{3.19}$$

Finite additivity then follows by taking $B_j = \emptyset$ *for all but finitely many* j*, noting that* $\mu_0(\emptyset) = 0$.

Let $B_j = \prod_{i=1}^n A_{ji}$ *where* $\{A_{ji}\}_{j=1}^{\infty} \subset \sigma(X_i)$ *for each* i *and assume that* $\bigcup_{j=1}^{\infty} B_j = \prod_{i=1}^n A_i$ *with* $A_i \in \sigma(X_i)$*. Then for all* $x = (x_1, x_2, ..., x_n) \in X$*:*

$$\prod_{i=1}^n \chi_{A_i}(x_i) = \sum_{j=1}^{\infty} \prod_{i=1}^n \chi_{A_{ji}}(x_i). \tag{1}$$

This follows because the left side product equals 1 *if and only if* $x \in \prod_{i=1}^n A_i = \bigcup_{j=1}^{\infty} \prod_{i=1}^n A_{ji}$*. Hence, such* $x \in B_j = \prod_{i=1}^n A_{ji}$ *for exactly one* j *by the disjointedness of* $\{B_j\}_{j=1}^{\infty}$*.*

Fixing $x_1, x_2, ..., x_{n-1}$*, consider the identity in* (1) *as a functional identity in* x_n*. Because* $\chi_{A_n}(x_n)$ *and each* $\chi_{A_{jn}}(x_n)$ *are* μ_n*-measurable functions, we can take* μ_n*-integrals of this identity and apply (3.17):*

$$\int \prod_{i=1}^{n-1} \chi_{A_i}(x_i)\chi_{A_n}(x_n)d\mu_n = \sum_{j=1}^{\infty} \int \left(\prod_{i=1}^{n-1} \chi_{A_{ji}}(x_i)\right) \chi_{A_{jn}}(x_n)d\mu_n.$$

Since $\prod_{i=1}^{n-1} \chi_{A_{ji}}(x_i)$ *and* $\prod_{i=1}^{n-1} \chi_{A_i}(x_i)$ *are constants relative to these* μ_n*-integrals, and* $\int \chi_A(x_n)d\mu_n = \mu_n(A)$ *for* $A = A_n$ *or* A_{nj}*, this obtains for all fixed* $x_1, x_2, ..., x_{n-1}$*:*

$$\prod_{i=1}^{n-1} \chi_{A_i}(x_i)\mu_n(A_n) = \sum_{j=1}^{\infty} \prod_{i=1}^{n-1} \chi_{A_{ji}}(x_i)\mu_n(A_{jn}). \tag{2}$$

Now fix $x_1, x_2, ..., x_{n-2}$ *and repeat the derivation to conclude that:*

$$\prod_{i=1}^{n-2} \chi_{A_i}(x_i)\mu_{n-1}(A_{n-1})\mu_n(A_n) = \sum_{j=1}^{\infty} \prod_{i=1}^{n-2} \chi_{A_{ji}}(x_i)\mu_{n-1}(A_{j,n-1})\mu_n(A_{jn}).$$

Continuing in this way produces:

$$\prod_{i=1}^n \mu_i(A_i) = \sum_{j=1}^{\infty} \prod_{i=1}^n \mu_i(A_{ji}),$$

which is (3.19) by the definitions of μ_0 *and the* A*-sets.* ■

Corollary I.7.3 proved that $\mathcal{A}'(\mathcal{A}_i')$, respectively $\mathcal{A}'(\mathcal{A}_i)$, were also semi-algebras on X:

Definition 3.35 ($\mathcal{A}'(\mathcal{A}_i')$ and $\mathcal{A}'(\mathcal{A}_i)$) $\mathcal{A}'(\cdot)$ *denotes the collection of measurable rectangles in* X *defined by* $A = \prod_{i=1}^n A_i$ *with:*

1. $\mathcal{A}'(\mathcal{A}_i') : A_i \in \mathcal{A}_i'$*, where* $\mathcal{A}_i' \subset \sigma(X_i)$ *is a semi-algebra;*
2. $\mathcal{A}'(\mathcal{A}_i) : A_i \in \mathcal{A}_i$*, where* $\mathcal{A}_i \subset \sigma(X_i)$ *is an algebra.*

By definition, when the algebra is generated by the semi-algebra:

$$\mathcal{A}'(\mathcal{A}_i') \subsetneq \mathcal{A}'(\mathcal{A}_i) \subsetneq \mathcal{A}'(\sigma(X_i)),$$

where $\mathcal{A}'(\sigma(X_i)) \equiv \mathcal{A}'$ is the semi-algebra above.

Corollary 3.36 (μ_0 is a pre-measure on $\mathcal{A}'(\cdot)$) *The product measure* μ_0 *is a pre-measure on* $\mathcal{A}'(\mathcal{A}_i')$*, respectively* $\mathcal{A}'(\mathcal{A}_i)$*.*

Proof. *The proof is identical since nothing but measurability of the component sets* $\{A_{ji}\}_{j=1}^{\infty} \subset \sigma(X_i)$ *and* $A_i \in \sigma(X_i)$ *was used.* ■

3.3 Integrating General Measurable Functions

The final step in the μ-integration sequence is to extend the definition from nonnegative to general μ-measurable functions, and this is now relatively easy to do. As was seen in the Lebesgue development, the key to this extension is to split a general function into positive and negative parts.

Definition 3.37 (Positive/negative parts of $f(x)$) *Given $f(x)$ defined on a measure space $(X, \sigma(X), \mu)$, the **positive part of** $f(x)$, denoted $f^+(x)$, is defined by:*

$$f^+(x) = \max\{f(x), 0\}, \tag{3.20}$$

*and the **negative part of** $f(x)$, denoted $f^-(x)$, is defined by:*

$$f^-(x) = \max\{-f(x), 0\}. \tag{3.21}$$

The positive and negative parts of a function are **both nonnegative functions**, so below we can apply the previous section's results to either part. The original function and its absolute value can then be recovered from these component functions:

$$\begin{aligned} f(x) &= f^+(x) - f^-(x), \\ |f(x)| &= f^+(x) + f^-(x), \end{aligned} \tag{3.22}$$

and these will provide a basis for the definitions of $\int_E f(x)dx$ and $\int_E |f(x)|\, dx$.

If $f(x)$ is μ-measurable, then by Proposition 2.19 so too is $f^+(x)$, $f^-(x)$ and $|f(x)| = \max\{f^+(x), f^-(x)\}$ as a maximum of measurable functions. Alternatively, if $A^+ \equiv \{x | f(x) > 0\}$ and $A^- \equiv \{x | f(x) < 0\}$, then:

$$f^+(x) = f(x)\chi_{A^+}(x), \quad f^-(x) = -f(x)\chi_{A^-}(x),$$

and $\chi_{A^\pm}(x)$ is μ-measurable since $A^\pm \in \sigma(X)$.

Recalling the Lebesgue development of Section III.2.5, the following will be no surprise.

Definition 3.38 (μ-integral of measurable $f(x)$) *A μ-measurable, extended real-valued function $f(x)$ defined on a measure space $(X, \sigma(X), \mu)$ is said to be **μ-integrable** over a μ-measurable set E if both $f^+(x)$ and $f^-(x)$ are μ-integrable over E, and in this case we define:*

$$\int_E f(x)d\mu = \int_E f^+(x)d\mu - \int_E f^-(x)d\mu. \tag{3.23}$$

Then $|f(x)|$ is also μ-integrable over E and we define:

$$\int_E |f(x)|\, d\mu = \int_E f^+(x)d\mu + \int_E f^-(x)d\mu. \tag{3.24}$$

*If one of the functions $f^+(x)$ and $f^-(x)$ is μ-integrable and one is not, then $f(x)$ is said to be **not μ-integrable,** although it is then common to define $\int_E f(x)d\mu = \infty$ or $\int_E f(x)d\mu = -\infty$ as appropriate. If neither function is μ-integrable, then $f(x)$ is said to be **not μ-integrable,** and $\int_E f(x)dx$, which formally equals the expression $\infty - \infty$, is undefined. But in these cases, we can again say that $\int_E |f(x)|\, d\mu = \infty$.*

Remark 3.39 (On two definitions of $\int_E |f(x)|\,d\mu$) *There is no conflict between the definition of $\int_E |f(x)|\,dx$ in (3.24) and that which would be produced directly by the prior section as applied to the nonnegative function $|f(x)|$.*

For measurable $f(x)$, if $|f(x)|$ is μ-integrable by the prior section, then both $f^+(x)$ and $f^-(x)$ are μ-integrable by Proposition 3.29 since $f^+(x) \leq |f(x)|$ and $f^-(x) \leq |f(x)|$. Then since $|f(x)| = f^+(x) + f^-(x)$, (3.24) will follow by linearity of the integral in Proposition 3.29.

Conversely by contradiction, if both $f^+(x)$ and $f^-(x)$ are μ-integrable, then so too will $|f(x)|$ be μ-integrable as noted above. Thus if $|f(x)|$ is not μ-integrable by the prior section, then at least one of $f^+(x)$ and $f^-(x)$ will not be μ-integrable.

While Definition 3.38 allows $f(x)$ to be an extended real-valued functions, **μ-integrable** functions are real-valued almost everywhere.

Proposition 3.40 (μ-integrable $\Rightarrow \mu(\{f(x) = \pm\infty\}) = 0$) *If $f(x)$ is μ-integrable, then:*

$$\mu(\{f(x) = \pm\infty\}) = 0.$$

Proof. *If $\mu(E \bigcup E') > 0$, where $E \equiv \{x|f(x) = \infty\}$ and $E' \equiv \{x|f(x) = -\infty\}$, then at least one of E or E' has positive μ-measure. Hence, at least one of $f^+(x)$ and $f^-(x)$ would not be μ-integrable by item 2 of Remark 3.12, and thus $f(x)$ would not be μ-integrable.* ■

As in the Lebesgue case with $f(x) = 1/x$ and $f(0) = \infty$, the assumption that $\mu(E) = \mu(E') = 0$ does not assure μ-integrability.

Example 3.41 ($\mu(\{f(x) = \pm\infty\}) = 0 \nRightarrow \mu$-integrability) *Recall Example 3.27 with $(X, \sigma(X), \mu) = ([0,\infty), \mathcal{M}_{\mu_F}([0,\infty)), \mu_F)$, the Borel measure space on $[0,\infty)$ defined with $F(x) = x^2$. We prove that $f(x) = x^2$ is not μ-integrable on $E = [0,\infty)$ by evaluating $\int_E f(x)d\mu_F$.*

Since E is an interval we again utilize a sequence of simple functions defined as step functions. As $f(x)$ is nonnegative, if $f(x)$ is the pointwise limit of an increasing sequence of simple functions, the integral of $f(x)$ can be evaluated by Lebesgue's monotone convergence theorem.

Given n, define a partition of $E_N \equiv [0, N]$ by $\{A_i\}_{i=1}^n$ with $A_i = [N(i-1)/n, Ni/n]$, and a simple function with $\varphi_n(x) \leq f(x)$ defined on E_N by:

$$\varphi_n(x) = \sum\nolimits_{i=1}^n \left[\frac{N(i-1)}{n}\right]^2 \chi_{A_i}(x).$$

To ensure that $\{\varphi_n(x)\}_{n=1}^\infty$ is an increasing sequence we again use interval bisection, meaning $n = 2m$ for positive integers m.

As in Example 3.27:

$$\int_{E_N} \chi_{A_i}(x)d\mu_F = \frac{N^2[2i-1]}{n^2},$$

and then by (3.1):

$$\int_{E_N} \varphi_n(x)d\mu_F = N^4 \sum\nolimits_{i=1}^n \frac{2i^3 - 5i^2 + 4i - 1}{n^4}.$$

Since $\sum_{i=1}^{n} i^k = O(n^{k+1})$*:*

$$\int_{E_N} \varphi_n(x) d\mu_F = 2N^4 \sum_{i=1}^{n} \frac{i^3}{n^4} + O(\frac{1}{n}),$$

and $\sum_{i=1}^{n} i^3 = \frac{1}{4} n^2 (n+1)^2$ *obtains:*

$$\int_0^\infty x^2 d\mu_F \geq \int_{E_N} \varphi_n(x) d\mu_F = \frac{N^4}{2}.$$

Hence, x^2 *is not* μ_F*-integrable on* E*.*

3.3.1 Properties of Integrals

The following proposition summarizes the essential properties of the μ-integral. We assume that both $f(x)$ and $g(x)$ are μ-integrable functions, and not just μ-measurable, to avoid definitional problems. For example, if stated assuming only μ-measurability, then for item 2, $f(x) + g(x)$ may be μ-integrable and indeed identically 0, with both $f(x)$ and $g(x)$ not μ-integrable. Similarly, if $g(x)$ is μ-integrable then μ-measurable $f(x)$ in item 4 need not be μ-integrable. Recall Definition 3.38.

On the other hand, if $g(x)$ is μ-integrable, then μ-measurable $f(x)$ in item 3 is also μ-integrable.

The reader is encouraged to fill in these details with counterexamples or proofs.

Proposition 3.42 (Properties of integrals of integrable functions) *If* $f(x)$ *and* $g(x)$ *are defined on a measure space* $(X, \sigma(X), \mu)$ *and* μ*-integrable on* $E \in \sigma(X)$*, then:*

1. *For any* $a \in \mathbb{R}$*,* $af(x)$ *is* μ*-integrable on* E *and:*

$$\int_E af(x) d\mu = a \int_E f(x) d\mu.$$

2. *Arbitrarily defining* $f(x) + g(x)$ *on the set of* μ*-measure* 0 *for which this sum is* $\infty - \infty$ *or* $-\infty + \infty$*,* $f(x) + g(x)$ *is* μ*-integrable on* E *and:*

$$\int_E [f(x) + g(x)] d\mu = \int_E f(x) dx + \int_E g(x) d\mu.$$

3. *If* $f(x) = g(x)$ *a.e.:*

$$\int_E f(x) d\mu = \int_E g(x) d\mu.$$

4. *If* $f(x) \leq g(x)$ *a.e.:*

$$\int_E f(x) d\mu \leq \int_E g(x) d\mu.$$

5. *If* $C_1 \leq f(x) \leq C_2$ μ*-a.e., and* $\mu(E) < \infty$*:*

$$C_1 \mu(E) \leq \int_E f(x) d\mu \leq C_2 \mu(E).$$

6. *If* $E = \bigcup_{i=1}^{n} E_i$, *a finite union of disjoint* μ*-measurable sets:*

$$\int_E f(x)d\mu = \sum\nolimits_{i=1}^{n} \int_{E_i} f(x)d\mu. \tag{3.25}$$

7. *The* ***triangle inequality:***

$$\left| \int_E f(x)d\mu \right| \leq \int_E |f(x)|\, d\mu. \tag{3.26}$$

8. *If* $\int_E f(x)d\mu = 0$ *for all measurable* E*, then* $f(x) = 0$ μ*-a.e.*

9. *If* $\mu(E) = 0$*, then:*

$$\int_E f(x)d\mu = 0.$$

Proof. *Taking these properties in turn:*

1. *This follows from Proposition 3.29 and (3.23) and is left as an exercise. Hint: Express* $[af(x)]^{\pm}$ *in terms of* $af^{\pm}(x)$*, noting that this expression depends on* $sgn(a)$*, the sign if* a.

2. *By the triangle inequality,* $|f(x) + g(x)| \leq |f(x)| + |g(x)|$*, so the assumed integrability of* $f(x)$ *and* $g(x)$ *and item 5 of Proposition 3.29 imply the integrability of* $|f(x) + g(x)|$*. This assures integrability of* $f(x) + g(x)$ *and this sum's positive and negative parts,* $[f(x) + g(x)]^{\pm}$*. The subtlety in this proof is that* $[f(x)+g(x)]^{\pm}$ *need not, and in general will not, equal* $f^{\pm}(x)+g^{\pm}(x)$*. So we must investigate the implication of splitting* $f(x)+g(x)$ *two ways into positive and negative parts.*

   By definition:

$$f(x) + g(x) = [f(x) + g(x)]^{+} - [f(x) + g(x)]^{-},$$

   and then by splitting $f(x)$ *and* $g(x)$ *separately:*

$$f(x) + g(x) = [f^{+}(x) + g^{+}(x)] - [f^{-}(x) + g^{-}(x)].$$

   These representations obtain:

$$[f(x) + g(x)]^{+} + [f^{-}(x) + g^{-}(x)] = [f^{+}(x) + g^{+}(x)] + [f(x) + g(x)]^{-}.$$

   All four terms are integrable by the integrability of $f(x)$*,* $g(x)$ *and the sum, so integrating and applying linearity from Proposition 3.29 obtains:*

$$\begin{aligned} &\int_E [f(x) + g(x)]^{+} d\mu + \int_E [f^{-}(x) + g^{-}(x)]d\mu \\ = \; &\int_E [f^{+}(x) + g^{+}(x)]d\mu + \int_E [f(x) + g(x)]^{-} d\mu. \end{aligned}$$

   Then by Definition 3.38:

$$\int_E [f(x) + g(x)]d\mu = \int_E [f^{+}(x) + g^{+}(x)]d\mu - \int_E [f^{-}(x) + g^{-}(x)]d\mu,$$

   and the final result follows from another application of Proposition 3.29, then Definition 3.38.

3.- 4. Let:

$$g(x) = f(x) + [g(x) - f(x)].$$

Since $g(x) - f(x) = 0$ μ-a.e. or $g(x) - f(x) \geq 0$ μ-a.e., respectively, the results follow from linearity of the integral by items 1 and 2, and then items 4 and 5 of Proposition 3.29.

5. *Both $f(x) - C_1\chi_E(x) \geq 0$ and $C_2\chi_E(x) - f(x) \geq 0$ μ-a.e., and thus have nonnegative integrals by item 4. The final result is completed by items 1 and 2 and (3.1).*

6. *Since $\sum_{i=1}^n \chi_{E_i}(x) = \chi_E(x)$ for disjoint subsets, Proposition 3.29 obtains:*

$$\begin{aligned} \int_E f(x)dx &\equiv \int_E f^+(x)d\mu - \int_E f^-(x)d\mu \\ &= \int \sum\nolimits_{i=1}^n f^+(x)\chi_{E_i}(x)d\mu - \int \sum\nolimits_{i=1}^n f^-(x)\chi_{E_i}(x)d\mu \\ &= \sum\nolimits_{i=1}^n \int_{E_i} f^+(x)d\mu - \sum\nolimits_{i=1}^n \int_{E_i} f^-(x)d\mu \\ &= \sum\nolimits_{i=1}^n \int_{E_i} f(x)d\mu. \end{aligned}$$

7. *This result follows from items 1 and 4 and the observation that $f(x) \leq |f(x)|$ and $-f(x) \leq |f(x)|$.*

8. *Assume that $\mu\{f(x) \neq 0\} > 0$. Then $\mu(E') > 0$ for at least one of $E' = \{f(x) > 0\}$ or $E' = \{f(x) < 0\}$. We prove the result for the former set and leave the latter set as an exercise.*

   If $\mu\{f(x) > 0\} > 0$, then there exists n so that $\mu(E'_n) > 0$ for $E'_n = \{f(x) > 1/n\}$. Otherwise, since $\bigcup_{n=1}^{\infty} E'_n = E'$ and $E'_n \subset E'_{n+1}$, continuity from below of μ would obtain that $\mu(E') = 0$, a contradiction. Given such n, measurability of E'_n and item 4 obtain that:

$$\int_{E'_n} f(x)d\mu > \mu(E'_n)/n > 0,$$

   a contradiction.

9. *Applying item 8 of Proposition 3.29 to $f^{\pm}(x)$, this result follows by Definition 3.38.*

■

Recall Corollary 3.30, which stated that every nonnegative μ-integrable function can be used as a **test function** to identify other nonnegative μ-integrable functions. To generalize to the current context, we need to be a little careful. If $g(x)$ is μ-integrable and $f(x) \leq g(x)$ μ-a.e., then we cannot be assured that $f(x)$ is μ-integrable since this inequality provides no lower bound on $f(x)$. However, absolute values provide the solution.

Corollary 3.43 (Integrability comparison test) *Let $f(x)$, $g(x)$ be μ-measurable functions defined on a measure space $(X, \sigma(X), \mu)$, and $E \in \sigma(X)$.*

1. ***If** $g(x)$ **is μ-integrable on** E, and $|f(x)| \leq |g(x)|$ μ-a.e., then $f(x)$ is μ-integrable on E with:*

$$\left|\int_E f(x)d\mu\right| \leq \int_E |f(x)|\, d\mu \leq \int_E |g(x)|\, d\mu.$$

2. ***If*** $g(x)$ ***is not*** μ***-integrable on*** E, *and* $|g(x)| \leq |f(x)|$ μ*-a.e., then* $f(x)$ *is not* μ*-integrable on* E.

Proof. *The assumed* μ*-integrability of* $g(x)$ *obtains* μ*-integrability of* $|g(x)|$ *by Definition 3.38. Since* $|f(x)|$ *and* $|g(x)|$ *are* μ*-measurable and nonnegative, the second integral bound in item 1 follows from part 5 of Proposition 3.29, and thus* $|f(x)|$ *is* μ*-integrable. This implies the integrability of* $f^{\pm}(x)$ *by Definition 3.38, which in turn implies integrability of* $f(x)$*, and then the first integral bound is the triangle inequality of Proposition 3.42.*

Item 2 is a logical restatement of item 1. If $f(x)$ *and thus* $|f(x)|$ *were* μ*-integrable, then integrability of* $g(x)$ *would follow from item 1, a contradiction.* ■

3.3.2 Beppo Levi's Theorem

For nonnegative measurable functions, **Fatou's lemma** bounds the μ-integral of the limit inferior of a μ-measurable function sequence with the limit inferior of the μ-integral sequence. Adding the assumption that this sequence is increasing, **Lebesgue's monotone convergence theorem** establishes equality between the limit of the μ-integral sequence and the integral of the corresponding limit function. Neither of these results requires μ-integrability of the functions in the sequence, but when so integrable, integrability of the limit function can often be concluded. In this and the next few sections, additional results relating to "integration to the limit" are developed.

The first result, **Beppo Levi's theorem,** is closely related to Lebesgue's monotone convergence theorem and is named for **Beppo Levi** (1875–1961). In contrast to the earlier result, the nonnegativity assumption is replaced by the assumptions that the functions in the sequence are in fact μ-integrable and that the associated integral values are bounded. This is then sufficient to guarantee that the limiting function is μ-integrable and to specify the value of its integral.

As was the case for earlier results, this result and that of the next section are sometimes expressed:

$$\int_E \lim_{n\to\infty} f_n(x)d\mu = \lim_{n\to\infty}\int_E f_n(x)d\mu,$$

to emphasize the interchanging of two limiting processes.

Proposition 3.44 (Beppo Levi's theorem) *Let* $\{f_n(x)\}_{n=1}^{\infty}$ *be an increasing sequence of* μ*-integrable functions defined on a measure space* $(X, \sigma(X), \mu)$ *that converges pointwise to a function* $f(x)$ *on* $E \in \sigma(X)$.

If for all n:

$$\int_E f_n(x)d\mu \leq K < \infty,$$

then $f(x)$ *is* μ*-integrable, and:*

$$\int_E f(x)d\mu = \lim_{n\to\infty}\int_E f_n(x)d\mu. \tag{3.27}$$

Proof. *Consider the nonnegative, increasing function sequence,* $\{f_n(x) - f_1(x)\}_{n\geq 2}$ *with pointwise limit* $f(x) - f_1(x)$*. By (3.16) of Lebesgue's monotone convergence theorem and the assumption of integrability of all* $f_n(x)$:

$$\int_E [f(x) - f_1(x)]\,d\mu = \lim_{n\to\infty}\int_E [f_n(x) - f_1(x)]\,d\mu = \lim_{n\to\infty}\int_E f_n(x)d\mu - \int_E f_1(x)d\mu.$$

The sequence of integrals on the right is increasing and bounded by K*, and so* $f(x) - f_1(x)$ *is integrable.*

The integrability of $f(x)$ and the identity in (3.27) now follow by linearity and then the addition of finite $\int_E f_1(x)d\mu$. ■

3.3.3 Lebesgue's Dominated Convergence Theorem

The next result is a cornerstone limit theorem for function sequences, called **Lebesgue's dominated convergence theorem,** and named for **Henri Léon Lebesgue** (1875–1941). By "dominated convergence" is meant that the functions in the convergent sequence are dominated by an integrable function. That is, $|f_n(x)| \leq g(x)$ for some integrable $g(x)$, which then implies that $|f(x)| \leq g(x)$, and thus all $f_n(x)$ and $f(x)$ are integrable. The conclusion is then that the integral sequence also converges and indeed converges to the integral of the limit function.

As corollaries to this result, we obtain results on integrating convergent function series and partitioning domains of integration, as well as derive the **bounded convergence theorem**.

Proposition 3.45 (Lebesgue's dominated convergence theorem) *Let $\{f_n(x)\}_{n=1}^{\infty}$ be a sequence of μ-measurable functions defined on a measure space $(X, \sigma(X), \mu)$ that converges pointwise to a function $f(x)$ on $E \in \sigma(X)$, or pointwise μ-a.e. to a measurable function $f(x)$ on E.*

If there exists a function $g(x)$, μ-integrable on E, so that for all n:

$$|f_n(x)| \leq g(x),$$

then $f(x)$ is μ-integrable on E and:

$$\int_E f(x)d\mu = \lim_{n\to\infty} \int_E f_n(x)d\mu. \tag{3.28}$$

Further,

$$\lim_{n\to\infty} \int_E |f_n(x) - f(x)|\, d\mu = 0. \tag{3.29}$$

Proof. *The limit function $f(x)$ is μ-measurable by Proposition 2.19 or by assumption. Since $|f_n(x)| \leq g(x)$ implies that $|f(x)| \leq g(x)$, all $|f_n(x)|$ and $|f(x)|$ are μ-integrable by Corollary 3.43, and then all $f_n(x)$ and $f(x)$ are μ-integrable by Definition 3.38.*

Assuming that (3.29) has been demonstrated, the triangle inequality in (3.26) obtains:

$$\left| \int_E [f_n(x) - f(x)]d\mu \right| \leq \int_E |f_n(x) - f(x)|\, d\mu,$$

and thus:

$$\int_E [f_n(x) - f(x)]d\mu \to 0.$$

Since $f(x)$ is μ-integrable, (3.28) is obtained by linearity of the integral and addition of finite $\displaystyle\int_E f(x)d\mu$.

For (3.29), $|f_m(x) - f(x)| \leq 2g(x)$ obtains that $2g(x) - |f_m(x) - f(x)|$ is nonnegative for all m and has pointwise limit $2g(x)$. Applying Fatou's lemma of Proposition 3.19:

$$\begin{aligned}\int_E 2g(x)dx &\leq \liminf \int_E [2g(x) - |f_m(x) - f(x)|]dx \\ &= \int_E 2g(x)dx + \liminf \int_E [-|f_m(x) - f(x)|]dx \\ &= \int_E 2g(x)dx - \limsup \int_E |f_m(x) - f(x)|\, dx,\end{aligned}$$

where the last step follows as in the proof of Proposition 3.22. Subtracting $\int_E 2g(x)dx$, which is finite by assumption, obtains:

$$\limsup \int_E |f_m(x) - f(x)|\, dx \leq 0. \tag{1}$$

Since each integral in (1) *is nonnegative, the limit superior of this integral sequence equals 0. Since* $\liminf \leq \limsup$ *for any sequence, the limit inferior of this sequence is also 0, and thus so too is the limit by Corollary I.3.46, proving (3.29).* ■

Integration to the limit results are conventionally stated in terms of a sequence of measurable functions $\{f_n(x)\}_{n=1}^{\infty}$, but these results can also be stated in terms of a continuum of functions $\{f(x,t)\}_{t\in[a,b]}$. Left as an exercise, the following version of the above result will be used for the Leibniz integral rule below.

Proposition 3.46 (Lebesgue's dominated convergence theorem for $\{f(x,t)\}_{t\in[a,b]}$) *Let $\{f(x,t)\}_{t\in[a,b]}$ be a collection of μ-measurable functions defined on a measure space $(X, \sigma(X), \mu)$ that converges pointwise to a function $f(x)$ on $E \in \sigma(X)$ in the sense that $f(x) = \lim_{t\to b} f(x,t)$.*

If there exists a μ-integrable function $g(x)$ so that for all $t \in [a,b]$:

$$|f(x,t)| \leq g(x),$$

then $f(x)$ is μ-integrable on E and:

$$\int_E f(x)d\mu = \lim_{t\to b} \int_E f(x,t)d\mu.$$

Further,

$$\lim_{t\to b} \int_E |f(x) - f(x,t)|\, d\mu = 0.$$

Proof. *Left as an exercise. Hint: Measurability follows for $f(x)$ since equivalently,* $\lim_{t_n\to b} f(x,t_n)$ *for any sequence. For the results, consider proof by contradiction.* ■

Corollary 3.47 (Lebesgue's dominated convergence theorem, μ-a.e) *Let $\{f_n(x)\}_{n=1}^{\infty}$ be a sequence of μ-measurable functions on a complete measure space $(X, \sigma(X), \mu)$ with $f(x) \equiv \lim_{n\to\infty} f_n(x)$ μ-a.e on $E \in \sigma(X)$.*

If there exists a μ-integrable function $g(x)$ so that for all n:

$$|f_n(x)| \leq g(x), \ \mu\text{-a.e.},$$

then the conclusions of Lebesgue's dominated convergence theorem remain true.

Proof. *Define* $\widetilde{f}_n(x)$ *by:*

$$\widetilde{f}_n(x) = \begin{cases} f_n(x), & \text{if } |f_n(x)| \leq g(x), \\ 0, & \text{otherwise.} \end{cases}$$

Then $\left\{\widetilde{f}_n(x)\right\}_{n=1}^{\infty}$ *is a* μ*-measurable sequence by Proposition 2.14 since* $\widetilde{f}_n(x) = f_n(x)$ μ*-a.e. and* $(X, \sigma(X), \mu)$ *is complete. Further,* $\left|\widetilde{f}_n(x)\right| \leq g(x)$ *for all* x*, and:*

$$\lim_{n\to\infty} \widetilde{f}_n(x) = \lim_{n\to\infty} f_n(x), \ \mu\text{-a.e.} \tag{1}$$

Define $\widetilde{f}(x) \equiv \lim_{n\to\infty} \widetilde{f}_n(x)$*, which is* μ*-measurable by Proposition 2.19. By Proposition 3.45:*

$$\int_E \widetilde{f}(x)d\mu = \lim_{n\to\infty} \int_E \widetilde{f}_n(x)d\mu, \tag{2}$$

and:

$$\int_E \left|\widetilde{f}_n(x) - \widetilde{f}(x)\right| d\mu \to 0, \ \text{as } n \to \infty. \tag{3}$$

Now $f(x)$ *is* μ*-measurable by Corollary 2.20 since* $(X, \sigma(X), \mu)$ *is complete, and* $f(x) = \widetilde{f}(x)$ μ*-a.e. by* (1)*. Further,* $|f_n(x) - f(x)| = \left|\widetilde{f}_n(x) - \widetilde{f}(x)\right|$, μ*-a.e. By item 3 of Proposition 3.42,* (2) *and* (3) *obtain the results in (3.28) and (3.29).* ■

We next turn to some useful corollaries of Lebesgue's result which generalize Corollaries 3.31 and 3.32 for nonnegative functions. We state the first result in the general case of almost everywhere convergence, thereby necessitating the assumption of completeness.

Another corollary will be found in the next section.

Corollary 3.48 (μ-integral of a series) *Let* $\{h_j(x)\}_{n=1}^{\infty}$ *be a sequence of* μ*-measurable functions on a* μ*-measurable set* E *of a complete measure space* $(X, \sigma(X), \mu)$ *and assume that* $f(x) \equiv \sum_{j=1}^{\infty} h_j(x)$, μ*-a.e.*

If there exists a μ*-integrable* $g(x)$ *on* E *so that for all* n*:*

$$\left|\sum_{j=1}^{n} h_j(x)\right| \leq g(x), \ \mu\text{-a.e.}$$

then $f(x)$ *is integrable on* E *and:*

$$\int_E f(x)d\mu = \sum_{j=1}^{\infty} \int_E h_j(x)d\mu. \tag{3.30}$$

Further, as $n \to \infty$,

$$\int_E \left|\sum_{j=n}^{\infty} h_j(x)\right| d\mu \to 0.$$

Proof. *Define* $f_n(x) = \sum_{j=1}^{n} h_j(x)$*. Then* $\{f_n(x)\}_{n=1}^{\infty}$ *satisfy the conditions of Corollary 3.47. The above results are then a restatement of (3.28) and (3.29), noting that by item 2 of Proposition 3.42:*

$$\int_E f_n(x)d\mu = \sum_{j=1}^{n} \int_E h_j(x)d\mu.$$

■

Corollary 3.49 (μ-Integral over a countable union) *Let $f(x)$ be a μ-integrable function on μ-measurable set E of a measure space $(X, \sigma(X), \mu)$, where $E = \bigcup_{n=1}^{\infty} E_n$, a countable union of disjoint μ-measurable sets. Then:*

$$\int_E f(x)d\mu = \sum\nolimits_{n=1}^{\infty} \int_{E_n} f(x)d\mu. \tag{3.31}$$

Proof. *Left as an exercise. Hint:* $\chi_E(x) = \lim_{m\to\infty} \sum\nolimits_{n=1}^{m} \chi_{E_n}(x)$. ■

3.3.4 Bounded Convergence Theorem

The next integration to the limit result is the **bounded convergence theorem.** In the Lebesgue theory of Book III, this was obtained in the study of the integrals of bounded functions over domains E with $m(E) < \infty$. We skipped this step in the general theory as noted in Remark 3.13 and instead can now present a one-sentence proof with the power of Lebesgue's result.

Proposition 3.50 (Bounded convergence theorem) *Let $\{f_n(x)\}_{n=1}^{\infty}$ be a sequence of μ-measurable functions with $f(x) \equiv \lim_{n\to\infty} f_n(x)$ on a μ-measurable set E with $\mu[E] < \infty$.*

If $|f_n(x)| \leq M < \infty$ on E, then:

$$\begin{aligned} \int_E f(x)d\mu &= \lim_{n\to\infty} \int_E f_n(x)d\mu, \\ \lim_{n\to\infty} \int_E |f_n(x) - f(x)|\, d\mu &= 0. \end{aligned} \tag{3.32}$$

If $f(x) \equiv \lim_{n\to\infty} f_n(x)$ μ-a.e. and $|f_n(x)| \leq M < \infty$ μ-a.e on E, then (3.32) remains true if the measure space $(X, \sigma(X), \mu)$ is complete.

Proof. *Let $g(x) = M\chi_E(x)$ in Lebesgue's dominated convergence theorem and corollary.* ■

Remark 3.51 (Evaluating μ-integrals) *The definition of the integral $\int_E f(x)d\mu$ in (3.23), as was the case for the definition in (3.9), potentially contemplates uncountably many simple functions in the supremum calculations. Analogous to the case of nonnegative functions addressed in Remark 3.26, Lebesgue's dominated convergence theorem gives a theoretical basis for evaluating the μ-integral of a general function defined on $(X, \sigma(X), \mu)$ which is useful in proofs.*

*Specifically, if $\{f_n(x)\}_{n=1}^{\infty}$ is any sequence of μ-**measurable functions** defined on a μ-measurable set E with $|f_n(x)| < g(x)$ for some μ-integrable $g(x)$, and $f_n(x) \to f(x)$ for $x \in E$, then $f(x)$ is μ-integrable and:*

$$\int_E f(x)d\mu = \lim_{n\to\infty} \int_E f_n(x)d\mu.$$

When $(X, \sigma(X), \mu)$ is complete, the same conclusion follows if only $f_n(x) \to f(x)$ for μ-almost all $x \in E$.

In theory, the μ-integral of $f(x)$ can be evaluated using any such sequence of functions, but by using simple functions the integrals in the sequence are then easily evaluated by (3.1).

For actual numerical evaluations of integrals, this result can sometimes be replaced by applications of monotone convergence. For example, when the domain E is relatively simple like an interval and splits:

$$E = E^+ \bigcup E^-,$$

where $E^{\pm}$ *are intervals or unions of such on which* $f^{\pm}(x) > 0$, *then:*

$$\int_E f(x)d\mu = \int_{E^+} f^+(x)d\mu - \int_{E^-} f^-(x)d\mu,$$

and each of these integrals can be evaluated as in Remark 3.26.

Exercise 3.52 **($\int_0^5 f(x)d\mu_F$)** *Recalling Example 3.27, let* $(X, \sigma(X), \mu) = ([0,\infty), \mathcal{M}_{\mu_F}([0,\infty)), \mu_F)$, *the Borel measure space on* $[0,\infty)$ *defined with* $F(x) = x^2$. *Evaluate* $\int_E f(x)d\mu_F$ *for* $E = [0,5]$ *and* $f(x) = x - 3$.

3.3.5 Uniform Integrability Convergence Theorem

The final integration to the limit theorem is "new" in the sense that it does not generalize a result developed for Lebesgue integration in Book III. There is a simple reason for this, and that is, that this result is applicable in finite measure spaces such as probability spaces, and thus the Lebesgue measure space does not qualify. On such spaces, this result is often useful when the function sequence converges pointwise or pointwise μ-a.e. but does not satisfy one of the conditions required for other integration to the limit results.

To summarize, each of the prior results is applicable to a sequence of μ-measurable functions $\{f_n(x)\}_{n=1}^{\infty}$ defined on a measure space $(X, \sigma(X), \mu)$. The most general result is:

1. **Fatou's lemma:** The sequence $\{f_n(x)\}_{n=1}^{\infty}$ must be nonnegative, but then this result "only" bounds the integral of $f(x) \equiv \liminf f_n(x)$ by $\liminf \int_E f_n(x)d\mu$.

Each of the other three results states that the integral of $f(x) \equiv \lim f_n(x)$ equals $\lim \int_E f_n(x)d\mu$ but correspondingly demands more of this sequence. The additional requirements are then:

2. **Lebesgue's monotone convergence theorem:** $\{f_n(x)\}_{n=1}^{\infty}$ must be nonnegative and monotonically increasing;

3. **Lebesgue's dominated convergence theorem:** $\{f_n(x)\}_{n=1}^{\infty}$ must be absolutely bounded by an integrable function $g(x)$;

4. **Bounded convergence theorem:** $\{f_n(x)\}_{n=1}^{\infty}$ must be absolutely bounded by a constant, and $\mu[E] < \infty$.

Each of these results has a version when the above criteria are only met μ-a.e., with the additional requirement that $(X, \sigma(X), \mu)$ be complete.

The next result uses the notion of **uniform integrability** introduced in Definition IV.4.66, but here adapted to the current notation:

Definition 3.53 (Uniformly integrable sequence) *Given a finite measure space* $(X, \sigma(X), \mu)$, *a sequence of measurable functions* $\{f_n(x)\}_{n=1}^{\infty}$ *is said to be* ***uniformly integrable*** *if:*

$$\lim_{N\to\infty} \sup_n \int_{|f_n|\geq N} |f_n(x)|\, d\mu = 0. \tag{3.33}$$

Remark 3.54 (On the measure space $(X, \sigma(X), \mu)$) *Given a* ***finite measure space*** *$(X, \sigma(X), \mu)$, meaning $\mu(X) < \infty$, the definition of uniform integrability of $\{f_n(x)\}_{n=1}^{\infty}$ assures that these functions are in fact integrable, and with integrals that are uniformly bounded. To see this, choose N so that $\sup_n \int_{|f_n|\geq N} |f_n(x)|\, d\mu \leq 1$, say. Then:*

$$\int_X |f_n(x)|\, d\mu \leq \int_{|f_n|\leq N} |f_n(x)|\, d\mu + \sup_n \int_{|f_m|\geq N} |f_n(x)| \leq N\mu(X) + 1.$$

If $f_n(x) \to f(x)$ μ-a.e., this will be enough below to assure that $f(x)$ is also μ-integrable and that the integrals converge.

If X is not a finite measure space, then (3.33) does not assure that $\left\{\int_X |f_n(x)|\, d\mu\right\}$ is bounded nor that $f(x)$ is integrable if $f_n(x) \to f(x)$. For a simple example, every uniformly bounded sequence of functions, $|f_n(x)| \leq M$ say, satisfies (3.33).

For another example, let $f_n(x) = x^{-1-1/n}$ and $f(x) = x^{-1}$ on $X = [1, \infty)$ as a Lebesgue measure space. Then $\{f_n(x)\}_{n=1}^{\infty}$ satisfy (3.33) and are uniformly bounded, $f_n(x) \to f(x)$ everywhere, yet $\int_X |f_n(x)|\, dm = n$ for all n and $f(x)$ is not integrable. If $f_n(x) = 1 + 1/n$ and $f(x) = 1$, then again $\{f_n(x)\}_{n=1}^{\infty}$ satisfy (3.33) and $f_n(x) \to f(x)$ everywhere, but none of these functions is integrable.

Hence, while the notion of uniform integrability has important applications in probability spaces and more generally in finite measure spaces, this is not the case in more general measure spaces.

We present this result in the more general case of almost everywhere converge, $f_n(x) \to f(x)$ μ-a.e. In this case, the open question is μ-measurability of the limit function $f(x)$, and this is assured below by completeness of the measure space. In the case of pointwise convergence, $f_n(x) \to f(x)$, measurability of $f(x)$ is assured by Proposition 2.19, and then the conclusions of this result follow without completeness.

Proposition 3.55 (Uniform integrability convergence theorem) *Let $\{f_n(x)\}_{n=1}^{\infty}$ be a uniformly integrable sequence of functions on a complete finite measure space $(X, \sigma(X), \mu)$. If $f_n(x) \to f(x)$ μ-a.e., then $f(x)$ is μ-integrable, and:*

$$\int_X f(x) d\mu = \lim_{n\to\infty} \int_X f_n(x) d\mu. \tag{3.34}$$

Proof. *By Corollary 2.20, $f(x)$ is μ-measurable. Then by Fatou's lemma of Corollary 3.21 and Remark 3.54, there exists N so that:*

$$\int_X |f(x)|\, d\mu \leq \liminf \int_X |f_n(x)|\, d\mu \leq N\mu(X) + 1.$$

Hence, $f(x)$ is μ-integrable.

Given arbitrary $N \in \mathbb{R}$, define $f_n^{(N)}(x) = f_n(x)$ for $|f_n(x)| < N$ and $f_n^{(N)}(x) = 0$ otherwise, and similarly $f^{(N)}(x) = f(x)$ for $|f(x)| < N$ and $f^{(N)}(x) = 0$ otherwise. To ensure that $f_n^{(N)}(x) \to f^{(N)}(x)$ μ-a.e., it must be verified that $\mu\{f(x) = N\} = 0$ since it is possible that $|f_n(x_0)| < N$ for all n, yet $f_n(x_0) \to f(x_0) = N$, and then $f_n^{(N)}(x_0) \nrightarrow f^{(N)}(x_0)$. But integrability of $f(x)$ assures that $\mu\{f(x) = N\} = 0$ for all but at most countably many N by Exercise 3.56.

Thus $f_n^{(N)}(x) \to f^{(N)}(x)$ μ-a.e. with at most countably many exceptions in N. As $\mu(X) < \infty$, the bounded convergence theorem of Proposition 3.50 applies to assure that for all but these exceptional values of N:

$$\int_X f_n^{(N)}(x) d\mu \to \int_X f^{(N)}(x) d\mu. \tag{1}$$

Now:

$$\int_X f_n(x)d\mu = \int_X f_n^{(N)}(x)d\mu + \int_{|f_n|\geq N} f_n(x)d\mu,$$

and similarly:

$$\int_X f(x)d\mu = \int_X f^{(N)}(x)d\mu + \int_{|f|\geq N} f(x)d\mu.$$

Using properties of Proposition 3.42:

$$\begin{aligned}\left|\int_X f_n(x)d\mu - \int_X f(x)d\mu\right| \leq & \left|\int_X f_n^{(N)}(x)d\mu - \int_X f^{(N)}(x)d\mu\right| \\ & + \int_{|f_n|\geq N} |f_n(x)|\, d\mu + \int_{|f|\geq N} |f(x)|\, d\mu.\end{aligned}$$

By the bounded convergence result in (1), *it follows that for all unexceptional* N :

$$\begin{aligned}\limsup_n \left|\int_X f_n(x)d\mu - \int_X f(x)d\mu\right| \leq & \limsup_n \int_{|f_n|\geq N} |f_n(x)|\, d\mu + \int_{|f|\geq N} |f(x)|\, d\mu \\ \leq & \sup_n \int_{|f_n|\geq N} |f_n(x)|\, d\mu + \int_{|f|\geq N} |f(x)|\, d\mu.\end{aligned}$$

Letting $N \to \infty$ *and avoiding the countably many values of* N *noted above, it follows that the right-hand side converges to* 0 *by uniform integrability of* $\{f_n(x)\}_{n=1}^{\infty}$ *and integrability of* $f(x)$. ■

Exercise 3.56 *Prove that if* $f(x)$ *is* μ*-integrable, then* $\mu\{f(x) = N\} = 0$ *for all but at most countably many* N*. Hint: It is enough by Definition 3.37 to prove this for a nonnegative function* $f(x)$*. If* $\mu\{f(x) = N\} > 0$ *for uncountable many* $N > 0$*, consider applying Definition 3.11 to evaluate* $\int_X f(x)d\mu$*.*

3.4 Leibniz Integral Rule

The **Leibniz integral rule,** named for **Gottfried Wilhelm Leibniz** (1646–1716), identifies conditions under which functions defined parametrically by integrals:

$$F(x) = \int f(x,t)dt,$$

can be differentiated by differentiating the integrand:

$$F'(x) = \int f_x(x,t)dt,$$

where $f_x(x,t) \equiv \frac{\partial f(x,t)}{\partial x}$.

There are many versions of this result which reflect the various ways in which this integral can be defined. In representations as above, the differential dt usually implies that this integral is defined in one-dimension as Riemann or Lebesgue integration, where in the latter case, $dt \equiv dm$. However, these then have multivariate versions, as well as versions for which this integral is defined as a Lebesgue-Stieltjes integral with differential $d\mu$. The proof of all such versions requires an "integration to the limit" result to reverse the limiting operations of integral and derivative.

To prove that such $F(x)$ is differentiable (or differentiable a.e.) requires that one establish convergence (or convergence a.e.):

$$\frac{F(x+\Delta x)-F(x)}{\Delta x} \to F'(x).$$

To justify that such a result can be obtained by differentiation under the integral uses Proposition 3.42 properties of such integrals, but then requires a proof of convergence (or convergence a.e.):

$$\int \frac{f(x+\Delta x,t)-f(x,t)}{\Delta x}dt \to \int f_x(x,t)dt.$$

When $f(x,t)$ is x-differentiable, these integrands converge by definition. But convergence of the integrals is a deeper question that is often addressed by an integration to the limit result. Using the integration to the limit results of the Lebesgue and Lebesgue-Stieltjes theories, one may only need that $f(x,t)$ is x-differentiable for m-a.e. or μ-a.e t.

The Riemann results are necessarily more demanding.

3.4.1 Riemann Integrals

One version of the Leibniz integral rule was introduced in Proposition III.1.40, where the above integral is a 1-dimensional Riemann integral. While the Riemann theory has integration to the limit tools in Propositions III.1.47 and III.1.48, these results are quite demanding, requiring both t-boundedness and Riemann t-integrability of $[f(x+\Delta x,t)-f(x,t)]/\Delta x$, and uniform convergence (in t) to $f_x(x,t)$.

Instead, the Book III version required that $f(x,t)$ be continuous and continuously x-differentiable (Definition III.1.29) and that the integral be defined over a bounded domain. Then continuity strengthens to uniform continuity by Exercise III.1.14, and the proof proceeded using the mean value theorem and properties of such functions.

We restate and generalize the Book III result here, adding the proof of continuity of $F'(x)$.

Proposition 3.57 (Leibniz integral rule: Riemann I) *Assume that $a(x)$ and $b(x)$ are continuously differentiable on $[x_0,x_1]$ with $a(x)<b(x)$ for all such x and that $f(x,t)$ and the x-partial derivative $f_x(x,t)\equiv\frac{\partial f(x,t)}{\partial x}$ are continuous for $t\in[a(x),b(x)]$ and $x_0\leq x\leq x_1$.*

Then defined as a Riemann integral:

$$F(x)\equiv(\mathcal{R})\int_{a(x)}^{b(x)} f(x,t)dt, \tag{3.35}$$

is continuously differentiable on $x_0\leq x\leq x_1$, with:

$$F'(x)=(\mathcal{R})\int_{a(x)}^{b(x)} f_x(x,t)dt+f(x,b\,(x))b'(x)-f(x,a\,(x))a'(x). \tag{3.36}$$

Proof. *See Proposition III.1.40 for details on the derivation of (3.36). Since the boundary terms are continuous by assumption, continuity of $F'(x)$ follows if:*

$$F(a,b,x)\equiv\int_a^b f_x(x,t)dt,$$

is continuous in a, b, *and* x, *since then* $F(a(x), b(x), x)$ *is continuous as a composition of continuous functions.*

For this, $F(a, b, x)$ *is differentiable in* a *and* b *by Proposition III.1.33 and Remark III.1.35 and thus continuous. For continuity in* $x \in [x_0, x_1)$, *there exists* $h > 0$ *so that* $f_x(x, t)$ *is continuous on* $[x, x+h] \times [a, b]$ *and then uniformly continuous on this rectangle by Exercise III.1.14. Thus given* $\epsilon > 0$, *there exists* $0 < \Delta x < h$ *so that for all* Δt *and* $j \leq (b-a)/\Delta t$:

$$|f_x(x, t + j\Delta t) - f_x(x + \Delta x, t + j\Delta t)| < \epsilon.$$

Using Riemann sums and letting $\Delta t \to 0$ *obtains that:*

$$\left| \int_a^b f_x(x + \Delta x, t)dt - \int_a^b f_x(x, t)dt \right| < \epsilon(b-a).$$

A similar construction applies for $x \in (x_0, x_1]$ *and some* $h < 0$, *and the proof of continuity is complete.* ■

This result can be generalized in various ways.

1. **Higher Order Derivatives:** Assume that $f(x,t)$ and the nth x-partial derivative $f^{(n)}(x,t) \equiv \frac{\partial^n f(x,t)}{\partial x^n}$ are continuous for $a(x) \leq t \leq b(x)$ and $x_0 \leq x \leq x_1$.
   When $a(x) = a$ and $b(x) = b$, then for $j \leq n$:
   $$F^{(j)}(x) = (\mathcal{R}) \int_a^b f^{(j)}(x,t)dt,$$
   and is continuous as above. This follows from (3.36) iteratively, since continuity of $f^{(n)}(x,t)$ assures continuity of $f^{(j)}(x,t)$ for $j < n$.
   For general $a(x)$ and $b(x)$, (3.36) generalizes to higher derivatives in the same way, assuming that these functions are n-times continuously differentiable. However, the formula gets complicated quickly. For example, even for $j = 2$:
   $$F^{(2)}(x) = \frac{\partial}{\partial x} \int_{a(x)}^{b(x)} f^{(1)}(x,t)dt + \frac{\partial}{\partial x} \left[f(x, b(x)) b^{(1)}(x) - f(x, a(x)) a^{(1)}(x) \right].$$

2. **Multivariate x, Partial Derivatives:** When $x = (x_1, ..., x_n)$, the formula in (3.36) readily adapts to first partial derivatives, whereas higher derivatives are possible but again more complicated.

Proposition 3.58 (Leibniz integral rule: Riemann II) *Assume that* $a(x)$ *and* $b(x)$ *are continuously differentiable on* $\prod_{j=1}^n \left[x_0^{(j)}, x_1^{(j)}\right]$ *with* $a(x) < b(x)$ *for all* x *and that* $f(x,t)$ *and the* x_k*-partial derivative* $f_{x_k}(x,t) \equiv \frac{\partial f(x,t)}{\partial x_k}$ *are continuous for* $t \in [a(x), b(x)]$ *and* $x \in \prod_{j=1}^n \left[x_0^{(j)}, x_1^{(j)}\right]$.

Then defined as a Riemann integral:

$$F(x) \equiv (\mathcal{R}) \int_{a(x)}^{b(x)} f(x,t)dt,$$

is continuously differentiable on $\prod_{j=1}^n \left[x_0^{(j)}, x_1^{(j)}\right]$, *with:*

$$F_{x_k}(x) = (\mathcal{R}) \int_{a(x)}^{b(x)} f_{x_k}(x,t)dt + f(x, b(x)) b_{x_k}(x) - f(x, a(x)) a_{x_k}(x). \tag{3.37}$$

Proof. *Fixing $k = 1$ for notational simplicity and $\widehat{x} \in \prod_{j=2}^{n}\left[x_0^{(j)}, x_1^{(j)}\right]$, let $\widetilde{a}(y) = a(y, \widehat{x})$, $\widetilde{b}(y) = b(y, \widehat{x})$, and $\widetilde{f}(y,t) = f(y, \widehat{x}, t)$. Then $\widetilde{a}(y)$ and $\widetilde{b}(y)$ are continuously differentiable on $\left[x_0^{(1)}, x_1^{(1)}\right]$, and $\widetilde{f}(y,t)$ and the y-partial of $\widetilde{f}(y,t)$ are continuous for $\widetilde{a}(y) \leq t \leq \widetilde{b}(y)$. Hence:*

$$\widetilde{F}(y) \equiv (\mathcal{R})\int_{\widetilde{a}(y)}^{\widetilde{b}(y)} \widetilde{f}(y,t)dt,$$

is continuously differentiable by (3.36), and the result follows by a notational change. ■

3. **Multivariate Riemann Integral:** In the most general case, $x = (x_1, ..., x_n)$, $t = (t_1, ..., t_m)$, and $F(x)$ is defined as a Riemann integral over a fixed t-rectangle $R = \prod_{i=1}^{m}[a_i, b_i]$, or a variable rectangle $R(x) = \prod_{i=1}^{m}[a_i(x), b_i(x)]$ defined on $x \in \prod_{j=1}^{n}\left[x_0^{(j)}, x_1^{(j)}\right]$ where we assume that $a_i(x) < b_i(x)$ for all i and all x.

   Suppressing the $(\mathcal{R})$ notation.

Proposition 3.59 (Leibniz integral rule: Riemann III) *Assume that $a_i(x)$ and $b_i(x)$ are continuously differentiable on $\prod_{j=1}^{n}\left[x_0^{(j)}, x_1^{(j)}\right]$ for $1 \leq i \leq m$, with $a_i(x) < b_i(x)$ for all x and that $f(x,t)$ and the x_k-partial derivative $f_{x_k}(x,t) \equiv \frac{\partial f(x,t)}{\partial x_k}$ are continuous for $t \in R(x) \equiv \prod_{i=1}^{m}[a_i(x), b_i(x)]$ and $x \in \prod_{j=1}^{n}\left[x_0^{(j)}, x_1^{(j)}\right]$.*

Then defined as a Riemann integral in $\mathbb{R}^m$ with $m > 1$:

$$F(x) \equiv \int_{R(x)} f(x,t)dt,$$

is continuously differentiable on $\prod_{j=1}^{n}\left[x_0^{(j)}, x_1^{(j)}\right]$ with:

$$\begin{aligned} F_{x_k}(x) &= \int_{R(x)} f_{x_k}(x,t)dt + \sum\nolimits_{i=1}^{m} \frac{\partial b_i(x)}{\partial x_k}\int_{R_i(x)} f(x, (\widehat{t}, b_i(x)))d\widehat{t} \\ &\quad - \sum\nolimits_{j=1}^{m} \frac{\partial a_i(x)}{\partial x_k}\int_{R_i(x)} f(x, (\widehat{t}, a_i(x)))d\widehat{t}, \end{aligned} \tag{3.38}$$

where $R_i(x) \equiv \prod_{l \neq i}[a_l(x), b_l(x)]$, $(\widehat{t}, b_i(x)) \equiv (t_1, .., t_{i-1}, b_i(x), t_{i+1}, .., t_m)$, and similarly $(\widehat{t}, a_i(x)) \equiv (t_1, .., t_{i-1}, a_i(x), t_{i+1}, .., t_m)$.

Proof. *Here and below, all integrals are Riemann.*

Assume that we have proved (3.38) for constant $R(u,v) = \prod_{i=1}^{m}[u_i, v_i]$ where $a_i(x) \leq u_i < v_i \leq b_i(x)$. That is, assume that:

$$F(x) \equiv \int_{R(u,v)} f(x,t)dt$$

is proved differentiable for $x \in \prod_{j=1}^{n}\left[x_0^{(j)}, x_1^{(j)}\right]$ and that for all k:

$$\frac{\partial F(x)}{\partial x_k} = \int_{R(u,v)} f_{x_k}(x,t)dt. \tag{1}$$

We claim that:

$$F(x, u_1, ..., u_m, v_1, ..., v_m) \equiv \int_{R(u,v)} f(x,t)dt,$$

is also differentiable in u_j *and* v_j.

Fixing $x \in \prod_{j=1}^{n}\left[x_0^{(j)}, x_1^{(j)}\right]$, *Proposition III.1.76 and the comments that followed obtain that for any* i:

$$F(x, u_1, ..., u_m, v_1, ..., v_m) = \int_{u_i}^{v_i} g(x, t_i)dt_i,$$

where $g(x, t_i)$ *is continuous on* $[u_i, v_i]$ *and defined by:*

$$g(x, t_i) = \int_{R_i(u,v)} f(x, (\widehat{t}, t_i))d\widehat{t}. \tag{2}$$

Here, $\widehat{t} = (t_1, ., t_{i-1}, t_{i+1}, .., t_m)$ *and* $R_i(u, v) = \prod_{l \neq i}[u_l, v_l]$. *Thus* $F(x, u_1, ..., v_m)$ *is differentiable in* u_i *and* v_i *by Proposition III.1.33 and Remark III.1.35, with:*

$$\begin{aligned} \tfrac{\partial F(x,u_1,...,v_m)}{\partial v_i} &= g(x, v_i), \\ \tfrac{\partial F(x,u_1,...,v_m)}{\partial u_i} &= -g(x, u_i). \end{aligned} \tag{3}$$

Since all $a_i(x)$ *and* $b_i(x)$ *are differentiable,* $F(x) \equiv F(x, a_1(x), ...b_m(x))$ *is differentiable in* x *by the chain rule, with:*

$$\begin{aligned} F_{x_k}(x) = & \left.\frac{\partial F(x, u_1, ..., v_m)}{\partial x_k}\right|_{u=a(x), v=b(x)} \\ & + \sum_{i=1}^{m} \left.\frac{\partial F(x, u_1, ..., v_m)}{\partial v_i}\right|_{u=a(x), v=b(x)} \frac{\partial b_i(x)}{\partial x_k} \\ & + \sum_{i=1}^{m} \left.\frac{\partial F(x, u_1, ..., v_m)}{\partial u_i}\right|_{u=a(x), v=b(x)} \frac{\partial a_i(x)}{\partial x_k}, \end{aligned}$$

where we simplify notation so that $u = a(x), v = b(x)$ *means that* $u_i = a_i(x)$ *and* $v_i = b_i(x)$ *for all* i. *Then with (1)–(3) this obtains (3.36).*

To prove the differentiability result of (1) *let* $F(x) \equiv \int_{R(u,v)} f(x, t)dt$, *where we assume that* $f(x, t)$ *and* $f_{x_k}(x, t)$ *are continuous in* $t \in R(u, v) \equiv \prod_{i=1}^{m}[u_i, v_i]$ *and* $x \in \prod_{j=1}^{n}\left[x_0^{(j)}, x_1^{(j)}\right]$. *For notational simplicity, let* $h_k \equiv (0, .., 0, h, 0, .., 0)$ *with* h *in the* k*th coordinate. Then* (1) *is proved if for* $x \in \prod_{j=1}^{n}\left[x_0^{(j)}, x_1^{(j)}\right]$:

$$\lim_{h \to 0}\left[\frac{F(x + h_k) - F(x)}{h} - \int_{R(u,v)} f_{x_k}(x, t)dt\right] = 0. \tag{4}$$

For $x = x_0^{(k)}$, *we restrict* $h > 0$, *and for* $x = x_1^{(k)}$, *we restrict* $h < 0$.

Now:

$$F(x + h_k) - F(x) = \int_{R(u,v)} [f(x + h_k, t) - f(x, t)]\, dt,$$

while by continuity of $f(x, t)$ *and the mean value theorem, there exists* $0 < \theta < 1$ *with* $\theta \equiv \theta(x, h, t)$ *so that:*

$$f(x + h_k, t) - f(x, t) = h f_{x_k}(x + \theta h_k, t).$$

Rewriting (4), *it must be proved that:*

$$\lim_{h \to 0} \int_{R(u,v)} [f_{x_k}(x + \theta h_k, t) - f_{x_k}(x, t)]\, dt = 0. \tag{5}$$

Since $f_{x_k}(x,t)$ *is continuous in* $\prod_{j=1}^n \left[x_0^{(j)}, x_1^{(j)}\right] \times \prod_{i=1}^m [u_i, v_i]$, *it is uniformly continuous by Exercise III.1.14. Thus given* $\epsilon > 0$, *there exists* δ *so that* $|f_{x_k}(x+\theta h_k, t) - f_x(x,t)| < \epsilon$ *if* $|(x+\theta h_k, t) - (x,t)| = \theta |h| < \delta$. *Taking* $|h| < \delta$ *accomplishes this, and thus by the triangle inequality of Proposition III.1.72, if* $|h| < \delta$:

$$\begin{aligned} \left| \int_{R(u,v)} [f_{x_k}(x+\theta h_k, t) - f_{x_k}(x,t)]\, dt \right| &\leq \int_{R(u,v)} |f_x(x+\theta h, t) - f_x(x,t)|\, dt \\ &< \epsilon m(R(u,v)), \end{aligned} \tag{6}$$

with $m(R(u,v))$ *the Lebesgue measure of* $R(u,v)$. *With* $X \equiv \prod_{j=1}^n \left[x_0^{(j)}, x_1^{(j)}\right]$:

$$m(R(u,v)) = \prod_{i=1}^m [v_i - u_i] \leq \max_{x \in X} \prod_{j=1}^m [b_i(x) - a_i(x)] \leq K, \tag{7}$$

where this is bounded by continuity of all $a_i(x)$ *and* $b_i(x)$ *and compactness of* X *(Exercise III.1.13). This and* (6) *prove* (5) *since* ϵ *is arbitrary,*

For continuity of $F_{x_k}(x)$ *in (3.38), it is enough to prove as in Proposition 3.57 that if* $g(x,t)$ *is continuous on* $\prod_{j=1}^n \left[x_0^{(j)}, x_1^{(j)}\right] \times R(u,v)$ *with* $R(u,v) \equiv \prod_{i=1}^{m'} [u_i, v_i]$ *and* $m' \leq m$, *then:*

$$F(x, u_1, ..., u_{m'}, v_1, ..., v_{m'}) \equiv \int_{R(u,v)} g(x,t)dt,$$

is continuous in all variables. Continuity in u, v *follows from differentiability, which was proved in the first paragraphs of this proof.*

Continuity in x *follows the template of Proposition 3.57, though more complex notationally. If* x, $x+h \in \prod_{j=1}^n [x_0^{(j)}, x_1^{(j)}]$ *with* $h \equiv (h_1, ..., h_n)$, *then* $g(x,t)$ *is uniformly continuous on* $\prod_{j=1}^n [x_j, x_j + h_j] \times R(u,v)$, *where* $[x_j, x_j+h_j] \equiv [x_j+h_j, x_j]$ *for* $h_j < 0$. *Thus given* $\epsilon > 0$, *there exists* $0 < \Delta x_j < h_j$ *(or* $h_j < \Delta x_j < 0$*) so that for all* Δt_i *and* $k_i \leq (v_i - u_i)/\Delta t_i$:

$$|g(x+\Delta x, t+k\Delta t) - g(x, t+k\Delta t)| < \epsilon,$$

where $\Delta x \equiv (\Delta x_1, ..., \Delta x_n)$ *and* $k\Delta t \equiv (k_1 \Delta t_1, ..., k_{m'} \Delta t_{m'})$. *Using Riemann sums and letting* $\Delta t_i \to 0$ *obtains:*

$$\left| \int_{R(u,v)} g(x+\Delta x, t)dt - \int_{R(u,v)} g(x,t)dt \right| < \epsilon m(R(u,v)),$$

where $m(R(u,v))$ *is bounded as in* (7). ■

3.4.2 Lebesgue/Lebesgue-Stieltjes Integrals

In this section, we develop the measure-theoretic versions of the Leibniz integral rule. By measure-theoretic is meant that the integral is defined on $(X, \sigma(X), \mu)$ in the Lebesgue-Stieltjes sense of Section 3.3, which includes the special case of the Lebesgue integral of Chapter I.2 with $(X, \sigma(X), \mu) = (\mathbb{R}^p, \mathcal{M}_L(\mathbb{R}^p), m)$. In contrast to the Riemann developments above, which rely on the tools of calculus and in particular properties of uniformly continuous functions, the developments below will use the integration to the limit results of these integration theories.

Also, these integrals will be defined over fixed measurable domains $E \in \sigma(X)$ rather than over domains that are parametrized by the independent variable x, such as with the rectangle $R(x) \equiv \prod_{i=1}^{m}[a_i(x), b_i(x)]$. One corollary of this approach is that the Leibniz integral rule is then equally true for "unbounded" domains, meaning where $\mu(E) = \infty$, a result not possible in the Riemann theory, at least not possible using the above proofs and uniform continuity.

We prove two versions of this result. The first is true for general $(X, \sigma(X), \mu)$ but requires assumptions to be made "everywhere." The second result allows the familiar "almost everywhere" assumptions and applies to complete measure spaces, including $(\mathbb{R}^p, \mathcal{M}_L(\mathbb{R}^p), m)$.

Proposition 3.60 (Leibniz integral rule: Lebesgue-Stieltjes) *Given the measure space $(X, \sigma(X), \mu)$, let $E \in \sigma(X)$ and $G \subset \mathbb{R}^n$ an open set, and let $f(x, y)$ and $f_{y_k}(x, y) \equiv \frac{\partial f(x, y_1, ..., y_n)}{\partial y_k}$ be defined on $E \times G$. Assume that for all $y \in G$ that:*

1. *$f(x, y)$ is μ-integrable over E;*
2. *$|f_{y_k}(x, y)| \leq g(x)$ where $g(x)$ is μ-integrable over E.*

Then defined on G:

$$F(y) \equiv \int_E f(x, y) d\mu,$$

is differentiable with respect to y_k with:

$$F_{y_k}(y) = \int_E f_{y_k}(x, y) d\mu. \tag{3.39}$$

Proof. *Given $y \in G$, wherever $F_{y_k}(y)$ exists Proposition 3.42 obtains:*

$$\begin{aligned} F_{y_k}(y) &\equiv \lim_{h \to 0} \frac{F(y + h_k) - F(y)}{h} \qquad (1) \\ &= \lim_{h \to 0} \int_E \frac{f(x, y + h_k) - f(x, y)}{h} d\mu, \end{aligned}$$

where $h_k = (0, .., h, 0, ...0)$ with h in the kth coordinate. Since G is open, there exists $\epsilon > 0$ so that $y + h_k \in G$ for $h < \epsilon$, and thus the integral in (1) is well-defined for such h.

By the mean value theorem:

$$\frac{f(x, y + h_k) - f(x, y)}{h} = f_{y_k}(x, y + \theta(x, y) h_k), \tag{2}$$

where $0 < \theta(x, y) < 1$. By definition in (2), $f_{y_k}(x, y + \theta(x, y) h_k)$ is μ-measurable for $0 < h < \epsilon$, and since $y + \theta(x, y) h_k \in G$:

$$|f_{y_k}(x, y + \theta(x, y) h_k)| \leq g(x). \tag{3}$$

Further:

$$\lim_{h \to 0} f_{y_k}(x, y + \theta(x, y) h_k) = f_{y_k}(x, y). \tag{4}$$

Hence, by (3) and (4), $f_{y_k}(x, y)$ is μ-integrable by Lebesgue's dominated convergence theorem of Proposition 3.46, and by (1) and (2):

$$\int_E f_{y_k}(x, y) d\mu = \lim_{h \to 0} \int_E f_{y_k}(x, y + \theta(x, y) h_k) d\mu = F_{y_k}(y).$$

■

When the measure space $(X,\sigma(X),\mu)$ is complete, for example, when $(X,\sigma(X),\mu) = (\mathbb{R}^p,\mathcal{M}_L(\mathbb{R}^p),m)$, it is enough for the requirements of Proposition 3.60 to be valid μ-a.e.

Proposition 3.61 (Leibniz integral rule: Lebesgue and complete Lebesgue-Stieltjes) *Given $(X,\sigma(X),\mu) = (\mathbb{R}^p,\mathcal{M}_L(\mathbb{R}^p),m)$, or more generally the complete measure space $(X,\sigma(X),\mu)$, let $E\in\sigma(X)$ and $G\subset\mathbb{R}^n$ an open set. Let $f(x,y)$ be defined on $E\times G$ and assume that for all $y\in G$ that:*

1. *$f(x,y)$ is μ-integrable over E;*
2. *$f_{y_k}(x,y)\equiv\frac{\partial f(x,y_1,\ldots,y_n)}{\partial y_k}$ exists μ-a.e.;*
3. *$|f_{y_k}(x,y)|\leq g(x)$ μ-a.e., where $g(x)$ is μ-integrable over E.*

Then defined on G:

$$F(y)\equiv\int_E f(x,y)d\mu,$$

is differentiable with respect to y_k μ-a.e., with:

$$F_{y_k}(y)=\int_E f_{y_k}(x,y)d\mu,\ \mu\text{-a.e.} \tag{3.40}$$

Proof. *Referencing the above proof, (1) remains valid by Proposition 3.42, and (2) is now valid μ-a.e. Similarly, (3) and (4) are valid μ-a.e. Hence, the result follows from Lebesgue's dominated convergence theorem of Corollary 3.47.* ■

3.5 Lebesgue-Stieltjes vs. Riemann-Stieltjes Integrals

Important special cases of the general integrals of this chapter are provided by the Borel measure space $(\mathbb{R},\mathcal{M}_{\mu_F}(\mathbb{R}),\mu_F)$, and the n-dimensional counterpart $(\mathbb{R}^n,\mathcal{M}_{\mu_F}(\mathbb{R}^n),\mu_F)$, summarized in Sections 1.1.3 and 1.3.2. These Borel measure spaces are often referred to as **Lebesgue-Stieltjes measure spaces** and μ_F **the Lebesgue-Stieltjes measure induced by the function** F. These are named for **Henri Lebesgue** (1875–1941) and **Thomas Stieltjes** (1856–1894), and integration on these measure spaces is then referred to as **Lebesgue-Stieltjes integration.** The associated **Lebesgue-Stieltjes integrals** are also called **Radon integrals,** after **Johann Radon** (1887–1956).

When $n=1$, the function $F(x)$ underlying μ_F was of necessity (Proposition I.5.7) increasing and right continuous, while in the general case (Propositions I.8.10 and I.8.12), such functions are of necessity continuous from above and n-increasing (Definition 1.24).

In this section, we prove that for continuous integrands and such $F(x)$, that the Lebesgue-Stieltjes integrals developed in this chapter agree with the Riemann-Stieltjes integrals of Chapter III.4.

3.5.1 Lebesgue-Stieltjes Integrals on $\mathbb{R}$

For a continuous and thus (Proposition I.3.13) Borel measurable function $g(x)$, the Lebesgue-Stieltjes integral:

$$\int_a^b g(x)d\mu_F,$$

can be approximated by the Riemann-Stieltjes sums of Definition III.4.3. It will then follow that this integral is closely related to the Riemann-Stieltjes integral of Section III.4.2:

$$\int_a^b g(x)dF.$$

Let $\Delta_n = \{x_i\}_{i=0}^n$ be a **partition** of $[a,b]$ with $a = x_0 < x_1 \cdots < x_n = b$, and define the simple function $g_n(x)$ on $[a,b]$ by:

$$g_n(x) = \sum\nolimits_{i=0}^{n-1} g(x_i')\chi_{(x_i,x_{i+1}]}(x) + g(a)\chi_{\{a\}}(x).$$

As in Definition 2.26, $\chi_A(x)$ denotes the **characteristic function** of A, defined to equal 1 on this set and 0 elsewhere, while $x_i' \in [x_i, x_{i+1}]$ is an arbitrary **interval tag**.

As $g_n(x)$ is a simple function, we have by (3.1):

$$\int_a^b g_n(x)d\mu_F = \sum\nolimits_{i=0}^{n-1} g(x_i')\left[F(x_{i+1}) - F(x_i)\right] + g(a)\left[F(a) - F(a^-)\right]. \tag{3.41}$$

Here, $\mu_F((x_i, x_{i+1}]) = F(x_{i+1}) - F(x_i)$ by (1.10), and $\mu_F(\{a\}) = F(a) - F(a^-)$ by I.(5.15) with $F(a^-)$ the left limit at a.

The summation in (3.41) is seen to be the Riemann-Stieltjes summation of Definition III.4.3.

Proposition 3.62 (Lebesgue-Stieltjes integrals by Riemann-Stieltjes sums) *Let $F(x)$ be an increasing, right continuous function and $g(x)$ a continuous function on $[a,b]$. Then for any sequence $\Delta_n = \{x_i\}_{i=0}^n$ of partitions of $[a,b]$ with **mesh size** $\delta_n \equiv \max_{1\leq i\leq n}\{x_i - x_{i-1}\} \to 0$, and arbitrary interval tags $\{x_i'\}_{i=0}^{n-1}$ with $x_i' \in [x_i, x_{i+1}]$:*

$$\sum\nolimits_{i=0}^{n-1} g(x_i')\left[F(x_{i+1}) - F(x_i)\right] \to \int_a^b g d\mu_F - g(a)\left[F(a) - F(a^-)\right]. \tag{3.42}$$

Proof. *By (3.41):*

$$\sum\nolimits_{i=0}^{n-1} g(x_i')\left[F(x_{i+1}) - F(x_i)\right] = \int_a^b g_n d\mu_F - g(a)\left[F(a) - F(a^-)\right],$$

and thus we must prove that as $\delta_n \to 0$:

$$\int_a^b g_n(x)d\mu_F \to \int_a^b g(x)d\mu_F, \tag{1}$$

for all selections of subinterval tags.

First, if $\int_a^b g_n(x)d\mu_F$ converges to a limit as $n \to \infty$ for some set of subinterval tags $\{x_i'\}_{i=0}^{n-1}$, this limit must be unique and independent of this choice. Since $g(x)$ is uniformly continuous on $[a,b]$ by Exercise III.1.14, if $\{x_i''\}_{i=0}^{n-1}$ is another collection of tags:

$$\left|\sum\nolimits_{i=0}^{n-1} g(x_i')\left[F(x_{i+1}) - F(x_i)\right] - \sum\nolimits_{i=0}^{n-1} g(x_i'')\left[F(x_{i+1}) - F(x_i)\right]\right|$$
$$\leq \sup\left|g(x_i') - g(x_i'')\right| \sum\nolimits_{i=0}^{n-1} \left|F(x_{i+1}) - F(x_i)\right|.$$

As F is increasing, this summation equals $F(b) - F(a)$, and so this difference converges to 0 as $\delta_n \to 0$ by uniform continuity of $g(x)$.

For the existence of the limit in (1), note that $g_n(x) \to g(x)$ pointwise for any choice of tags if $\delta_n \to 0$ as $n \to \infty$. This follows since for all $x \in [a,b]$:

$$|g(x) - g_n(x)| \leq \sup_{J_i} |g(x) - g(x')|,$$

where x, $x' \in J_i \equiv [x_i, x_{i+1}]$ is a subinterval of maximum length δ_n. Since $g(x)$ is uniformly continuous on $[a,b]$, $\sup_{J_i} |g(x) - g(x')| \to 0$ and this obtains that $g_n(x) \to g(x)$ for all x as $\delta_n \to 0$.

By continuity of $g(x)$ and Exercise III.1.13, there exists $K < \infty$ so that $|g(x)| \leq K$ on $[a,b]$. Thus by Proposition 3.29:

$$\int_a^b |g(x)|\, d\mu_F \leq K\, [F(b) - F(a)],$$

and $g(x)$ is μ_F-integrable.

Now define $x'_i \in J_i \equiv [x_i, x_{i+1}]$ so that $g(x'_i) = \inf_{x \in J_i} g(x)$. That x'_i exists follows by continuity of $g(x)$ and Exercise III.1.13. Define $g_n(x)$ by:

$$g_n(x) = \sum\nolimits_{i=0}^{n-1} g(x'_i)\chi_{(x_i, x_{i+1}]}(x) + g(a)\chi_{\{a\}}(x).$$

Then $g_n(x) \leq g(x)$ for all x and $g_n(x) \to g(x)$ pointwise as $\delta_n \to 0$. Hence, by Lebesgue's dominated convergence theorem:

$$\int_a^b g_n(x) d\mu_F \to \int_a^b g(x) d\mu_F,$$

and the result follows by (3.41). ■

To induce a Borel measure μ_F on $\mathbb{R}$, it is necessary and sufficient for the function $F(x)$ to be increasing and right continuous by Propositions I.5.7 and I.5.23. For the **Riemann-Stieltjes integrals** of Section III.4.2 of $g(x)$ over $[a,b]$, denoted $\int_a^b g(x)dF$, one existence result of Proposition III.4.17 stated that if the integrator $F(x)$ is increasing and $g(x)$ is continuous, then this integral is well-defined.

The next result addresses the equivalence of Lebesgue-Stieltjes and Riemann-Stieltjes integrals and generalizes the equality of Riemann and Lebesgue integrals noted in Propositions III.2.18 and III.2.56. These earlier results were quite general, while for the current context we focus on a more specific result.

Proposition 3.63 (Lebesgue-Stieltjes vs. Riemann-Stieltjes integrals) *If $F(x)$ is increasing and right continuous, and $g(x)$ is continuous on $[a,b]$, then the associated Riemann-Stieltjes and Lebesgue-Stieltjes integrals over $[a,b]$ are related by:*

$$\int_a^b g(x) d\mu_F = \int_a^b g(x) dF + g(a)\left[F(a) - F(a^-)\right]. \tag{3.43}$$

Thus if $F(x)$ is continuous at a, these integrals agree.

Proof. *By Propositions III.4.17 and III.4.12, the Riemann-Stieltjes summation of (3.42) converges to the integral $\int_a^b g(x)dF$, and the result follows by Proposition 3.62.* ■

It may well have occurred to the reader that the slight asymmetry in the conclusion in (3.43) is because in a Riemann-Stieltjes integral, the F-measure of an interval $[a,b]$ is defined by (1.10) as $F(b) - F(a)$, but for the Lebesgue-Stieltjes integral, this is the μ_F-measure of the interval $(a,b]$. Thus this asymmetry is eliminated by defining the Lebesgue-Stieltjes integral over $(a,b]$.

Exercise 3.64 (Lebesgue-Stieltjes vs. Riemann-Stieltjes integrals) *Prove that if $F(x)$ is increasing and right continuous, and $g(x)$ is continuous on $[a,b]$, then the associated Riemann-Stieltjes and Lebesgue-Stieltjes integrals are related by:*

$$\int_{(a,b]} g d\mu_F = \int_a^b g dF. \tag{3.44}$$

Hint: Redefine the approximating simple function sequence.

3.5.2 Lebesgue-Stieltjes Integrals on $\mathbb{R}^n$

This section will largely follow the template for $n = 1$ with appropriate generalizations of notation and references to earlier results. However, the asymmetry noted in (3.43) becomes increasingly complicated in higher dimensions, so we will first follow the insight of Exercise 3.64, and then discuss the general result.

In more detail, for n-increasing and continuous from above $F(x)$, we first compare the Riemann-Stieltjes integral:

$$\int_{\bar{R}} g(x)dF,$$

where $\bar{R} = \prod_{j=1}^n [a_j, b_j]$ is a bounded, closed rectangle in $\mathbb{R}^n$ and $g(x)$ is continuous on $\bar{R}$, to the Lebesgue-Stieltjes integral:

$$\int_R g(x)d\mu_F,$$

with $R = \prod_{j=1}^n (a_j, b_j]$. We will then discuss the needed adjustments to this result when the latter integral is defined over $\bar{R}$.

To partition R into subrectangles, recall the notation from Section III.4.4. Let each interval $[a_j, b_j]$ be partitioned:

$$a_j = x_{j,0} < x_{j,1} < \cdots < x_{j,m_j-1} < x_{j,m_j} = b_j,$$

with the **mesh size** δ of this collection of partitions defined by:

$$\delta \equiv \max\{x_{j,i} - x_{j,i-1}\}.$$

The maximum here is defined over $1 \leq i \leq m_j$ and $1 \leq j \leq n$.

These interval partitions lead to a partition of R into disjoint, right semi-closed subrectangles:

$$\bigcup\nolimits_{J\in I} R_J = R,$$

where $R_J \equiv \prod_{j=1}^n R_{j,i_j}$ with $R_{j,i_j} \equiv (x_{j,i_j-1}, x_{j,i_j}]$. The index set $I = \{(i_1, i_2, ..., i_n) | 1 \leq i_j \leq m_j\}$ identifies the $\prod_{j=1}^n m_j$ disjoint subrectangles that are defined by this partition.

To approximate $g(x)$ on R by simple functions, we assume that $m_j = m$ for all j to simplify notation and define the functions $g_m(x)$ on R by:

$$g_m(x) = \sum\nolimits_{J\in I} g(x'_J)\chi_{R_J}(x).$$

The subrectangles $\{R_J\}_{J\in I}$ are defined as above with all $m_j = m$, and $x'_J \in \bar{R}_J \equiv \prod_{j=1}^n [x_{j,i_j-1}, x_{j,i_j}]$, the closure of R_J.

As $g_m(x)$ is a simple function, we have by (3.1):

$$\int_R g_m(x)d\mu_F = \sum\nolimits_{J\in I} g(x'_J)\mu_F[R_J].$$

The expression on the right is a Riemann-Stieltjes summation of Definition III.4.80 since by (1.21):

$$\mu_F\left[R_J\right] = \Delta F_J.$$

Here ΔF_J denotes the F-content of $\bar{R}_J$ of Definition III.4.63, defined as the summation in (1.21). Thus:

$$\int_R g_m(x)d\mu_F = \sum\nolimits_{J\in I} g(x'_J)\Delta F_J. \tag{3.45}$$

Proposition 3.65 (Lebesgue-Stieltjes integrals by Riemann-Stieltjes sums) *Let $F(x)$ be a continuous from above and n-increasing function on $\mathbb{R}^n$, and $g(x)$ a continuous function on the closure $\bar{R} = \prod_{j=1}^n [a_j, b_j]$ of the right semi-closed rectangle $R = \prod_{j=1}^n (a_j, b_j]$. Then for any sequence of partitions of R:*

$$R = \bigcup\nolimits_{J\in I} R_J,$$

with $\delta \equiv \max\{x_{j,i} - x_{j,i-1}\} \to 0$ as $m \to \infty$, and arbitrary $\{x'_J\}_{J\in I}$ with $x'_J \in \bar{R}_J$:

$$\sum\nolimits_{J\in I} g(x'_J)\Delta F_J \to \int_R g(x)d\mu_F. \tag{3.46}$$

Proof. *If $\int_R g_m(x)d\mu_F$ in (3.45) converges to a limit as $m \to \infty$, or equivalently as $\delta \to 0$, for any collections of tags $\{x'_J\}_{J\in I}$, this limit must be unique. If $\{x''_J\}_{J\in I}$ is another selection, then by uniform continuity of $g(x)$ by Exercise III.1.14, and that $\Delta F_J \geq 0$ for all J by definition of n-increasing:*

$$\left|\sum\nolimits_{J\in I} g(x'_J)\Delta F_J - \sum\nolimits_{J\in I} g(x''_J)\Delta F_J\right| \leq \sup\left|g(x'_J) - g(x''_J)\right| \sum\nolimits_{J\in I} \Delta F_J.$$

By Proposition III.4.67:

$$\sum\nolimits_{J\in I} \Delta F_J = \Delta F, \tag{3.47}$$

where ΔF denotes the F-content of $\bar{R}$, and thus this difference converges to 0 as $\delta \to 0$ by uniform continuity of $g(x)$.

For the existence of this limit, first note that for $x \in R$:

$$|g(x) - g_m(x)| \leq \sup_{x,x'\in \bar{R}_J} |g(x) - g(x')|.$$

Since $g(x)$ is uniformly continuous on $\bar{R}$ by Exercise III.1.14, it follows that $g_m(x) \to g(x)$ as $\delta \to 0$ for any choice of tags.

By continuity of $g(x)$ and Exercise III.1.13, $|g(x)| \leq K < \infty$ on $\bar{R}$. Thus for n-increasing $F(x)$, Proposition 3.29 obtains:

$$\int_R |g(x)|\, d\mu_F \leq K \int_R d\mu_F = K\mu_F(R),$$

so $g(x)$ is μ_F-integrable.

Now define $x'_J \in \bar{R}_J$ so that $g(x'_J) = \inf_{\bar{R}_J} g(x)$, noting that x'_J exists by continuity of $g(x)$ and Exercise III.1.13. Define $g_m(x)$ by:

$$g_m(x) = \sum\nolimits_{J\in I} g(x'_J)\chi_{R_J}(x).$$

Then $g_m(x) \leq g(x)$ for all x, and $g_m(x) \to g(x)$ pointwise as $m \to \infty$.

By Lebesgue's dominated convergence theorem:

$$\int_R g_m(x)d\mu_F \to \int_R g(x)d\mu_F,$$

and the result follows by (3.45). ■

To induce a Borel measure μ_F on $\mathbb{R}^n$, it is sufficient by Proposition I.8.16, and necessary by Propositions I.8.10 and I.8.12, for the function $F(x)$ to be n-increasing and continuous from above. For the **Riemann-Stieltjes integral** of Section III.4.4 of $g(x)$ over $\bar{R} = \prod_{j=1}^{n}[a_j, b_j]$, denoted $\int_{\bar{R}} g(x)dF$, one existence result of Corollary III.4.90 provides that if the integrator $F(x)$ is n-increasing and $g(x)$ is continuous, then this integral is well-defined.

Proposition 3.66 (Lebesgue-Stieltjes vs. Riemann-Stieltjes integrals) *If $F(x)$ is n-increasing and continuous from above and $g(x)$ is continuous on $\bar{R} = \prod_{j=1}^{n}[a_j, b_j]$, the closure of $R = \prod_{j=1}^{n}(a_j, b_j]$, then the associated Riemann-Stieltjes and Lebesgue-Stieltjes integrals are related by:*

$$\int_{R} g(x)d\mu_F = \int_{\bar{R}} g(x)dF. \tag{3.48}$$

Proof. *By Corollary III.4.90 and Proposition III.4.88, the Riemann-Stieltjes summation of (3.46) converges to the integral $\int_{\bar{R}} g(x)dF$, and the result follows by Proposition 3.65.* ■

It is natural to wonder how the result of Proposition 3.63 generalizes to $\mathbb{R}^n$, where the Lebesgue-Stieltjes integral is defined over the closed set $\bar{R}$. To investigate this requires a more general simple function sequence defined over a partition of $\bar{R}$. To this end, again assuming that $m_j = m$ for all j:

$$\begin{aligned}\bar{R} &= \prod\nolimits_{j=1}^{n}[a_j, b_j] \\ &= \prod\nolimits_{j=1}^{n}\left[\{a_j\} \bigcup \bigcup\nolimits_{i_j=1}^{m} R_{j,i_j}\right],\end{aligned}$$

where $R_{j,i_j} \equiv (x_{j,i_j-1}, x_{j,i_j}]$ as above. This can be expanded as a union of $(m+1)^n$ rectangles, but we require more notation.

With $R_J \equiv \prod_{j=1}^{n} R_{j,i_j}$ as above, let $R_J(a_k) \equiv \prod_{j=1}^{k-1} R_{j,i_j} \times \{a_k\} \times \prod_{j=k+1}^{n} R_{j,i_j}$ and note that there are nm^{n-1} such sets. Similarly, defining $R_J(a_k, a_l)$ with the convention that $k < l$, there are then $\binom{n}{2}m^{n-2}$ such sets. Continuing in this way, $R_J(a_1, ..., a_n) = \prod_{j=1}^{n}\{a_j\}$ is independent of $J \in I$.

Each derived R_J-set is μ_F-measurable. For example:

$$R_J(a_k) = \bigcap\nolimits_{r=1}^{\infty}\left[\prod\nolimits_{j=1}^{k-1} R_{j,i_j} \times \left(a_k - \frac{1}{r}, a_k\right] \times \prod\nolimits_{j=k+1}^{n} R_{j,i_j}\right],$$

with similar expressions for other boundary partition sets. Further, as Borel measures are finite on compact sets by Definition 1.8, and each $R_J()$-subset is contained in a compact set, each such set has finite measure by monotonicity of μ_F.

Thus continuity from above of Proposition 1.29 applies, and with (1.21):

$$\mu_F\left[R_J(a_k)\right] = \lim_{r\to\infty} \sum\nolimits_{x_r} sgn(x_r)F(x_r). \tag{3.49}$$

Here, each $x_r = (x_{r,1}, ..., x_{r,n})$ in the summation is one of the 2^n vertices of the rectangle defined with r in the above intersection, and $sgn(x_r)$ equals -1 if there are an odd number of components of x_r equal to the subinterval left endpoints, and equals $+1$ otherwise. Similar expressions hold for all such derived sets.

Example 3.67 **($F(x_1, ..., x_n) = \prod_{j=1}^{n} F_j(x_j)$)** *Recalling Exercise 1.25, if $F(x_1, ..., x_n) = \prod_{j=1}^{n} F_j(x_j)$ with increasing, right continuous $\{F_j\}_{j=1}^{n}$ and $R = \prod_{j=1}^{n}(a_i, b_i]$, then:*

$$\mu_F[R] = \prod\nolimits_{j=1}^{n}\left[F_j(b_j) - F_j(a_j)\right].$$

Defining $R(a_k)$ as above:

$$\begin{aligned}\mu_F\left[R(a_k)\right] &= \prod\nolimits_{j\neq k}\left[F_j(b_j)-F_j(a_j)\right]\lim_{r\to\infty}\left[F_{k_1}(a_k)-F_k(a_k-\frac{1}{r})\right] \\ &= \left[F_k(a_k)-F_k(a_k^-)\right]\prod\nolimits_{j\neq k}\left[F_j(b_j)-F_j(a_j)\right].\end{aligned}$$

Similarly:

$$\mu_F\left[R(a_k,a_l)\right]=\left[F_k(a_k)-F_k(a_k^-)\right]\left[F_l(a_l)-F_l(a_l^-)\right]\prod\nolimits_{j\neq k,l}\left[F_j(b_j)-F_j(a_j)\right],$$

and so forth, with:

$$\mu_F\left[R(a_1,...,a_n)\right]=\prod\nolimits_{j=1}^n\left[F_j(a_j)-F_j(a_j^-)\right].$$

It follows from this that such measures will be zero when the associated F_j is continuous at a_j.

Given the above partition:

$$\bar{R}=\bigcup\nolimits_J R_J\cup\bigcup\nolimits_k\bigcup\nolimits_J R_J(a_k)\cup\bigcup\nolimits_{k<l}\bigcup\nolimits_J R_J(a_k,a_l)\cup...\cup\prod\nolimits_{j=1}^n\{a_j\}, \tag{1}$$

and we can define simple functions $g_m(x)$ on $\bar{R}$ by:

$$\begin{aligned}g_m(x) &= \sum\nolimits_J g(x'_J)\chi_{R_J}(x)+\sum\nolimits_{k=1}^n\sum\nolimits_J g(x'_J(a_k))\chi_{R_J(a_k)}(x) \\ &\quad+\sum\nolimits_{k<l}\sum\nolimits_J g(x'_J(a_k,a_l))\chi_{R_J(a_k,a_l)}(x)+ \\ &\quad ...+g(a_1,...,a_n)\chi_{(a_1,...,a_n)}(x),\end{aligned}$$

where for each rectangle, $x'_J(\cdot)\in\bar{R}_J(\cdot)$, the closure of $R_J(\cdot)$.
Then by (3.1):

$$\begin{aligned}\int_{\bar{R}}g_m(x)d\mu_F &= \sum\nolimits_J g(x'_J)\mu_F(R_J)+\sum\nolimits_{k=1}^n\sum\nolimits_J g(x'_J(a_k))\mu_F(R_J(a_k)) \\ &\quad+\sum\nolimits_{k<l}\sum\nolimits_J g(x'_J(a_k,a_l))\mu_F(R_J(a_k,a_l))+ \\ &\quad ...+g(a_1,...,a_n)\mu_F((a_1,...,a_n)).\end{aligned} \tag{2}$$

By uniform continuity of $g(x)$, each $x'_J(\cdot)$ can be chosen as above so that:

$$g(x'_J(\cdot))=\inf_{x\in\bar{R}_J(\cdot)}g(x),$$

and thus $g_m(x)\leq g(x)$ for all m and $x\in\bar{R}$.
With details left as an exercise, Lebesgue's dominated convergence theorem assures that:

$$\int_{\bar{R}}g_m(x)d\mu_F\to\int_{\bar{R}}g(x)d\mu_F.$$

On the other hand, the first summation on the right in (2) converges as above to the Riemann-Stieltjes integral:

$$\sum\nolimits_J g(x'_J)\mu_F(R_J)\to\int_{\bar{R}}g(x)dF.$$

The other summations in general converge to various Lebesgue-Stieltjes integrals defined on the lower dimensional boundary sets defined by the $R_J(\cdot)$-sets. Finally, there is the constant $g(a_1,...,a_n)\mu_F((a_1,...,a_n))$ as in (3.43).

If we use Lebesgue's dominated convergence theorem for every term on the right in (2), this decomposition of a Lebesgue-Stieltjes integral over $\bar{R}$ into a summation of such integrals over sets in (1) which union to $\bar{R}$, obtains a special case of the result in (3.25).

For these latter integrals from (2), the function $g(x)$ is effectively restricted by the a_k-components that define $R_J(\cdot)$. For example, when $k=1$ in the second summation:

$$\sum\nolimits_J g(x'_J(a_1))\mu_F(R_J(a_1)) \to \int_{R(a_1)} g(a_1,x_2,...,x_n)d\mu_F, \tag{3}$$

where $R(a_1) \equiv \{a_1\}\times\prod_{j=2}^n(a_j,b_j]$. In some cases, these latter integrals will also be Riemann-Stieltjes.

Example 3.68 ($F(x_1,...,x_n)=\prod_{j=1}^n F_j(x_j)$) *Continuing Example 3.67, if* $k=1$ *in the second summation in* (2)*:*

$$\begin{aligned}&\sum\nolimits_J g(x'_J(a_1))\mu_F(R_J(a_1))\\ &= \left[F_1(a_1)-F_1(a_1^-)\right]\sum\nolimits_J g(a_1,x'_2,...,x'_n)\prod\nolimits_{j=2}^n\left[F_j(x_{j,i_j})-F_j(x_{j,i_j-1})\right].\end{aligned}$$

Now $\prod_{j=2}^n\left[F_j(x_{j,i_j})-F_j(x_{j,i_j-1})\right]$ *is the* $\tilde{F}$*-content of* $\tilde{R}_J\equiv\prod_{j=2}^n[x_{j,i_j-1},x_{j,i_j}]$ *with* $\tilde{F}(x_2,...,x_n)\equiv\prod_{j=2}^n F_j(x_j)$ *and denoted* $\Delta\tilde{F}(\tilde{R}_J)$ *in Definition III.4.80. Since* $g(a_1,x_2,...,x_n)$ *is continuous in* $(x_2,...,x_n)$*, this summation converges to a Riemann-Stieltjes integral by Corollary III.4.90, and thus with* $\tilde{R}\equiv\prod_{j=2}^n[a_j,b_j]$*:*

$$\sum\nolimits_J g(x'_J(a_1))\mu_F(R_J(a_1)) \to \left[F_1(a_1)-F_1(a_1^-)\right]\int_{\tilde{R}} g(a_1,x_2,...,x_n)d\tilde{F}. \tag{4}$$

An analogous conclusion follows for all the boundary summations in (2)*, and thus in this case,* $\int_{\bar{R}} g(x)d\mu_F$ *can be expressed as a sum of Riemann-Stieltjes integrals, plus the constant* $g(a_1,...,a_n)\prod_{j=1}^n\left[F_j(a_j)-F_j(a_j^-)\right]$.

From the expression in (4), it is apparent that if $F(x_1,...,x_n)=\prod_{j=1}^n F_j(x_j)$ and each $\{F_j\}_{j=1}^n$ is continuous, then $F_j(a_j)=F_j(a_j^-)$ for all j and thus from (2):

$$\sum\nolimits_{k=1}^n\sum\nolimits_J g(x'_J(a_k))\mu_F(R_J(a_k)) \to 0.$$

This is also true for the other "boundary" summations, and thus we have the following:

Proposition 3.69 ($F(x_1,...,x_n)=\prod_{j=1}^n F_j(x_j)$) *If* $F(x_1,...,x_n)=\prod_{j=1}^n F_j(x_j)$ *with* $\{F_j\}_{j=1}^n$ *continuous,* $\bar{R}=\prod_{j=1}^n[a_j,b_j]$*, and* $g(x)$ *continuous on* $\bar{R}$*:*

$$\int_{\bar{R}} g(x)d\mu_F = \int_{\bar{R}} g(x)dF. \tag{3.50}$$

This generalizes to continuous $F(x_1,...,x_n)$.

Exercise 3.70 (Continuous $F(x_1,...,x_n)$**)** *Prove that if* $F(x_1,...,x_n)$ *is continuous, then Proposition 3.69 is true for all* $\bar{R}=\prod_{j=1}^n[a_j,b_j]$*. Hint: From* (2) *and the above discussion, it must be proved that all the boundary integrals are again zero. For this, it is enough to prove that* $\mu_F(R_J(a_k))=0$ *for all* k*, since then by monotonicity of measures,* $\mu_F(R_J(a_k,a_l))=0$ *for all* $k<l$*, and so forth. For* $\mu_F(R_J(a_k))=0$*, use (3.49) and split the summation into two sums, one with all* $x_{r,k}=a_k$*, the other with all* $x_{r,k}=a_k-1/r$*. Recall the* $sgn(x_r)$ *convention.*

4

Change of Variables

It is a common exercise in Riemann integration to use the "method of substitution" or "change of variables" in a Riemann integral to simplify its evaluation.

Example 4.1 (Riemann method of substitution) *Assume that we seek to evaluate $\int_0^\infty h(x)dx$ for $h(x) = x\exp(-x^2)$. The method of substitution defines a new "variable" $y \equiv g(x) = x^2$ and then defines $dy \equiv g'(x)dx = 2xdx$ to obtain:*

$$\begin{aligned}\int_0^\infty x\exp(-x^2)dx &= 0.5\int_{g(0)}^{g(\infty)} \exp(-y)dy \\ &= 0.5.\end{aligned}$$

While grounded in calculus theory, this manipulation can also be interpreted as a ***trompe l'oeil,*** *French for "deceive the eye." The goal of this substitution is to simplify the application of the fundamental theorem of calculus (Proposition III.1.31) by making it easier to identify the antiderivative of the integrand. Specifically, to identify a function $H(x)$ with $H'(x) = x\exp(-x^2)$. The function $H(x)$ is often called the "antiderivative" of $h(x)$, and applying the fundamental theorem:*

$$\int_0^\infty x\exp(-x^2)dx \equiv \lim_{b\to\infty}\int_0^b H'(x)dx = H(\infty) - H(0), \tag{1}$$

assuming $H(\infty) \equiv \lim_{b\to\infty} H(b)$ exists.

The method of substitution attempts to represent the integrand $H'(x)$ as the derivative of a composite function $H(x) = f(g(x))$, and so:

$$H'(x) = f'(g(x))\, g'(x).$$

Thus the goal is to find functions f and g with:

$$x\exp(-x^2) = f'(g(x))\, g'(x),$$

so that (1) becomes:

$$\int_0^\infty x\exp(-x^2)dx = f(g(\infty)) - f(g(0)). \tag{2}$$

The result in (2) is seen to be the same as that obtained by the fundamental theorem for the integral:

$$\int_{g(0)}^{g(\infty)} f'(y)dy,$$

when $f(g(\infty))$ exists in the limit. The substitution in the first paragraph simply suppresses the details associated with explicitly identifying $g(x)$ and $f(y)$ so that $x\exp(-x^2) = f'(g(x))\, g'(x)$. Guessing $g(x) = x^2$ above leads to:

$$x\exp(-x^2) = 2xf'(x^2),$$

and thus $f'(y) = 0.5\exp(-y)$ as above.

DOI: 10.1201/9781003264576-4

The method of substitution simplifies trial and error in determining the most appropriate variable $y = g(x)$ and also supports sequential substitutions to ultimately reduce a complicated integrand to something manageable, if indeed this is possible.

In this chapter, we will see that this change of variable formula also has an interpretation from a measure-theoretic perspective, and this has important applications in probability theory in Book VI.

We begin with results on the evaluation of Lebesgue-Stieltjes and more general integrals when the measure in the integral is of a certain type. We then turn to a general investigation into the transformation of measures between measure spaces and the implications for integrals in the domain space.

The final section addresses three applications of these results to change of variables in Riemann and Lebesgue integrals.

4.1 Change of Measure: A Special Case

This section introduces a special class of measures that can be defined on a given measure space using the earlier integration theory. It then investigates how integrals defined relative to these new measures compare with integrals with respect to the original measures on this space.

These results will look familiar from their probability theory applications and will also be reminiscent of Propositions III.4.28 and III.4.97 in the Riemann-Stieltjes development.

4.1.1 Measures Defined by Integrals

Given the Borel measure space $(\mathbb{R}^n, \mathcal{B}(\mathbb{R}^n), \mu)$, assume that there is a nonnegative Lebesgue measurable function $f(x)$ defined on $\mathbb{R}^n$ so that for any Borel set $A \in \mathcal{B}(\mathbb{R}^n)$:

$$\mu(A) = (\mathcal{L})\int_A f(x)dx. \tag{4.1}$$

Here, $x \equiv (x_1, ..., x_n)$ and dx with the $(\mathcal{L})$-qualifier denotes Lebesgue measure. This integral can also be denoted with dm and without the $(\mathcal{L})$-qualifier.

The representation in (4.1) may look familiar from the study of probability theory.

Example 4.2 (Probability theory) *Let X be a random variable defined on a probability space $(\mathcal{S}, \mathcal{E}, \lambda)$, which means by Definition II.3.1 and Exercise II.3.3 that $X : (\mathcal{S}, \mathcal{E}) \to (\mathbb{R}, \mathcal{B}(\mathbb{R}))$ is measurable, so $X^{-1}(A) \in \mathcal{E}$ for all $A \in \mathcal{B}(\mathbb{R})$. With X measurable, the* ***distribution function*** *$F(x)$ of X is well-defined by:*

$$F(x) = \lambda\left[X^{-1}(-\infty, x]\right].$$

As summarized in Notation II.3.5, the function $F(x)$ is increasing and right continuous by Proposition I.3.60 and Remark I.3.61.

Now assume that there exists a ***density function for*** *$F(x)$, which is a nonnegative Lebesgue integrable $f(x)$ with $(\mathcal{L})\int_{-\infty}^{\infty} f(y)dy = 1$, so that:*

$$F(x) = (\mathcal{L})\int_{-\infty}^{x} f(y)dy.$$

By Proposition III.3.62, such $f(x)$ exists when $F(x)$ is absolutely continuous (Definition III.3.54), and then this representation is satisfied if $f(x) = F'(x)$ m-a.e., with m Lebesgue measure.

By Proposition III.3.62, the set function:

$$\mu_F((a,b]) \equiv F(b) - F(a) = (\mathcal{L})\int_a^b f(y)dy,$$

and thus this set function is well-defined on the semi-algebra of right semi-closed intervals $\mathcal{A}'$. It is then well-defined on the associated algebra $\mathcal{A}$ of finite disjoint unions of $\mathcal{A}'$-sets (Proposition I.5.11) and can be extended by Proposition I.5.23 to a measure on $\mathcal{B}(\mathbb{R})$.

Since $\mu_F = \mu$ on the semi-algebra $\mathcal{A}'$ and thus also on $\mathcal{A}$, it follows from Proposition I.6.14 that $\mu_F = \mu$ on $\sigma(\mathcal{A})$, the smallest sigma algebra that contains $\mathcal{A}$. But $\sigma(\mathcal{A}) = \mathcal{B}(\mathbb{R})$ by Proposition I.8.1, and thus μ_F is given by (4.1) for all $A \in \mathcal{B}(\mathbb{R})$.

More generally, we have the following. Recall Definition 1.7.

Proposition 4.3 (μ is a measure on $\mathcal{B}(\mathbb{R}^n)$) *If $f(x)$ is a nonnegative Lebesgue measurable function on $\mathbb{R}^n$, then the set function μ in (4.1) defines a measure on $\mathcal{B}(\mathbb{R}^n)$.*

If $f(x)$ is bounded, then μ is a Borel measure. If $f(x)$ integrable, then μ is a finite Borel measure.

Proof. *First, note that μ is well-defined on $\mathcal{B}(\mathbb{R}^n)$. Given $A \in \mathcal{B}(\mathbb{R}^n)$, let $A_m \equiv \{|x| \leq m\} \bigcap A$ and define:*

$$f_m(x) = f(x)\chi_{\{f\leq m\}}(x)\chi_{A_m}(x).$$

Recall that the characteristic function $\chi_B(x)$ is defined to equal 1 for $x \in B$ and 0 otherwise. Then $\{f_m(x)\}_{m=1}^{\infty}$ is an increasing sequence of functions with $f_m(x) \to f(x)$ pointwise on A.

Each $f_m(x)$ is Lebesgue measurable and bounded, and $\int_A f_m(x)dx$ exists since by Definition III.2.9:

$$\begin{aligned} \int_A f_m(x)dx &= \int f(x)\chi_{\{f\leq m\}}(x)\chi_{A_m}(x)\chi_A(x)dx \\ &= \int_{A_m} f(x)\chi_{\{f\leq m\}}(x)dx. \end{aligned}$$

Since $f(x)\chi_{\{f\leq m\}}(x)$ is bounded and $m(A_m) < \infty$, Proposition III.2.15 applies and this integral is well-defined by measurability of the integrand. Further, this integral is nonnegative by Definition III.2.26.

Lebesgue's monotone convergence theorem of Proposition III.2.37 then obtains that:

$$\int_A f(x)dx = \lim_{m\to\infty}\int_A f_m(x)dx \geq 0, \tag{1}$$

and thus $\mu(A)$ is well-defined on $\mathcal{B}(\mathbb{R}^n)$.

For countable additivity of μ, assume that $\{A_j\}_{j=1}^{\infty} \subset \mathcal{B}(\mathbb{R}^n)$ are disjoint and let $A = \bigcup_{j=1}^{\infty} A_j$. If $f(x)$ is integrable over A then countable additivity follows from Corollary III.2.40. If $f(x)$ is not integrable over A, then $\mu(A) = \infty$ by nonnegativity. If there exists j with $f(x)$ also not integrable over A_j, then $\mu(A_j) = \infty$ and countable additivity is again satisfied.

So, assume that $f(x)$ is integrable over all A_j, and thus $\mu(A_j) < \infty$ for all j. Define $C_m \equiv \bigcup_{j=1}^{m} A_j$ and $f_m(x) \equiv f(x)\chi_{C_m}(x)$. Then $f_m(x)$ is nonnegative and Lebesgue integrable, and by Definition III.2.9 and Proposition III.2.31:

$$\int f_m(x)dx \equiv \int_{C_m} f(x)dx = \sum\nolimits_{j=1}^{m}\int_{A_j} f(x)dx = \sum\nolimits_{j=1}^{m}\mu(A_j). \tag{2}$$

Now, $\{f_m\}_{j=1}^{\infty}$ is a monotone sequence of nonnegative functions with $f_m(x) \to f(x)\chi_A(x)$. Thus by Lebesgue's monotone convergence theorem:

$$\int f_m(x)dx \to \int_A f(x)dx.$$

Combining with (2):

$$\sum\nolimits_{j=1}^{m} \mu(A_j) \to \int_A f(x)dx = \infty.$$

This proves countable additivity and thus μ is a measure.

If A is compact, then $A \subset R = \prod_{j=1}^{n}[a_j, b_j]$ by the Heine-Borel theorem of Proposition I.2.27. Thus by item 5 of Proposition III.2.31, if $f(x)$ is bounded by M say:

$$\mu(A) \leq M \prod\nolimits_{j=1}^{n} (b_j - a_j) < \infty,$$

and μ is a Borel measure.

For integrable $f(x)$, $\mu(\mathbb{R}^n) = \int_{\mathbb{R}^n} f(x)dx < \infty$, and thus μ is a finite Borel measure since measures of compact sets are finite by monotonicity of μ. ■

Remark 4.4 (μ is a measure on $\mathcal{M}_L(\mathbb{R}^n)$) *Recall that $\mathcal{B}(\mathbb{R}^n) \subset \mathcal{M}_L(\mathbb{R}^n)$, the complete Lebesgue sigma algebra of Section I.7.6. Using the same proof, μ also defines a measure on $\mathcal{M}_L(\mathbb{R}^n)$ since the integral in (4.1) is well-defined for $A \in \mathcal{M}_L(\mathbb{R}^n)$ by the same argument.*

This observation then extends to Proposition 4.8, in that the results are valid for Lebesgue measurable $g(x)$. However, for results in Sections 4.3.2 and 4.3.3 on change of variables in Lebesgue integrals, we will focus on Borel measurable g for reasons to be clarified there. Hint: Proposition I.3.33.

The result of Proposition 4.3 can be generalized beyond Lebesgue and the special Borel measures on $\mathbb{R}^n$ defined by (4.1).

Let $(X, \sigma(X), \nu)$ be a σ-finite measure space, recalling Definition 1.21, and $f(x)$ a nonnegative $\sigma(X)$-measurable function,

$$f : (X, \sigma(X), \nu) \to (\mathbb{R}, \mathcal{B}(\mathbb{R}), m),$$

meaning $f^{-1}(A) \in \sigma(X)$ for all $A \in \mathcal{B}(\mathbb{R})$. Define a set function on $\sigma(X)$ by:

$$\mu(A) \equiv \int_A f(x)d\nu. \tag{4.2}$$

Proposition 4.5 (μ is a measure on $\sigma(X)$) *If $f(x)$ is a nonnegative $\sigma(X)$-measurable function on σ-finite $(X, \sigma(X), \nu)$, then the set function μ in (4.2) defines a measure on $\sigma(X)$.*

If $f(x)$ is bounded, then μ is a σ-finite measure. If $f(x)$ is ν-integrable, then μ is a finite measure.

Proof. *The proof of Proposition 4.3 must be adapted for σ-finite $(X, \sigma(X), \nu)$ and to reflect properties of integrals developed in Chapter 3. We include the details for completeness.*

First, μ is well-defined on $\sigma(X)$. Let $\{B_j\}_{j=1}^{\infty} \subset \sigma(X)$ be the sets of Definition 1.21, with $\nu(B_j) < \infty$ for all j and $X = \bigcup_{j=1}^{\infty} B_j$. Given $A \in \mathcal{B}(\mathbb{R}^n)$, let $A_m \equiv \bigcup_{j=1}^{m} B_j \bigcap A$ and define:

$$f_m(x) = f(x)\chi_{\{f \leq m\}}(x)\chi_{A_m}(x).$$

Then $\{f_m(x)\}_{m=1}^{\infty}$ is an increasing sequence of functions with $f_m(x) \to f(x)$ pointwise on A.

As each $f_m(x)$ is ν-measurable and bounded, $\int_A f_m(x)dx$ exists since by Definition 3.7:

$$\begin{aligned}\int_A f_m(x)d\nu &\equiv \int f(x)\chi_{\{f\leq m\}}(x)\chi_{A_m}(x)\chi_A(x)d\nu \\ &= \int_{A_m} f(x)\chi_{\{f\leq m\}}(x)d\nu.\end{aligned}$$

This integral is nonnegative and finite by item 6 of Proposition 3.29 since $f(x)\chi_{\{f\leq m\}}(x)$ is nonnegative and bounded, and $\nu(A_m) < \infty$.

Lebesgue's monotone convergence theorem of Proposition 3.23 then obtains that:

$$\int_A f(x)dx = \lim_{m\to\infty}\int_A f_m(x)dx \geq 0, \tag{1}$$

and thus $\mu(A)$ is well-defined on $\sigma(X)$.

For countable additivity, assume $\{A_j\}_{j=1}^{\infty} \subset \sigma(X)$ are disjoint and let $A = \bigcup_{j=1}^{\infty} A_j$. If $f(x)$ is integrable over A, then countable additivity follows from Corollary 3.32. If $f(x)$ is not integrable over A, this implies that $\mu(A) = \infty$ by nonnegativity. Thus if there exists j with $f(x)$ also not integrable over A_j, then $\mu(A_j) = \infty$ and countable additivity is satisfied.

So, assume that $\mu(A) = \infty$ and that $f(x)$ is integrable over all A_j, and thus $\mu(A_j) < \infty$ for all j. Define $C_m \equiv \bigcup_{j=1}^{m} A_j$ and $f_m(x) \equiv f(x)\chi_{C_m}(x)$. Then $f_m(x)$ is nonnegative and integrable, and by Definition 3.7 and Proposition 3.29:

$$\int f_m(x)dx \equiv \int_{C_m} f(x)dx = \sum\nolimits_{j=1}^{m}\int_{A_j} f(x)dx = \sum\nolimits_{j=1}^{m}\mu(A_j). \tag{2}$$

By definition, $\{f_m\}_{j=1}^{\infty}$ is a monotone sequence of nonnegative functions with $f_m(x) \to f(x)\chi_A(x)$. Thus by Lebesgue's monotone convergence theorem:

$$\int f_m(x)dx \to \int_A f(x)dx.$$

Combining with (2):

$$\sum\nolimits_{j=1}^{m}\mu(A_j) \to \int_A f(x)dx = \infty.$$

This proves countable additivity in this case, and thus μ is a measure.

For bounded $f(x)$, given $\{B_j\}_{j=1}^{\infty} \subset \sigma(X)$ with $\nu(B_j) < \infty$ for all j and $X = \bigcup_{j=1}^{\infty} B_j$, it follows from item 6 of Proposition 3.29 that $\nu(B_j) < \infty$ for all j, and hence ν is σ-finite.

For integrable $f(x)$, $\mu(X) = \int_X f(x)d\nu < \infty$, and thus μ is a finite measure. ■

Remark 4.6 (On $\mu \ll m$ and $\mu \ll \nu$) *The measure μ defined relative to Lebesgue measure m in (4.1) has a special property, independent of which $f(x)$ is used in this definition, and that is, for all $A \in \mathcal{B}(\mathbb{R}^n)$:*

$$m(A) = 0 \Rightarrow \mu(A) = 0.$$

Similarly, the measure μ defined relative to σ-finite measure ν in (4.2) has a special property, that for all $A \in \sigma(X)$:

$$\nu(A) = 0 \Rightarrow \mu(A) = 0.$$

In Definition 8.3, this property will be called ***absolute continuity,*** *and specifically, it will be said that:*

- *μ in (4.1) is absolutely continuous with respect to m, and denoted $\mu \ll m$;*
- *μ in (4.2) is absolutely continuous with respect to ν, and denoted $\mu \ll \nu$.*

Quite remarkably, not only are the measures defined as μ in (4.1) and (4.2) absolutely continuous with respect to the underlying measures in these spaces, but we will see that this is essentially the only way absolutely continuous measures can arise.

The ***Radon-Nikodým theorem*** *of Proposition 8.23, named for* ***Johann Radon*** *(1887–1956), who proved this result on $\mathbb{R}^n$, and* ***Otto Nikodým*** *(1887–1974), who generalized Radon's result to σ-finite measure spaces, states the following:*

Radon-Nikodým theorem: *Given σ-finite $(X, \sigma(X), \nu)$, if μ is a σ-finite measure on $\sigma(X)$ with $\mu \ll \nu$, then there exists a ν-measurable $f(x)$ so that (4.2) is satisfied. Further, $f(x)$ is unique ν-a.e.*

Remark 4.7 (On absolute continuity) *The reader may recall that the notion of absolute continuity was introduced in Definition III.3.54 as a property of a function. The connection between these notions will be addressed in Proposition 8.18.*

4.1.2 Integrals and Change of Measure

In this section, we investigate integrals defined with respect to the derived measures μ of (4.1) and (4.2) and prove that these can be related to integrals defined with respect to the measures of the underlying spaces.

Starting with the Lebesgue-Stieltjes case, if the measure μ is defined as in (4.1), let $g(x)$ be a Borel measurable function defined on $(\mathbb{R}^n, \mathcal{B}(\mathbb{R}^n), \mu)$:

$$g : (\mathbb{R}^n, \mathcal{B}(\mathbb{R}^n), \mu) \to (\mathbb{R}, \mathcal{B}(\mathbb{R}), m),$$

meaning that $g^{-1}(A) \in \mathcal{B}(\mathbb{R}^n)$ for all $A \in \mathcal{B}(\mathbb{R})$. As $(\mathbb{R}^n, \mathcal{B}(\mathbb{R}^n), \mu)$ is a measure space and g is measurable, the μ-integral of $g(x)$ over any Borel set $A \in \mathcal{B}(\mathbb{R}^n)$:

$$\int_A g(x) d\mu,$$

is defined as a Lebesgue-Stieltjes integral by Definition 3.38, though need not be finite.

The following result relates the μ-integrability of $g(x)$ and the value of this Lebesgue-Stieltjes integral to the Lebesgue integrability of $f(x)g(x)$ and the value of the associated Lebesgue integral. This result substantially generalizes the result of Proposition III.4.97, which required continuity of $f(x)$ and $g(x)$ to be able to utilize the framework of Riemann and Riemann-Stieltjes integrals. On the other hand, the earlier result also applied to $f(x)$ that was not nonnegative, as long as the associated $F(x)$ was of Vitali bounded variation. See Section 6.1 for a generalization for so-called "signed measures."

Proposition 4.8 ($\int_A g(x)d\mu$ as a Lebesgue integral; $\mu(A) = (\mathcal{L}) \int_A f(x)dx$) *Given a nonnegative Lebesgue measurable function $f : \mathbb{R}^n \to \mathbb{R}$, let μ be defined on $\mathcal{B}(\mathbb{R}^n)$ as in (4.1).*

For any Borel measurable function $g : (\mathbb{R}^n, \mathcal{B}(\mathbb{R}^n)) \to (\mathbb{R}, \mathcal{B}(\mathbb{R}))$:

$$\int_{\mathbb{R}^n} g(x) d\mu = (\mathcal{L}) \int_{\mathbb{R}^n} g(x) f(x) dx, \tag{4.3}$$

though both integrals may be infinite.

In addition, $g(x)$ is μ-integrable if and only if $f(x)g(x)$ is Lebesgue integrable, and when integrable:

$$\int_A g(x) d\mu = (\mathcal{L}) \int_A g(x) f(x) dx, \tag{4.4}$$

for all $A \in \mathcal{B}(\mathbb{R}^n)$.

Proof. *Given $A \in \mathcal{B}(\mathbb{R}^n)$, let $g(x) = \chi_A(x)$. Then (4.3) is satisfied because by (3.1):*

$$\int_{\mathbb{R}^n} g(x)d\mu = \mu(A),$$

while by Definition III.2.9:

$$(\mathcal{L})\int_{\mathbb{R}} g(x)f(x)dx \equiv (\mathcal{L})\int_A f(x)dx = \mu(A).$$

If $g(x)$ is a simple function, then (4.3) again follows by linearity of both integrals by Propositions 3.42 and III.2.49.

For general nonnegative $g(x)$, let $\{g_m(x)\}_{m=1}^{\infty}$ be an increasing sequence of nonnegative simple functions given by Proposition 2.28 with $g_m(x) \to g(x)$ for all x. Since $f(x)$ is nonnegative and Lebesgue measurable, $\{g_m(x)f(x)\}_{m=1}^{\infty}$ is an increasing sequence of nonnegative Lebesgue measurable functions and $g_m(x)f(x) \to g(x)f(x)$.

By Lebesgue's monotone convergence theorem of Propositions 3.23 and III.2.37:

$$\int_{\mathbb{R}^n} g_m(x)d\mu \to \int_{\mathbb{R}^n} g(x)d\mu,$$

$$(\mathcal{L})\int_{\mathbb{R}^n} g_m(x)f(x)dx \to (\mathcal{L})\int_{\mathbb{R}^n} g(x)f(x)dx,$$

and (4.3) follows since this identity is satisfied for all $g_m(x)$.

For general measurable $g(x)$, the conclusion of (4.3) now applies to $g^+(x)$ and $g^-(x)$, the positive and negative parts of $g(x)$ of (3.22), and to $|g(x)|$:

$$\int_{\mathbb{R}^n} |g(x)|\, d\mu = (\mathcal{L})\int_{\mathbb{R}^n} f(x)\, |g(x)|\, dx.$$

Hence, $g(x)$ is μ-integrable if and only if $f(x)g(x)$ is Lebesgue integrable.

Finally, if $g(x)$ is μ-integrable and $A \in \mathcal{B}(\mathbb{R}^n)$, then $g(x)\chi_A(x)$ is μ-integrable since by Proposition 3.42:

$$\left|\int_{\mathbb{R}^n} g(x)\chi_A(x)d\mu\right| \leq \int_{\mathbb{R}^n} |g(x)|\, \chi_A(x)d\mu \leq \int_{\mathbb{R}^n} |g(x)|\, d\mu.$$

Hence, applying (4.3) to $g(x)\chi_A(x)$ over $\mathbb{R}^n$ obtains (4.4). ■

Example 4.9 (Expectations of random variables 1) *Let X be a random variable defined on a probability space $(\mathcal{S}, \mathcal{E}, \lambda)$ with distribution function $F(x)$ as in Example 4.2. Let $(\mathbb{R}, \mathcal{B}(\mathbb{R}), \mu_F)$ be a Borel measure space induced by this increasing, right continuous, and bounded function $F(x)$ as in the construction of Section 1.1.3.*

If $F(x)$ is absolutely continuous, then $F'(x) = f(x)$ exists almost everywhere and is Lebesgue integrable on bounded intervals by Proposition III.3.59, and since $F(-\infty) = 0$, it follows from Proposition III.3.62 that:

$$F(x) = (\mathcal{L})\int_{-\infty}^{x} f(y)dy.$$

It then follows as in Example 4.2 that μ_F is given as in (4.1) for all $A \in \mathcal{B}(\mathbb{R})$.

If $g : (\mathbb{R}, \mathcal{B}(\mathbb{R}), \mu_F) \to (\mathbb{R}, \mathcal{B}(\mathbb{R}), m)$ is μ_F-measurable, then the Lebesgue-Stieltjes integral of $g(x)$ over $A \in \mathcal{B}(\mathbb{R})$ can be evaluated as a Lebesgue integral by (4.4):

$$\int_A g(x)d\mu_F = (\mathcal{L})\int_A g(x)f(x)dx.$$

In addition, $g(x)$ will be μ_F-integrable if and only if $g(x)f(x)$ is Lebesgue integrable.

Since μ_F is a probability measure and thus $(\mathbb{R}, \mathcal{B}(\mathbb{R}), \mu_F)$ is a probability space, the μ_F-measurable function $g(x)$ is a **random variable** *and as in Example 4.2, usually denoted by an upper case letter such as X, Y, etc. Then by IV.(4.7), the* **expectation of such** $Y \equiv g(x)$, *denoted $E[Y]$, is defined:*

$$E[Y] \equiv \int_{\mathbb{R}} Y d\mu_F.$$

When $F(x)$ is absolutely continuous and hence $f(x) = F'(x)$ m-a.e. by Proposition III.3.62, (4.4) obtains:

$$E[Y] = (\mathcal{L}) \int_{\mathbb{R}} Y(x)f(x)dx.$$

It is also the case that the original random variable X defined on $(\mathcal{S}, \mathcal{E}, \lambda)$ has an expectation analogously defined by:

$$E[X] = \int_{\mathcal{S}} X d\lambda.$$

To evaluate such an integral requires more general results on transformations of measures below. See Example 4.21 and Section IV.4.1.2 for a summary of these results.

Proposition 4.8 can be generalized to the context of Proposition 4.5.

Proposition 4.10 **($\int_A g(x)d\mu$ as a ν-integral; $\mu(A) = \int_A f(x)d\nu$)** *Given σ-finite $(X, \sigma(X), \nu)$ and a nonnegative $\sigma(X)$-measurable function $f : (X, \sigma(X), \nu) \to (\mathbb{R}, \mathcal{B}(\mathbb{R}), m)$, let μ be defined on $\sigma(X)$ as in (4.2).*

For any measurable function $g : (X, \sigma(X)) \to (\mathbb{R}, \mathcal{B}(\mathbb{R}))$:

$$\int_X g(x)d\mu = \int_X g(x)f(x)d\nu, \tag{4.5}$$

though both integrals may be infinite.

In addition, $g(x)$ is μ-integrable if and only if $f(x)g(x)$ is ν-integrable, and when integrable,

$$\int_A g(x)d\mu = \int_A g(x)f(x)d\nu, \tag{4.6}$$

for all $A \in \sigma(X)$.

Proof. *The above proof works with minor changes, but we include the details for completeness.*

Given $A \in \sigma(X)$, let $g(x) = \chi_A(x)$. Then (4.5) is satisfied because by (3.1):

$$\int_X g(x)d\mu = \mu(A),$$

while by Definition 3.7:

$$\int_X g(x)f(x)d\nu \equiv \int_A f(x)d\nu = \mu(A).$$

If $g(x)$ is a simple function, then (4.5) follows by linearity of both integrals by Proposition 3.42.

For general nonnegative $g(x)$, let $\{g_m(x)\}_{m=1}^{\infty}$ be an increasing sequence of nonnegative simple functions given by Proposition 2.28 with $g_m(x) \to g(x)$ for all x. Since $f(x)$ is nonnegative and ν-measurable, $\{g_m(x)f(x)\}_{m=1}^{\infty}$ is an increasing sequence of nonnegative ν-measurable functions with $g_m(x)f(x) \to g(x)f(x)$.

By Lebesgue's monotone convergence theorem of Proposition 3.23:

$$\int_X g_m(x)d\mu \to \int_X g(x)d\mu,$$

$$\int_X g_m(x)f(x)d\nu \to \int_X g(x)f(x)d\nu,$$

and (4.5) follows since this identity is satisfied for all $g_m(x)$.

For general measurable $g(x)$*, the conclusion of (4.5) now applies to* $g^+(x)$ *and* $g^-(x)$*, the positive and negative parts of* $g(x)$ *of (3.22), and to* $|g(x)|$:

$$\int_X |g(x)|\, d\mu = \int_X f(x)\, |g(x)|\, d\nu.$$

Hence, $g(x)$ *is* μ*-integrable if and only if* $f(x)g(x)$ *is* ν*-integrable.*

Finally, if $g(x)$ *is* μ*-integrable and* $A \in \sigma(X)$*, then* $g(x)\chi_A(x)$ *is* μ*-integrable since by Proposition 3.42:*

$$\left| \int_{\mathbb{R}^n} g(x)\chi_A(x)d\mu \right| \leq \int_{\mathbb{R}^n} |g(x)|\, \chi_A(x)d\mu \leq \int_{\mathbb{R}^n} |g(x)|\, d\mu.$$

Hence, applying (4.5) to $g(x)\chi_A(x)$ *over* $\mathbb{R}^n$ *obtains (4.6).* ■

Remark 4.11 (Radon-Nikodým theorem) *The reader may find it of interest to look ahead to Corollary 8.24 of the Radon-Nikodým theorem that frames the prior result from a different perspective. But be forewarned that there is a shift in notation between the discussions here and there.*

We summarize results in the current notation:

In the above development, we are given a σ*-finite measure space* $(X, \sigma(X), \nu)$ *and a nonnegative measurable* $f : (X, \sigma(X), \nu) \to (\mathbb{R}, \mathcal{B}(\mathbb{R}), m)$*, and then* ***defined*** *a new measure* μ *on* $(X, \sigma(X))$*, which provided the above conclusions. Thus the collection of such measurable functions* f *determines the class of measures* μ *for which the above integral identities are valid.*

But how would we identify if a given measure μ *was in this latter class?*

With the power of the Radon-Nikodým theorem, we are again given a σ*-finite measure space* $(X, \sigma(X), \nu)$ *and can answer this question. The defining property of a measure* μ *that ensures the existence of such* f *and the validity of the above identities is that* $\mu \ll \nu$*, meaning that* μ *is absolutely continuous with respect to* ν *by Definition 8.3.*

4.2 Transformations and Change of Measure

In this section, we study how a measurable transformation and the measure on the domain space induce a new measure on the range space, and then investigate how integrals on the domain space transform relative to this new measure. In effect, this transformation reflects a "change of variables" from the domain space to the range space.

4.2.1 Measures Induced by Transformations

Given sets X and X', a **transformation** $T : X \to X'$, and sometimes called a **map** or **mapping,** is simply a rule under which for each $x \in X$, there exists a unique $Tx \in X'$.

Since this is virtually unusable as a general notion, transformations of interest almost always have additional properties. As an example, linear and bounded linear transformations were introduced in Section III.4.3 and will be investigated in a new context in Section 9.4.

The subject of this section, **measurable transformations,** is defined next. For this definition, recall that if $A' \subset X'$, the inverse of T is defined as a set function:

$$T^{-1}(A') = \{x \in X | Tx \in A'\}. \tag{4.7}$$

The notion of an induced measure may well appear familiar to the reader. Before looking at Example 4.14, it is worth a moment of thought to identify where this has appeared in earlier books.

Definition 4.12 (Measurable transformations; Induced measures) *Given measure spaces $(X, \sigma(X), \mu)$ and $(X', \sigma(X'), \mu')$, a transformation $T : X \to X'$ is **measurable,** and sometimes $\sigma(X)/\sigma(X')$-**measurable,** if:*

$$T^{-1}(A') \in \sigma(X) \text{ for all } A' \in \sigma(X'), \tag{4.8}$$

or more economically:

$$T^{-1}[\sigma(X')] \subset \sigma(X).$$

*If $(X', \sigma(X'), \mu') = (\mathbb{R}, \mathcal{B}(\mathbb{R}), \mu')$ for a Borel measure μ', then T is called a **measurable function.***

*A measurable transformation/function induces a measure μ_T on the range space X', and hence induces a new measure space denoted $(X', \sigma(X'), \mu_T)$. The measure μ_T is called **the measure induced by** T and defined on $A' \in \sigma(X')$ by*

$$\mu_T(A') = \mu\left[T^{-1}(A')\right]. \tag{4.9}$$

Since (4.9) only defines a set function on $\sigma(X')$, it must be checked that μ_T is indeed a measure on $(X', \sigma(X'))$.

Proposition 4.13 (μ_T is a measure) *The set function μ_T defined in (4.9) is a measure on $(X', \sigma(X'))$.*

Proof. *First, μ_T is a well-defined set function on $\sigma(X')$ due to the measurability of T, while $\mu_T(A') \geq 0$ for all such A' because μ is a measure. Since $\mu_T(\emptyset) = 0$ follows from $T^{-1}(\emptyset) = \emptyset$, the last property to check is countable additivity.*

If $\{A'_j\}_{j=1}^{\infty} \subset \sigma(X')$ are disjoint, then $\{T^{-1}(A'_j)\}_{j=1}^{\infty} \subset \sigma(X)$ are also disjoint by definition of a transformation. By (4.7):

$$T^{-1}\left(\bigcup\nolimits_{j=1}^{\infty} A'_j\right) = \bigcup\nolimits_{j=1}^{\infty} T^{-1}(A'_j),$$

and this obtains by countable additivity of μ:

$$\mu_T\left[\bigcup\nolimits_{j=1}^{\infty} A'_j\right] = \mu\left[\bigcup\nolimits_{j=1}^{\infty} T^{-1}(A'_j)\right] = \sum\nolimits_{j=1}^{\infty} \mu_T\left[A'_j\right].$$

■

The measure μ_T is defined on a set $A' \in \sigma(X')$ to equal the μ-measure of the pre-image set $T^{-1}(A')$. This may well sound like a familiar idea, as we have encountered this already in earlier books. In particular, the next example reflects the development of Proposition II.6.3.

Example 4.14 **($\mu_X \equiv \mu_F$ for a random variable X)** *Let $(\mathcal{S}, \mathcal{E}, \mu)$ be a probability space and X a* ***random variable (r.v.)****:*

$$X : \mathcal{S} \longrightarrow \mathbb{R}.$$

By Definition II.3.1 and Exercise II.3.3 this means that $X : (\mathcal{S}, \mathcal{E}) \to (\mathbb{R}, \mathcal{B}(\mathbb{R}))$ is a measurable function, so $X^{-1}(A) \in \mathcal{E}$ for all $A \in \mathcal{B}(\mathbb{R})$.

The ***distribution function (d.f.)*** *or* ***cumulative distribution function (c.d.f.)*** *associated with X, denoted F or F_X, is then defined on $\mathbb{R}$ by:*

$$F(x) \equiv \mu[X^{-1}(-\infty, x]]. \tag{1}$$

This definition is identical with that of μ_X as given in (4.9) for sets $A' = (-\infty, x]$, since:

$$F(x) = \mu_X\left[(-\infty, x]\right],$$

and thus the random variable X induces the measure space $(\mathbb{R}, \mathcal{B}(\mathbb{R}), \mu_X)$ by the above definition. By finite additivity, for any right semi-closed interval $(a, b]$:

$$\mu_X\left[(a, b]\right] = F(b) - F(a). \tag{2}$$

The collection of such sets forms a semi-algebra $\mathcal{A}'$, and the collection of all finite unions of $\mathcal{A}'$-sets is an algebra denoted $\mathcal{A}$ (Example I.2.4).

Since $\mu_F = \mu_X$ on $\mathcal{A}'$ and thus also on $\mathcal{A}$, this identity extends by Proposition I.6.14 to the smallest sigma algebra $\sigma(\mathcal{A})$ that contains $\mathcal{A}$. By Proposition I.8.1, $\sigma(\mathcal{A}) = \mathcal{B}(\mathbb{R})$. Thus:

$$\mu_X \equiv \mu_F, \text{ on } \mathcal{B}(\mathbb{R}).$$

Not surprisingly, this example generalizes, again using earlier results.

Proposition 4.15 **($\mu_X \equiv \mu_F$ for a random vector X)** *Let $(\mathcal{S}, \mathcal{E}, \mu)$ be a probability space and X a* ***random vector,*** *$X : \mathcal{S} \longrightarrow \mathbb{R}^n$, with joint distribution function $F(x)$ with $x = (x_1, ..., x_n)$. Then:*

$$\mu_X \equiv \mu_F, \text{ on } \mathcal{B}(\mathbb{R}^n), \tag{4.10}$$

where μ_X is the induced measure of Proposition 4.13, and μ_F is the Borel measure induced by $F(x)$ of Section 1.3.2.

Proof. *By Proposition II.6.9, $\mu_F(A) = \mu\left[X^{-1}(A)\right]$ for all $A \in \mathcal{B}(\mathbb{R}^n)$, which is the definition of μ_X in (4.9).* ■

Although μ_T is defined on the same space and sigma algebra as was the original measure μ', the induced measure and original measure can be closely related or quite different.

Example 4.16 **(μ_T vs. μ')** *We provide three examples. See Chapter 8 for more results on measure relationships.*

1. *$\mu_T \perp \mu'$: Let the random variable X_1 be defined on a probability space, $X_1 : (\mathcal{S}, \mathcal{E}, \mu) \to (\mathbb{R}, \mathcal{B}(\mathbb{R}), m)$, with range $\mathbb{N} = \{0, 1, 2, 3...\}$. Assume that for some $\lambda > 0$, that for any $j \in \mathbb{N}$:*

   $$\mu\left[X_1^{-1}(j)\right] = e^{-\lambda}\lambda^j/j!,$$

   and hence:

   $$F_1(x) = \mu[X_1^{-1}(-\infty, x]] = e^{-\lambda}\sum\nolimits_{j \leq x} \lambda^j/j!.$$

The induced measure μ_{X_1} on $\mathbb{R}$ is the ***Poisson probability measure*** *μ_P introduced in Section II.1.3, and so in the various notations:*

$$\mu_{F_1} \equiv \mu_{X_1} = \mu_P.$$

This induced measure μ_{X_1} is quite different from the original Lebesgue measure of the range space, in that m assigns measure 0 to the set $\mathbb{N}$ on which μ_{X_1} assigns a total measure of 1.

In Chapter 8, we will say that m and μ_{X_1} are ***mutually singular****, denoted $m \perp \mu_{X_1}$, and define this to mean that there is a set $A \in \mathcal{B}(\mathbb{R})$ so that $m(A) = \mu_{X_1}(\tilde{A}) = 0$. In this example, $A = \mathbb{N}$ and $\tilde{A} = \mathbb{R} - \mathbb{N}$.*

2. *$\mu_T \ll \mu'$: Let X_2 be defined on a probability space, $X_2 : (\mathcal{S}, \mathcal{E}, \mu) \to (\mathbb{R}, \mathcal{B}(\mathbb{R}), m)$, with range $\mathbb{R}^+ \equiv \{x \in \mathbb{R} | x \geq 0\}$. Assume that for some $\lambda > 0$, that for $x \geq 0$:*

$$F_2(x) = \mu[X_2^{-1}(-\infty, x]] = 1 - e^{-\lambda x},$$

and $F_2(x) = 0$ for $x < 0$. The induced measure μ_{X_2} on $\mathbb{R}$ is the ***exponential probability measure*** *μ_E introduced in Section II.1.3, and so in the various notations:*

$$\mu_{F_2} \equiv \mu_{X_2} = \mu_E.$$

In this case, the induced measure μ_{X_2} is closely related to the original Lebesgue measure m. In particular, with the associated density function $f_E(x)$ defined on $x \geq 0$ by:

$$f_E(x) = \lambda e^{-\lambda x},$$

and $f_E(x) = 0$ elsewhere, we have that for $A \in \mathcal{B}(\mathbb{R})$:

$$\mu_{X_2}(A) = (\mathcal{L}) \int_A f_E(x) dx.$$

Hence, for any set $A \in \mathcal{B}(\mathbb{R})$ with $m(A) = 0$, it must be the case that $\mu_{X_2}(A) = 0$.

As in noted in Remark 4.6, in Chapter 8 we will say that μ_{X_2} is ***absolutely continuous*** *with respect to m, denoted $\mu_{X_2} \ll m$, where this notation means that if $m(A) = 0$ for $A \in \mathcal{B}(\mathbb{R})$, then $\mu_{X_1}(A) = 0$.*

3. *$\mu_T = \mu'$: See Proposition 4.31, which proves that given a linear transformation $T : (\mathbb{R}^n, \mathcal{B}(\mathbb{R}^n), \mu) \to (\mathbb{R}^n, \mathcal{B}(\mathbb{R}^n), m)$ and that μ can be defined so that $\mu_T = m$.*

4.2.2 Change of Variables under Transformations

The goal of this section is to relate integrals defined on the induced measure space $(X', \sigma(X'), \mu_T)$ with integrals defined on the domain space $(X, \sigma(X), \mu)$. Recall that a **measurable transformation** $T : X \to X'$ induces the measure μ_T on the range space X' that is defined on $A' \in \sigma(X')$ by (4.9):

$$\mu_T(A') = \mu\left[T^{-1}(A')\right].$$

Consider a $\sigma(X')$-measurable function $g(x')$ defined on $(X', \sigma(X'), \mu')$ with range in $(\mathbb{R}, \mathcal{B}(\mathbb{R}), m)$. By definition, $g(x')$ is also $\sigma(X')$-measurable on $(X', \sigma(X'), \mu_T)$ since the definition of measurability depends only on the sigma algebra $\sigma(X')$ and not on the measure defined on this sigma algebra. Diagrammatically:

$$(X, \sigma(X), \mu) \quad \xrightarrow{T} \quad (X', \sigma(X'), \mu_T) \quad \xrightarrow{g} \quad (\mathbb{R}, \mathcal{B}(\mathbb{R}), m).$$

As $(X', \sigma(X'), \mu_T)$ is a measure space and g is a measurable function:

$$\int_{A'} g(x')d\mu_T,$$

is definable for any measurable set $A' \in \sigma(X')$ by the development in Chapter 3, though such integrals need not be finite.

Measurability of $g(x')$ as a function on $(X', \sigma(X'), \mu_T)$ ensures that the composite function $g \circ T(x)$ is measurable as a function on $(X, \sigma(X), \mu)$. This follows because if $B \in \mathcal{B}(\mathbb{R})$, then:

$$[g \circ T]^{-1}(B) = T^{-1}\left[g^{-1}(B)\right].$$

Measurability of $g(x')$ assures that $g^{-1}(B) \in \sigma(X')$, and this assures that $T^{-1}\left[g^{-1}(A)\right] \in \sigma(X)$ by the measurability of T.

Hence, as $(X, \sigma(X), \mu)$ is a measure space and $g \circ T \equiv g(T)$ is a measurable function:

$$\int_A g(Tx)d\mu,$$

is again definable for any measurable set $A \in \sigma(X)$.

The goal of the next proposition is to relate the value of these integrals.

Remark 4.17 (On $\int_{A'} g(x')d\mu'$) *Although the μ'-integral $\int_{A'} g(x')d\mu'$ can also be defined on the original range space $(X', \sigma(X'), \mu')$ for $A' \in \sigma(X')$, the next proposition is silent on its value. And indeed it must be silent since, in general, the measure μ' need not be related in any predictable way to the domain measure μ.*

Proposition 4.18 ($\int_{A'} g(x')d\mu_T$ as a μ-integral) *Let $T : (X, \sigma(X), \mu) \rightarrow (X', \sigma(X'), \mu')$ be a measurable transformation and μ_T defined on $\sigma(X')$ by (4.9).*

Then for any measurable function $g : (X', \sigma(X')) \rightarrow (\mathbb{R}, \mathcal{B}(\mathbb{R}))$:

$$\int_{X'} g(x')d\mu_T = \int_X g(Tx)d\mu, \tag{4.11}$$

though both integrals may be infinite.

More generally, a measurable function $g(x')$ is μ_T-integrable if and only if $g(Tx)$ is μ-integrable, and when integrable:

$$\int_{A'} g(x')d\mu_T = \int_{T^{-1}A'} g(Tx)d\mu \tag{4.12}$$

for all $A' \in \sigma(X')$.

Proof. *Given $A' \in \sigma(X')$, let $g(x') = \chi_{A'}(x')$. Then by (3.1):*

$$\int_{X'} g(x')d\mu_T = \mu_T\left[A'\right].$$

Since $g(Tx) = \chi_{T^{-1}A'}(x)$, it follows from (4.9) that:

$$\int_X g(Tx)d\mu = \mu\left[T^{-1}(A')\right] \equiv \mu_T\left[A'\right].$$

This proves (4.11) for characteristic functions of sets in $\sigma(X')$, and then also for all simple functions on X' by linearity of the integral in Proposition 3.42.

For general nonnegative $g(x')$, let $\{g_n(x')\}_{n=1}^{\infty}$ be an increasing sequence of nonnegative simple functions given by Proposition 2.28, so that $g_n(x') \to g(x')$ for all x'. Then also $\{g_n(Tx)\}_{n=1}^{\infty}$ is an increasing sequence of simple functions since if:

$$g_n(x') = \sum\nolimits_{j=1}^{m} a_j \chi_{A_j}(x'),$$

then defined on X:

$$g_n(Tx) = \sum\nolimits_{j=1}^{m} a_j \chi_{T^{-1}(A_j)}(x),$$

noting that $T^{-1}(A_j) \in \sigma(X)$ by measurability. In addition, $g_n(Tx) \to g(Tx)$ for all x.

By Lebesgue's monotone convergence theorem of Proposition 3.23:

$$\int_X g_n(Tx)d\mu \to \int_X g(Tx)d\mu,$$

$$\int_{X'} g_n(x')d\mu_T \to \int_{X'} g(x')d\mu_T,$$

and (4.11) follows since this identity is satisfied for these sequences of simple functions.

Applying this conclusion to $|g(x')|$ for a general measurable function, it follows that:

$$\int_X |g(Tx)|\, d\mu = \int_{X'} |g(x')|\, d\mu_T,$$

and hence $g(x')$ is μ_T-integrable if and only if $g(Tx)$ is μ-integrable.

Finally, if $g(x')$ is μ_T-integrable, then so too is $g(x')\chi_{A'}(x')$ for $A' \in \sigma(X')$ since by Proposition 3.42:

$$\left| \int_{X'} g(x')\chi_{A'}(x')d\mu_T \right| \leq \int_{X'} |g(x')|\, d\mu_T.$$

Thus applying (4.11) to $g^{\pm}(x')$ of Definition 3.38 and noting that $\chi_{A'}(Tx) = \chi_{T^{-1}A'}(x)$ obtains:

$$\int_{X'} g(x')\chi_{A'}(x')d\mu_T = \int_X g(Tx)\chi_{T^{-1}A'}(x)d\mu.$$

By Definition 3.7, this proves (4.12):

$$\int_{A'} g(x')d\mu_T = \int_{T^{-1}A'} g(Tx)d\mu.$$

■

Remark 4.19 (On Proposition 4.18) *The reader is encouraged to return to Propositions 4.8 and 4.10. Note that when converting from a Lebesgue-Stieltjes integral to a Lebesgue integral in (4.4), that the domain of integration $A \in \mathcal{B}(\mathbb{R})$ is not changed. The same is true of (4.6) for $A \in \sigma(X)$. These results reflected the special nature of the induced measures μ of (4.1) and (4.2) and that both measures in these results are defined on the same space.*

For the general integral induced by a transformation T as in (4.12), these integrals are defined on different measure spaces. Thus the domain of integration in the range space must be transformed into its pre-image for the integral in the domain space.

It is tempting to rewrite (4.12) in terms of μ-integrals over sets $A \in \sigma(X)$ in the domain space, and μ_T-integrals over $T(A)$, as this initially appears to be merely a change of notation. But as will be seen in Example 4.23, while $T^{-1}A' \in \sigma(X)$ for all $A' \in \sigma(X')$, there may well exist $A \in \sigma(X)$, which cannot be so expressed. Measurability of T requires only that $T^{-1}[\sigma(X')] \subset \sigma(X)$, and it need not be the case that $T^{-1}[\sigma(X')] = \sigma(X)$.

Given this remark, the above result provides the following corollary.

Corollary 4.20 ($\int_{T(A)} g(x')d\mu_T$ as a μ-integral) *Let $T : (X, \sigma(X), \mu) \to (X', \sigma(X'), \mu')$ be a measurable transformation and μ_T defined on $\sigma(X')$ by (4.9).*

A measurable function $g : (X', \sigma(X')) \to (\mathbb{R}, \mathcal{B}(\mathbb{R}))$ is μ_T-integrable if and only if $g(Tx)$ is μ-integrable. When so integrable, for any $A \in T^{-1}[\sigma(X')]$:

$$\int_{T(A)} g(x')d\mu_T = \int_A g(Tx)d\mu. \tag{4.13}$$

Proof. *This follows from (4.12) since if $A \in T^{-1}[\sigma(X')]$, then by definition $A = T^{-1}[A']$ for some $A' \in \sigma(X')$.* ■

Example 4.21 (Expectations of random variables 2) *Let $X : (\mathcal{S}, \mathcal{E}, \mu) \to (\mathbb{R}, \mathcal{B}(\mathbb{R}), m)$ be a random variable defined on a probability space. As noted in Example 4.14, the induced measure μ_X on $(\mathbb{R}, \mathcal{B}(\mathbb{R}))$ equals μ_F, the Borel measure induced by $F(x)$, the distribution function of X:*

$$\mu_X \equiv \mu_F.$$

Hence, if $g(x)$ is a measurable function defined on $(\mathbb{R}, \mathcal{B}(\mathbb{R}), m)$, it is measurable as above when defined on $(\mathbb{R}, \mathcal{B}(\mathbb{R}), \mu_F)$, and (4.11) becomes:

$$\int_{\mathcal{S}} g(X(s))d\mu = \int_{\mathbb{R}} g(x)d\mu_F.$$

Thus the "expectation" of $g(X)$:

$$E[g(X)] \equiv \int_{\mathcal{S}} g(X(s))d\mu,$$

can be evaluated as a Lebesgue-Stieltjes integral on $\mathbb{R}$ with the Borel measure μ_F induced by the distribution function F.

If $F(x)$ is absolutely continuous and thus has an associated density function $f(x)$, this Lebesgue-Stieltjes integral can then be evaluated as a Lebesgue integral by (4.4):

$$E[g(X)] = (\mathcal{L}) \int_{\mathbb{R}} g(x)f(x)dx.$$

See Section IV.4.1.2 for a summary of these transformations, and Book VI for more details.

Not surprisingly, this example generalizes.

Corollary 4.22 (On $E[g(X)] \equiv \int_{\mathcal{S}} g(X(s))d\mu$) *Let $X : (\mathcal{S}, \mathcal{E}, \mu) \to (\mathbb{R}^n, \mathcal{B}(\mathbb{R}^n), m)$ be a random vector defined on a probability space with joint distribution function $F(x)$, and $g : (\mathbb{R}^n, \mathcal{B}(\mathbb{R}^n), m) \to (\mathbb{R}, \mathcal{B}(\mathbb{R}), m)$ a measurable function. Then:*

$$E[g(X)] \equiv \int_{\mathcal{S}} g(X(s))d\mu = \int_{\mathbb{R}^n} g(x)d\mu_F, \tag{4.14}$$

where μ_F is the Borel measure induced by $F(x)$ of Section 1.3.2.
Proof. *This is a restatement of (4.11), applying the result of Proposition 4.15 that $\mu_X = \mu_F$.* ■

4.3 Special Cases of Change of Variables

In this section, we develop three applications of the general framework of Proposition 4.18. Returning to the Riemann change of variables example in the introduction, we investigate this example as a change of variables in the associated Lebesgue integral and then generalize. The next two sections apply the general framework to transformations $T : \mathbb{R}^n \to \mathbb{R}^n$, first considering the case of invertible linear transformations, and then generalizing to invertible, continuously differentiable transformations.

4.3.1 Lebesgue Integrals on $\mathbb{R}$

Revisiting Example 4.1, we can apply the results from Propositions 4.8 and 4.18 to provide a measure-theoretic framework for this familiar Riemann procedure. While this Riemann integral is more easily analyzed with the calculus methods of Example 4.1, it also provides an accessible Lebesgue example to which the above methods can be applied, and then generalized in Proposition 4.24.

Example 4.23 (Example 4.1 cont'd) *We begin with a simple set-up which leads to a false start and then refine our approach.*

1. ***False Start:*** *The goal is to evaluate the Riemann integral* $(\mathcal{R})\int_0^\infty x\exp(-x^2)dx$. *Since* $x\exp(-x^2)$ *is absolutely Riemann integrable on this interval, it is also Lebesgue integrable by Proposition III.2.56, and the integrals agree. Thus this can also be considered an exercise to evaluate* $(\mathcal{L})\int_0^\infty x\exp(-x^2)dx$ *by applying (4.12):*

$$\int_{A'} g(x')d\mu_T = \int_{T^{-1}A'} g(Tx)d\mu. \tag{1}$$

Initially, it may seem natural to regard $(\mathcal{L})\int_0^\infty x\exp(-x^2)dx$ *as the integral on the left in* (1), *with* $g(x) = x\exp(-x^2)$ *and* $d\mu_T = dx$ *for Lebesgue measure. But this would require that we identify a domain space measure* μ *and a transformation* T *so that the induced measure* $\mu_T = m$, *Lebesgue measure.*

Thus we identify this integral with the integral on the right in (1), *and so* $d\mu = dx$, *and the domain space* $(X, \sigma(X), \mu) = (\mathbb{R}, \mathcal{B}(\mathbb{R}), m)$. *We then seek to determine* g *and* T *so that:*

$$g(Tx) = x\exp(-x^2).$$

The simplest solution is then:

$$Tx = x^2; \quad g(y) = \sqrt{y}\exp(-y).$$

Since $Tx \geq 0$, *our model becomes:*

$$\begin{aligned} T &: (\mathbb{R}, \mathcal{B}(\mathbb{R}), m) \to (\mathbb{R}^+, \mathcal{B}(\mathbb{R}^+), m_T), \\ g &: (\mathbb{R}^+, \mathcal{B}(\mathbb{R}^+), m_T) \to (\mathbb{R}^+, \mathcal{B}(\mathbb{R}^+), m), \end{aligned} \tag{2}$$

where $\mathbb{R}^+ = [0,\infty)$ *and* $\mathcal{B}(\mathbb{R}^+) \equiv \mathcal{B}(\mathbb{R})\bigcap[0,\infty)$. *The induced measure* m_T *in the range space of* T *is given by (4.9). Then by* (1):

$$(\mathcal{L})\int_{T^{-1}A'} x\exp(-x^2)dx = \int_{A'} \sqrt{x'}\exp(-x')dm_T.$$

Unfortunately, there is no $A' \in \mathcal{B}(\mathbb{R}^+)$ *with* $T^{-1}A' = [0,\infty)$. *More generally, this is an example noted in Remark 4.19 for which* $T^{-1}\left[\mathcal{B}(\mathbb{R}^+)\right] \subsetneqq \mathcal{B}(\mathbb{R})$.

2. ***Refined Approach:*** *Redefining the domain of* T *in* (2)*:*

$$T : (\mathbb{R}^+, \mathcal{B}(\mathbb{R}^+), m) \to (\mathbb{R}^+, \mathcal{B}(\mathbb{R}^+), m_T),$$

and applying (1) *with* $A' = [0,\infty)$ *obtains:*

$$(\mathcal{L})\int_0^\infty x\exp(-x^2)dx = \int_0^\infty \sqrt{x'}\exp(-x')dm_T. \tag{3}$$

The integrand on the right seems equally intractable as that on the left, in fact perhaps more so since a somewhat familiar Lebesgue integral has been replaced by a Lebesgue-Stieltjes integral with measure m_T.

But the measure m_T *is of the special type in (4.1) and addressed in Proposition 4.8. For any set* $(a,b] \in \mathcal{A}'(\mathbb{R}^+) \equiv \mathcal{A}' \bigcap [0,\infty)$:

$$m_T((a,b]) \equiv m\left(T^{-1}(a,b]\right) = \sqrt{b} - \sqrt{a}.$$

This also follows for sets of the form $[0,b]$ *for* $b > 0$. *Thus with* $f(x) = 0.5x^{-1/2}$ *for* $x > 0$ *and* 0 *otherwise:*

$$m_T((a,b]) = (\mathcal{L})\int_a^b f(x)dx.$$

Hence, Proposition 4.8 allows the Lebesgue-Stieltjes integral in (3) *to be restated in terms of a Lebesgue integral in which the original integrand is multiplied by* $f(x)$. *The domain of integration is kept the same as seen in (4.4):*

$$\int_0^\infty \sqrt{x'}\exp(-x')dm_T = (\mathcal{L})\int_0^\infty 0.5\exp(-x')dx'. \tag{4}$$

Putting the results of (3) *and* (4) *together, we have with the benefit of two measure-theoretic change of variable formulas:*

$$(\mathcal{L})\int_0^\infty x\exp(-x^2)dx = (\mathcal{L})\int_0^\infty (0.5)\exp(-x')dx'.$$

Since both $x\exp(-x^2)$ *and* $0.5\exp(-x')$ *are continuous and Lebesgue integrable, these integrals again equal the corresponding Riemann integrals, and hence:*

$$(\mathcal{R})\int_0^\infty x\exp(-x^2)dx = (\mathcal{R})\int_0^\infty (0.5)\exp(-x')dx'.$$

The following proposition generalizes this example, relabeling the transformation h. It requires the assumption of absolute continuity for the inverse transformation h^{-1}, recalling Definition III.3.54. By Remark III.3.56, it is sufficient to assume that h^{-1} is continuously differentiable.

Note that since the transformation $h(x)$ is continuous and thus Borel measurable by Proposition I.3.11, the function $f(h(x))$ is Borel measurable by Proposition I.3.33 when $f(x)$ is Borel measurable. As noted in Remark I.3.31, if $f(x)$ is Lebesgue measurable, then the measurability of $f(h(x))$ is not generally predictable.

Proposition 4.24 (Lebesgue substitution) *Let $h : \mathbb{R} \to \mathbb{R}$ be a continuously differentiable function with $h'(x) \neq 0$ and h^{-1} absolutely continuous, and let $[a, b]$ be a finite or infinite interval with $[a, b] \subset Rng[h]$, the range of h.*

If $f(x)$ is Borel integrable on $[a, b]$:

$$(\mathcal{L}) \int_a^b f(y)dy = (\mathcal{L}) \int_{h^{-1}(a)}^{h^{-1}(b)} f(h(x))h'(x)dx. \tag{4.15}$$

Proof. *As noted above, $h : (\mathbb{R}, \mathcal{B}(\mathbb{R}), m) \to (\mathbb{R}, \mathcal{B}(\mathbb{R}), m)$ is Borel measurable, while $g(y) \equiv f(y)h'(h^{-1}(y))$ is Borel measurable since $f(x)$ has this property by assumption, and $h'(h^{-1}(y))$ is continuous and thus Borel measurable. Further, $h'(x) \neq 0$ assures h is one-to-one, and this obtains that $g(h(x)) = f(h(x))h'(x)$.*

With $A' = [a, b]$, it follows from (4.12) that:

$$\int_{h^{-1}([a,b])} f(h(x))h'(x)dx = \int_a^b f(y)h'(h^{-1}(y))dm_h, \tag{1}$$

where m_h is the Borel measure induced by $h(x)$.

Since h is continuously differentiable with $h'(x) \neq 0$ and $[a, b] \subset Rng[h]$:

$$h^{-1}([a,b]) = \begin{cases} [h^{-1}(a), h^{-1}(b)], & h' > 0, \\ [h^{-1}(b), h^{-1}(a)], & h' < 0. \end{cases} \tag{2}$$

Suppressing $(\mathcal{L})$ and recalling Remark 3.8, the identity in (1) can be written:

$$sgn(h') \int_{h^{-1}(a)}^{h^{-1}(b)} f(h(x))h'(x)dx = \int_a^b f(y)h'(h^{-1}(y))dm_h, \tag{3}$$

where $sgn(h')$ is the sign of $h'(x)$.

Also by (2), the induced measure m_h in (4.9) is defined on the semi-algebra $\mathcal{A}' \equiv \{(c, d]\}$ by:

$$m_h((c,d]) = sgn(h') \left[h^{-1}(d) - h^{-1}(c)\right].$$

Absolute continuity of h^{-1} and Proposition III.3.59 assures the existence of $\frac{dh^{-1}}{dy}$ m-a.e., and by Proposition III.3.62, the measure $m_h((c, d])$ can be expressed on $\mathcal{A}'$ as a Lebesgue integral:

$$m_h((c,d]) = sgn(h') \int_c^d \frac{dh^{-1}}{dy} dy.$$

It follows from $h'(x) \neq 0$ that $\frac{dh^{-1}}{dy} = 1/h'(h^{-1}(y))$ m-a.e., and thus for all $(c, d] \in \mathcal{A}'$:

$$m_h((c,d]) = sgn(h') \int_c^d 1/h'(h^{-1}(y))dy. \tag{4}$$

Now Lebesgue measurability of $1/h'(h^{-1}(y))$ assures that the integral on the right in (4) defines a measure on $\mathcal{B}(\mathbb{R}^n)$ by Proposition 4.3, and since this integral agrees with m_h on $\mathcal{A}'$, it follows from uniqueness of extensions by Proposition 1.22 that for all $A \in \mathcal{B}(\mathbb{R}^n)$:

$$m_h(A) = sgn(h') \int_A 1/h'(h^{-1}(y))dy. \tag{5}$$

The result in (4.15) now follows from (3) and (5) by Proposition 4.8. ■

4.3.2 Linear Transformations on $\mathbb{R}^n$

In this section, we provide a more detailed change of variables result in the special case where T is an invertible linear transformation, $T : \mathbb{R}^n \rightarrow \mathbb{R}^n$. This result will be generalized to continuously differentiable and invertible transformations on $\mathbb{R}^n$ in the next section. See Section III.4.3.1 for an introduction to vector spaces and linear transformations. More generally, see **Strang** (2009) for a modern development of linear algebra, as well as **Edwards** (1973) for much of the linear algebra needed below.

By Definition III.4.55, that $T : \mathbb{R}^n \rightarrow \mathbb{R}^n$ is a **linear transformation** means that if $x, y \in \mathbb{R}^n$, then for all $a, b \in \mathbb{R}$:

$$T[ax + by] = aTx + bTy. \tag{4.16}$$

By invertible is meant that there exists a transformation, denoted T^{-1}: $\mathbb{R}^n \rightarrow \mathbb{R}^n$, so that for all x :

$$T\left(T^{-1}x\right) = T^{-1}\left(Tx\right) = x. \tag{4.17}$$

Exercise 4.25 (On T^{-1}) *Prove that if $T : \mathbb{R}^n \rightarrow \mathbb{R}^n$ is an invertible linear transformation, then T^{-1} is an invertible linear transformation with $(T^{-1})^{-1} = T$.*

Let $\{e_j\}_{j=1}^n$ denote the standard basis for $\mathbb{R}^n$, which is sometimes called the **canonical basis**, and so $e_j \equiv (0, ..., 1, 0, ..., 0)$ with 1 in the jth coordinate. Then if $x \equiv (x_1, x_2, ..., x_n) \in \mathbb{R}^n$:

$$x = \sum\nolimits_{j=1}^{n} x_j e_j.$$

Thus if $T : \mathbb{R}^n \rightarrow \mathbb{R}^n$ is a linear transformation, then by (4.16):

$$Tx = \sum\nolimits_{j=1}^{n} x_j Te_j, \tag{4.18}$$

and so Tx is completely determined by the collection of vectors $\{Te_j\}_{j=1}^n \subset \mathbb{R}^n$.

Given this basis, the transformation T has a matrix representation as $A = (a_{ij})_{i,j=1}^n$:

$$A \equiv \begin{pmatrix} a_{11} & a_{12} & \cdots & a_{1n} \\ a_{21} & a_{22} & \cdots & a_{2n} \\ \vdots & \vdots & \vdots & \vdots \\ a_{n1} & a_{n2} & \cdots & a_{nn} \end{pmatrix},$$

where the columns of this matrix are the vectors $\{Te_j\}_{j=1}^n$, so:

$$Te_j \equiv \begin{pmatrix} a_{1j} \\ a_{2j} \\ \vdots \\ a_{nj} \end{pmatrix}.$$

It then follows from (4.18) that:

$$Tx = \begin{pmatrix} \sum\nolimits_{j=1}^{n} x_j a_{1j} \\ \sum\nolimits_{j=1}^{n} x_j a_{2j} \\ \vdots \\ \sum\nolimits_{j=1}^{n} x_j a_{nj} \end{pmatrix}. \tag{4.19}$$

The reader may well recognize the expression on the right in (4.19) as Ax, the product of the matrix A with the vector x, where x is conventionally identified with a column vector (or matrix):

$$Ax = \begin{pmatrix} a_{11} & a_{12} & \cdots & a_{1n} \\ a_{21} & a_{22} & \cdots & a_{2n} \\ \vdots & \vdots & \vdots & \vdots \\ a_{n1} & a_{n2} & \cdots & a_{nn} \end{pmatrix} \begin{pmatrix} x_1 \\ x_2 \\ \vdots \\ x_n \end{pmatrix}.$$

If $B = (b_{ij})_{i=1,j=1}^{n,m}$ denotes a general $n \times m$ matrix with n rows and m columns, which includes row $(1 \times m)$ and column $(n \times 1)$ matrices, the **transpose** $B^t = \left(b'_{ij}\right)_{i=1,j=1}^{m,n}$ of B is an $m \times n$ matrix, with m rows and n columns, and is defined:

$$b'_{ij} = b_{ji}. \tag{4.20}$$

With this notation, (4.19) is more simply expressed notationally:

$$Tx = Ax \equiv \left(\sum\nolimits_{j=1}^{n} a_{1j}x_j, \sum\nolimits_{j=1}^{n} a_{2j}x_j, \cdots, \sum\nolimits_{j=1}^{n} a_{nj}x_j\right)^t. \tag{4.21}$$

Notation 4.26 *For most expressions involving vectors, there is no reason to "identify" vectors as column vectors. In other words, stating that* $x^t \equiv (x_1, x_2, ..., x_n)$ *or* $x \equiv (x_1, x_2, ..., x_n)$ *conveys exactly the same information, and it is common to drop the transpose designation unless required for well-definedness.*

For example, this notation is essential for expressions involving matrices. If A *is an* $n \times n$ *matrix and* x *is an* n*-vector, then* xAx *is not defined, although it could be argued that this expression is uniquely definable as* x^tAx*. On the other hand,* xx *is also not defined, nor is there a unique interpretation as both* x^tx *and* xx^t *are well-defined but very different. The reader should verify that the first item is a* 1×1 *matrix, while the second is an* $n \times n$ *matrix.*

Remark 4.27 (On $\{e_j\}_{j=1}^n$ and other bases) *The reader may have surmised, if not known from other contexts, that there is nothing special about the basis* $\{e_j\}_{j=1}^n$*, other than that it is the most natural in the Euclidean space* $\mathbb{R}^n$*. This space is an example of a vector space over the real field* $\mathbb{R}$*.*

In general, a collection of vectors $\{f_j\}_{j=1}^m$ *is a* ***basis*** *for a vector space* V *over a field* $\mathcal{F}$*, which is always* $\mathbb{R}$ *or* $\mathbb{C}$ *in these books:*

1. ***Linearly Independent:*** *If* $\sum\nolimits_{j=1}^{m} \alpha_j f_j = \theta$ *for* $\{\alpha_j\}_{j=1}^m \subset \mathcal{F}$*, then* $\alpha_j = 0$ *for all* j*.*

2. ***Spanning:*** *If* $x \in V$*, then there exists* $\{\alpha_j\}_{j=1}^m \subset \mathcal{F}$ *so that* $x = \sum\nolimits_{j=1}^{m} \alpha_j f_j$*.*

In item 1, "θ*" denotes the "zero vector" in* V*, so* $x + \theta = x$ *for all* $x \in V$*. For* $V = \mathbb{R}^n$*, which is a vector space over the real field* $\mathcal{F} = \mathbb{R}$*, the zero vector is, as expected, just* $\theta = (0, ..., 0)$*. When* $\mathcal{F}$ *is not one of* $\mathbb{R}$ *or* $\mathbb{C}$*, then "*$\alpha_j = 0$*" in item 1 means that all* α_j *equal the additive unit in* $\mathcal{F}$*, meaning that* $\alpha + 0 = \alpha$ *for all* $\alpha \in \mathcal{F}$*.*

It is a small exercise to check that the canonical basis $\{e_j\}_{j=1}^n$ *as defined above is indeed a basis for* $\mathbb{R}^n$*. This is the most natural basis for* $\mathbb{R}^n$ *since in natural units, if* $x \equiv (x_1, x_2, ..., x_n) \in \mathbb{R}^n$*, then* $x = \sum\nolimits_{j=1}^{n} x_j e_j$*. It can then be proved that every basis in* $\mathbb{R}^n$ *has* n *vectors.*

Given a linear transformation T*, the matrix representation for* T*, denoted* A *above, fundamentally reflects the choice of basis. This follows exactly as above. If* $\{f_j\}_{j=1}^n$ *is any basis, then the columns of the associated matrix are* $\{Tf_j\}_{j=1}^n$*.*

If T is invertible, then in any basis the associated matrix A has an $n \times n$ inverse matrix, denoted A^{-1}, for which by (4.17):

$$AA^{-1} = A^{-1}A = I, \tag{4.22}$$

where $I = (c_{ij})_{i,j=1}^{n}$ is the identity matrix, with $c_{jj} = 1$ and all other $c_{ij} = 0$.

Exercise 4.28 (On transformation of bases) *Show that if $T : \mathbb{R}^n \to \mathbb{R}^n$ is an invertible linear transformation and $\{f_j\}_{j=1}^{n}$ is any basis for $\mathbb{R}^n$, then $\{Tf_j\}_{j=1}^{n}$ is also a basis. By Remark 4.27, this implies that the columns of the associated matrix A for T in this basis form a basis. Since T is invertible if and only if T^{-1} is invertible, it follows that $\{T^{-1}f_j\}_{j=1}^{n}$, the columns of the matrix A^{-1}, also form a basis.*

Given T as above with associated matrix A in the standard basis, define a measure μ on the complete Lebesgue sigma algebra $\mathcal{M}_L(\mathbb{R}^n)$, and thus also on $\mathcal{B}(\mathbb{R}^n)$, by:

$$\mu(B) = |\det(A)|\, m(B). \tag{4.23}$$

Here, m is Lebesgue measure and $\det(A)$ is the **determinant of the invertible matrix** A **induced by** T.

Remark 4.29 (On $\det(A)$**)** *There is no efficient way to explicitly derive the various properties of the determinant of an $n\times n$ matrix A needed here. To avoid an extended development that will lead us astray, we simply identify some of these properties and direct the reader to the above noted and other references for details.*

Below, A and B are arbitrary $n\times n$ matrices. Such matrices need not be explicitly induced by an identified linear transformation T as above, where the columns of A are $\{Tf_j\}_{j=1}^{n}$ relative to a given basis $\{f_j\}_{j=1}^{n}$. But note that given an arbitrary basis, any $n \times n$ matrix can be deemed to have been induced by such a linear transformation T on $\mathbb{R}^n$. Given a basis $\{f_j\}_{j=1}^{n}$, define the associated transformation T so that Tf_j equals the jth column of A and then extend T to all x as in (4.18). Then by construction, A is the matrix induced by T in this basis.

Properties of determinants that will be needed:

1. $\det(I) = 1$, *where I is the identity matrix. More generally,* $\det(A) = \prod_{j=1}^{n} a_{jj}$ *if A is a diagonal matrix, with $a_{ij} = 0$ for $i \neq j$.*

2. $\det(A) = 0$ *if and only if A has linearly dependent column vectors, if and only if (by item 3) A has linearly dependent row vectors. The vectors $\{f_j\}_{j=1}^{m}$ are* ***linearly dependent*** *if there exist $\{\alpha_j\}_{j=1}^{m} \subset \mathcal{F}$, not all zero, so that $\sum_{j=1}^{m} \alpha_j f_j = \theta$.*

3. $\det(A) = \det(A^t)$, *where A^t is the transpose of A.*

4. $\det(AB) = \det(A)\det(B)$.

5. $\det(A^{-1}) = 1/\det(A)$ *by items 1 and 4 if A^{-1} exists.*

6. *Given a rectangle $R = \prod_{j=1}^{n}[a_j, b_j]$, then with m Lebesgue measure:*

$$m\left[A(R)\right] = |\det(A)|\, m[R] = |\det(A)| \prod_{j=1}^{n}[b_j - a_j], \tag{4.24}$$

where $A(R) = \{Ax | x \in R\}$. The same result holds if R is defined to be an open or semi-closed rectangle.

Exercise 4.30 (Item 6 special case) *Derive (4.24) when A is a diagonal matrix as defined in item 1. Hint: $R = a + R'$ where $a = (a_1, ..., a_n)$ and $R' = \prod_{j=1}^{n}[0, b_j - a_j]$. Use linearity of A and properties of m to conclude that $m[A(R)] = m[A(R')]$. Evaluate $A(R')$ and $m[A(R')]$ explicitly, noting* $\det(A)$ *as in item 1.*

Consider next the above linear transformation T to be defined between **measure spaces:**

$$T : (\mathbb{R}^n, \mathcal{B}(\mathbb{R}^n), \mu) \to (\mathbb{R}^n, \mathcal{B}(\mathbb{R}^n), \mu_T). \tag{4.25}$$

Here, μ denotes the measure defined in (4.23), and μ_T is the measure induced by T in (4.9). Since T is apparently continuous by the expression in (4.21), it follows that T is a measurable transformation by Proposition I.3.13, and thus μ_T is well-defined.

Proposition 4.31 ($\mu_T = m$) *With μ defined in (4.23), the measure μ_T induced by T in (4.25) is Lebesgue measure:*

$$\mu_T = m. \tag{4.26}$$

Proof. *If $B \in \mathcal{B}(\mathbb{R}^n)$, then by (4.9) and (4.25):*

$$\mu_T(B) \equiv \mu\left[T^{-1}(B)\right] \equiv |\det(A)|\, m(T^{-1}(B)). \tag{1}$$

If $B = \prod_{j=1}^{n}(a_j, b_j]$ is a right semi-closed rectangle, then since T^{-1} is again a linear transformation by Exercise 4.25 with matrix A^{-1}, (4.24) and item 5 of Remark 4.29 obtain:

$$m(T^{-1}(B)) = \left|\det(A^{-1})\right| m[B] = m[B]/\left|\det(A)\right|.$$

This and (1) *prove (4.26) for all $B \in \mathcal{A}'$, the semi-algebra of right semi-closed rectangles, $\left\{\prod_{j=1}^{n}(a_j, b_j]\right\}$.*

Equality of these measures then extends to the associated algebra $\mathcal{A}$ of all finite disjoint unions of $\mathcal{A}'$-sets by the Carathéodory extension theorem 2 of Proposition 1.20, since each is extended additively. Then by Proposition 1.22, this identity extends to the sigma algebra $\sigma(\mathcal{A})$, the smallest sigma algebra that contains $\mathcal{A}$. The proof is complete by the result of Proposition I.8.1 that $\sigma(\mathcal{A}) = \mathcal{B}(\mathbb{R}^n)$. ■

Given Proposition 4.31, the results of Section 4.2.2 can now be applied to address invertible linear transformations in Lebesgue integrals. These results will be generalized in the next section to continuously differentiable invertible transformations $T : \mathbb{R}^n \to \mathbb{R}^n$, where the absolute value of the **Jacobian determinant of** T will take the place of $|\det(A)|$ below.

Note that since linear transformations are continuous and thus Borel measurable by Proposition I.3.13, the function $g(Ax)$ is Borel measurable as in Proposition I.3.33 when $g(x)$ is Borel measurable. As noted in Remark I.3.31, if $g(x)$ is Lebesgue measurable, then the measurability of $g(Ax)$ is not generally predictable.

Proposition 4.32 (Linear transformation in Lebesgue integrals) *Let $T : \mathbb{R}^n \to \mathbb{R}^n$ be an invertible linear transformation, and $|\det(A)|$ the absolute value of the determinant of the matrix A induced by T. Then for any Borel measurable function $g : \mathbb{R}^n \to \mathbb{R}$:*

$$\int_{\mathbb{R}^n} g(Ax)\, |\det(A)|\, dm = \int_{\mathbb{R}^n} g(y) dm, \tag{4.27}$$

though both integrals may be infinite.

More generally, a Borel measurable function $g(x)$ is Lebesgue integrable if and only if $g(Ax)\,|\det(A)|$ is Lebesgue integrable, and when integrable:

$$\int_{T^{-1}B} g(Ax)\,|\det(A)|\,dm = \int_B g(y)dm, \tag{4.28}$$

for all $B \in \mathcal{B}(\mathbb{R}^n)$.

Proof. *As $\mu \equiv |\det(A)|\,m$ by definition, and the induced $\mu_T = m$ by Proposition 4.31, the result follows from Proposition 4.18 by substitution.* ■

This result obtains a generalization of (4.24).

Corollary 4.33 **($m\,[T\,(B)]\,,\,B \in \mathcal{B}(\mathbb{R}^n)$)** *If $T : \mathbb{R}^n \to \mathbb{R}^n$ is an invertible linear transformation, then for all $B \in \mathcal{B}(\mathbb{R}^n)$:*

$$m\,[T\,(B)] = |\det(A)|\,m\,[B]\,. \tag{4.29}$$

Proof. *Letting $g(x) = 1$, (4.28) implies that for any $B' \in \mathcal{B}(\mathbb{R}^n)$, $m\,[B'] = |\det(A)|\,m\,[T^{-1}\,(B')]$. As T and T^{-1} are continuous, $B' \in \mathcal{B}(\mathbb{R}^n)$ if and only if $T^{-1}\,(B') \in \mathcal{B}(\mathbb{R}^n)$, so given B, let $B' = T\,(B)$ in this identity.* ■

Remark 4.34 (On $|\det(A)|$) *It should be noted that in contrast to the expression in (4.28), $h'(x)$ in (4.15) appears without absolute values. However, $h(x) = ax$ is a linear transformation on $\mathbb{R}$ to which either result applies. Thus in (4.28) we would have $|\det(A)| = |h'(x)| = |a|$, while in (4.15), $h'(x) = a$.*

Below, we generalize (4.28) to continuously differentiable transformations and again an absolute value will be seen in the term that generalizes the role of $\det(A)$, so the same problem is apparently observed for more general $h(x)$.

The explanation for this is that over any interval $[a, b]$ for which $h'(x) < 0$, it would follow that $h^{-1}(b) < h^{-1}(a)$ and hence by Remark 3.8:

$$\int_{h^{-1}(a)}^{h^{-1}(b)} f(h(x))h'(x)dx = \int_{h^{-1}(b)}^{h^{-1}(a)} f(h(x))\,|h'(x)|\,dx = \int_{h^{-1}([a,b])} f(h(x))\,|h'(x)|\,dx.$$

For subintervals on which $h'(x) > 0$, $h^{-1}(a) < h^{-1}(b)$ and so:

$$\int_{h^{-1}(a)}^{h^{-1}(b)} f(h(x))h'(x)dx = \int_{h^{-1}(a)}^{h^{-1}(b)} f(h(x))\,|h'(x)|\,dx = \int_{h^{-1}([a,b])} f(h(x))\,|h'(x)|\,dx.$$

In either case, integrating with $|h'(x)|$ over the interval $h^{-1}\,([a, b])$ as in (4.28) produces the same result as integrating with $h'(x)$ from $h^{-1}(a)$ to $h^{-1}(b)$ as in (4.15).

4.3.3 Differentiable Transformations on $\mathbb{R}^n$

In this section, we generalize Proposition 4.32 to invertible, continuously differentiable transformations, a result that also generalizes (4.15) to multivariate Lebesgue integrals. Recall the set-up in Section 4.2.2, but here modified to:

$$(\mathbb{R}^n, \mathcal{B}(\mathbb{R}^n), \mu) \quad \underset{\longrightarrow}{T} \quad (\mathbb{R}^n, \mathcal{B}(\mathbb{R}^n), m) \quad \underset{\longrightarrow}{g} \quad (\mathbb{R}, \mathcal{B}(\mathbb{R}), m).$$

Thus T is a transformation between an n-dimensional Borel measure space with measure μ, to be defined below, and a Borel measure space with Lebesgue measure, while $g(x)$ is a Borel measurable function on this latter space.

For $x = (x_1, x_2, ..., x_n)$,

$$T(x) = (t_1(x), t_2(x), \cdots, t_n(x)),$$

where $t_i : \mathbb{R}^n \to \mathbb{R}$ for all i. We will say that T is continuous, differentiable, continuously differentiable, etc., if the component functions $\{t_i\}_{i=1}^n$ have such properties.

When this transformation has differentiable component functions, the **Jacobian matrix** associated with T, denoted $T'(x)$, is defined as:

$$T'(x) \equiv \begin{pmatrix} \frac{\partial t_1}{\partial x_1} & \frac{\partial t_1}{\partial x_2} & \cdots & \frac{\partial t_1}{\partial x_n} \\ \frac{\partial t_2}{\partial x_1} & \frac{\partial t_2}{\partial x_2} & \cdots & \frac{\partial t_2}{\partial x_n} \\ \vdots & \vdots & \vdots & \vdots \\ \frac{\partial t_n}{\partial x_1} & \frac{\partial t_n}{\partial x_2} & \cdots & \frac{\partial t_n}{\partial x_n} \end{pmatrix}. \tag{4.30}$$

The **Jacobian determinant** associated with T is defined as the determinant of $T'(x)$ and thus denoted $\det(T'(x))$. This matrix and its determinant are named for **Carl Gustav Jacob Jacobi** (1804–1851), an early developer of determinants and their applications in analysis.

Notation 4.35 **($T'(x)$)** *The Jacobian matrix is denoted in many ways:*

$$T'(x) \equiv \frac{\partial(t_1, t_2, ..., t_n)}{\partial(x_1, x_2, ..., x_n)} = \left(\frac{\partial(t_1, t_2, ..., t_n)}{\partial(x_1, x_2, ..., x_n)}\right) = \left(\frac{\partial T}{\partial x}\right),$$

and the Jacobian determinant is sometimes expressed:

$$\det T'(x) \equiv \left|\frac{\partial(t_1, t_2, ..., t_n)}{\partial(x_1, x_2, ..., x_n)}\right|.$$

Absolute values are commonly used with a matrix to denote determinant, but it can be ambiguous in the context of the current development where we use the absolute value of the Jacobian determinant, $|\det T'(x)|$. Thus we will avoid this alternative notation.

Remark 4.36 (Multivariate calculus) *We cannot review multivariate calculus in any detail in these books, but there are several properties of the Jacobian matrix which are used below. The reader can find such results in* ***Edwards*** *(1973) and elsewhere.*

1. ***Linear Approximations for Continuously Differentiable Transformations:*** *If the transformation $T : \mathbb{R}^n \to \mathbb{R}^n$ is continuously differentiable at x, then:*

$$T(y) = T(x) + T'(x)(y - x) + o(|y - x|), \tag{4.31}$$

where $|y - x|$ denotes the ***standard Euclidean norm*** *on $\mathbb{R}^n$. In the terminology of Remark 4.43 and Example 9.13, this is also called the l_2-norm of $x \equiv (x_1, x_2, ..., x_n)$ and defined by:*

$$|(x_1, x_2, ..., x_n)| = \sqrt{\sum_{j=1}^{n} x_j^2}. \tag{4.32}$$

*The "****little*** *o" error term means that:*

$$\frac{o(|y - x|)}{|y - x|} \to 0 \text{ as } |y - x| \to 0. \tag{4.33}$$

The term $T'(x)(y-x)$ is a standard matrix product as in (4.21) of the Jacobian matrix $T'(x)$ and the vector $(y-x)$. Thus $T'(x)$ serves the same role in linearly approximating T as does $f'(x)$ in the usual Taylor series approximation for a continuously differentiable function:

$$f(y) = f(x) + f'(x)(y - x) + o(|y - x|).$$

2. ***Bounds for Linear Transformations:*** *A linear transformation $T : \mathbb{R}^n \to \mathbb{R}^n$ is continuous by (4.21), and thus by Proposition III.4.58 is bounded in the sense of Definition III.4.55. In detail, if $A = (a_{ij})_{i,j=1}^n$ is the matrix representation of T in a given basis, then for all x:*

$$|Ax| \leq \sqrt{\sum\nolimits_{i,j=1}^{n} a_{ij}^2}\, |x|, \tag{4.34}$$

where the norms $|Ax|$ and $|x|$ are defined in (4.32). In the notation of Definition III.4.55, the norm of this transformation, denoted $\|T\|$, satisfies:

$$\|T\| \leq \sqrt{\sum\nolimits_{i,j=1}^{n} a_{ij}^2}.$$

To prove (4.34), it follows from (4.21) and (4.32) that:

$$|Ax|^2 = \sum\nolimits_{i=1}^{n}\left(\sum\nolimits_{j=1}^{n} a_{ij}x_j\right)^2.$$

Then by the Cauchy-Schwarz inequality of (9.2):

$$\left(\sum\nolimits_{j=1}^{n} a_{ij}x_j\right)^2 \leq |x|^2 \sum\nolimits_{j=1}^{n} a_{ij}^2.$$

3. ***Inverse Function Theorem:*** *Recall the one variable result that if $f'(x)$ is continuously differentiable and $f'(x_0) \neq 0$, then there are open intervals U with $x_0 \in U$ and V with $f(x_0) \in V$ so that $f^{-1} : V \to U$ is continuously differentiable, and for $v \in V$:*

$$\left(f^{-1}\right)'(v) = 1/f'\left[f^{-1}(v)\right].$$

The inverse function theorem generalizes this result to transformations.

Proposition 4.37 (Inverse function theorem) *If $T : \mathbb{R}^n \to \mathbb{R}^n$ is continuously differentiable at x_0 and $\det T'(x_0) \neq 0$, then there are open sets U with $x_0 \in U$ and V with $T(x_0) \in V$ so that $T^{-1} : V \to U$ is continuously differentiable, and for $v \in V$:*

$$\left(T^{-1}\right)'(v) = \left(T'\left(T^{-1}(v)\right)\right)^{-1}. \tag{4.35}$$

In words, the Jacobian matrix of T^{-1} at v is the inverse of the Jacobian matrix of T at $T^{-1}(v)$.

While the inverse function theorem identifies the behavior of T^{-1} locally, meaning on some open set V around every point $T(x_0)$ where $\det T'(x_0) \neq 0$, it also provides a global statement on T^{-1}.

Proposition 4.38 (Inverse function theorem - global) *Let $T : \mathbb{R}^n \to \mathbb{R}^n$ be a one-to-one and continuously differentiable transformation with $\det T'(x) \neq 0$ for all x. Given $A = \prod_{i=1}^n [a_i, b_i]$, a closed rectangle in $\mathbb{R}^n$, there exists an open set V with $T(A) \subset V$ so that $T^{-1} : V \to \mathbb{R}^n$ is continuously differentiable and (4.35) is satisfied for all $v \in V$.*

Proof. *For any $x \in A$, there exists U_x and V_x with $x \in U_x$ and $T(x) \in V_x$, so that $T_x^{-1} : V \to U$ is continuously differentiable and (4.35) is satisfied. Here, we denote this inverse by T_x^{-1} since it is defined locally in an open set about $T(x)$. Now $U' \equiv \bigcup_x U_x$ is* *an open set by Exercise III.4.46 with $A \subset U'$, and similarly $V' \equiv \bigcup_x V_x$ is an open set with $T(A) \subset V'$.*

Since A is closed and bounded, it is compact by the Heine-Borel theorem of Proposition I.2.27, and thus there exists a finite collection $\{U_{x_j}\}_{j=1}^m$ so that $A \subset U \equiv \bigcup_{j=1}^m U_{x_j}$, and thus so too $T(A) \subset V \equiv \bigcup_{j=1}^m V_{x_j}$. Now if $v \in V_{x_j} \bigcap V_{x_k}$, then $T_{x_j}^{-1}(v) = T_{x_k}^{-1}(v)$ since T is one-to-one, and hence T^{-1} is well-defined on V.

Continuous differentiability and (4.35) then follow from Proposition 4.37. ■

The next result proves that the error in the linear approximation to T in item 1 can be explicitly bounded in terms of the errors in the component functions $\{t_i(x)\}_{i=1}^n$. The next proposition identifies this connection and will be applied below. For notational simplicity, we denote $\frac{\partial t_i}{\partial x_j}$ by $\partial_j t_i$.

Proposition 4.39 (Bounding $|T(y) - T(x) - T'(x)(y - x)|$) *Let $T : \mathbb{R}^n \to \mathbb{R}^n$ be a continuously differentiable transformation with $T(x) = (t_1(x), t_2(x), \cdots, t_n(x))$, and $A = \prod_{i=1}^n [a_i, b_i]$ a closed rectangle in $\mathbb{R}^n$. If for all i, j, and for all $x, y \in A$:*

$$|\partial_j t_i(y) - \partial_j t_i(x)| \leq \beta,$$

then:

$$|T(y) - T(x) - T'(x)(y - x)| \leq \beta n^2 |y - x|. \tag{4.36}$$

Proof. *If $x \equiv (x_1, x_2, ..., x_n)$, then:*

$$|x| \leq \sum_{j=1}^n |x_i|,$$

as is confirmed by squaring both sides, where $|x|$ is defined as in (4.32) and $|x_i|$ is defined as an absolute value. Hence:

$$|T(y) - T(x) - T'(x)(y - x)| \leq \sum_{i=1}^n \left| t_i(y) - t_i(x) - \sum_{j=1}^n \partial_j t_i(x)(y_j - x_j) \right|.$$

As each t_i is a continuously differentiable function, the mean value theorem obtains that for some $0 < \lambda_i < 1$:

$$t_i(y) - t_i(x) = \sum_{j=1}^n \partial_j t_i(x + \lambda_i(y - x))(y_j - x_j).$$

Thus since $|y_j - x_j| \leq |y - x|$ for all j:

$$\begin{aligned} & |T(y) - T(x) - T'(x)(y - x)| \\ \leq\ & \sum_{i=1}^n \left| \sum_{j=1}^n [\partial_j t_i(x + \lambda_i(y - x)) - \partial_j t_i(x)] (y_j - x_j) \right| \\ \leq\ & |y - x| \sum_{i=1}^n \sum_{j=1}^n |\partial_j t_i(x + \lambda_i(y - x)) - \partial_j t_i(x)|. \end{aligned}$$

The result now follows since if $x, y \in A$, then $x + \lambda_i(y - x) \in A$. ■

Remark 4.40 ($O(|y - x|)$ vs. $o(|y - x|)$) *While (4.36) implies that the error term in $T(y) - T(x) - T'(x)(y - x)$ is only bounded by $O(|y - x|)$, meaning that:*

$$\frac{O(|y - x|)}{|y - x|} \to C \text{ as } |y - x| \to 0,$$

this error is in fact $o(|y - x|)$ as noted in item 1 of Remark 4.36. By continuity of $\partial_j t_i$, it follows that $\beta \to 0$ as $|y - x| \to 0$.

Generalizing the approach in (4.23) where T was a linear transformation, we define a measure μ on $\mathcal{B}(\mathbb{R}^n)$ or $\mathcal{M}_L(\mathbb{R}^n)$ by:

$$\mu(B) = \int_B |\det(T'(x))|\, dm, \tag{4.37}$$

where m is Lebesgue measure on $\mathbb{R}^n$ and $\det(T'(x))$ is the Jacobian determinant of T. When T is continuously differentiable, this integral is well-defined and finite for bounded B, since continuous $|\det(T'(x))|$ is bounded on such B. This follows from Exercise III.1.13, since by boundedness, $B \subset \prod_{i=1}^{n} [a_i, b_i]$ for some closed rectangle.

Proposition 4.41 (μ in (4.37) is a measure) *If T is a continuously differentiable transformation on $\mathbb{R}^n$, then μ in (4.37) defines a measure on $\mathcal{B}(\mathbb{R}^n)$ or $\mathcal{M}_L(\mathbb{R}^n)$.*
Proof. *This is a corollary to Proposition 4.3 since $|\det(T'(x))|$ is nonnegative, and the assumption of continuity assures Lebesgue measurability by Proposition I.3.13.* ■

This definition of μ generalizes that in (4.23) for a linear transformation. If $x = (x_1, x_2, ..., x_n)$ and T is a linear transformation as in (4.21), then $T'(x)$ is constant matrix given by the matrix of coefficients:

$$T'(x) = A \equiv \begin{pmatrix} a_{11} & a_{12} & \cdots & a_{1n} \\ a_{21} & a_{22} & \cdots & a_{2n} \\ \vdots & \vdots & \vdots & \vdots \\ a_{n1} & a_{n2} & \cdots & a_{nn} \end{pmatrix}.$$

Hence, for linear T, (4.37) becomes by Definitions II.2.4 and III.2.9:

$$\mu(B) = |\det(A)|\, m(B),$$

which is (4.23).

To simplify the following derivation, we first restate Proposition 4.8 and the results of (4.3) and (4.4) in the current context.

Proposition 4.42 ($\int_A g(x)d\mu$ as a Lebesgue integral) *Let $T : \mathbb{R}^n \to \mathbb{R}^n$ be a continuously differentiable, one-to-one, and satisfy $\det(T'(x)) \neq 0$ for all x. With μ as in (4.37), then for any Borel measurable function $g(x)$ defined on $\mathbb{R}^n$:*

$$\int_{\mathbb{R}^n} g(x)d\mu = \int_{\mathbb{R}^n} g(x)\, |\det(T'(x))|\, dm, \tag{4.38}$$

although both integrals may be infinite.

In addition, $g(x)$ is μ-integrable if and only if $g(x)\, |\det(T'(x))|$ is Lebesgue integrable, and when integrable:

$$\int_A g(x)d\mu = \int_A g(x)\, |\det(T'(x))|\, dm, \tag{4.39}$$

for all $A \in \mathcal{B}(\mathbb{R}^n)$.
Proof. *This is immediate from Proposition 4.8, since $f(x) \equiv |\det(T'(x))|$ is nonnegative and Lebesgue measurable.* ■

The result on change of variables in Lebesgue integrals in Proposition 4.44 is technically a corollary of the general results of Propositions 4.8 and 4.18. But to apply the earlier results, we must demonstrate that if $T : \mathbb{R}^n \to \mathbb{R}^n$ is a continuously differentiable, one-to-one transformation with $\det(T'(x)) \neq 0$, then for all $A \in \mathcal{B}(\mathbb{R}^n)$:

$$\mu_T(A) = m(A). \tag{4.40}$$

Here, μ_T denotes the measure on the range space induced by T, where μ is defined in (4.37) on the domain space. The demonstration of this identity is the essence of this result's rather long proof.

Since differentiable transformations are continuous and thus Borel measurable by Proposition I.3.13, the function $g(Tx)$ is Borel measurable as in Proposition I.3.33 when $g(x)$ is Borel measurable. This is the assumption made below. As noted in Remark I.3.31, if $g(x)$ is Lebesgue measure, then the measurability of $g(Tx)$ is not generally predictable.

Remark 4.43 (On norms and norm equivalence) *There is a small but important detail in the following proof that requires a short discussion on norms on $\mathbb{R}^n$, and in particular, norm equivalence. For additional details, see Section 3.2 in* **Reitano** *(2010), as well as other books in analysis noted in the References.*

The standard Euclidean norm was defined in (4.32) and is sometimes called an l_2-norm and denoted $\|x\|_2$. For any p with $1 \leq p \leq \infty$, there is an associated l_p-norm on $\mathbb{R}^n$, which is a real vector space by Definition 9.2, and defined by:

$$\|x\|_p \equiv \begin{cases} \left(\sum_{j=1}^{n} |x_j|^p\right)^{1/p}, & 1 \leq p < \infty, \\ \sup_j |x_j|, & p = \infty. \end{cases} \tag{4.41}$$

All such norms satisfy the defining properties of Definition 9.5, and moreover, all are equivalent in the following sense.

Named for ***Rudolf Lipschitz*** *(1832–1903), two norms $\|x\|$ and $\|x\|'$ are said to be* ***Lipschitz equivalent*** *if there exists constants λ_1 and λ_2 so that for all x:*

$$\lambda_1 \|x\|' \leq \|x\| \leq \lambda_2 \|x\|'.$$

An important corollary to the equivalence of the l_p-norms is that if $\|x\|_p \to 0$ for any p, then $\|x\|_{p'} \to 0$ for all p'. As noted in the Reitano reference, for $1 \leq p, p' < \infty$:

$$\begin{aligned} n^{-1/p'} \|x\|_{p'} \leq \|x\|_p \leq n^{1/p} \|x\|_{p'}, \\ \|x\|_\infty \leq \|x\|_p \leq n^{1/p} \|x\|_\infty. \end{aligned} \tag{4.42}$$

The significance of norm equivalence below is that while continuity of a function or a transformation is typically defined with respect to the l_2-norm, it can be equivalently defined with respect to any of the above norms. Put another way, a continuous transformation satisfies the continuity definition relative to any of the above norms, and for the construction below, we will want to use the l_∞-norm.

Proposition 4.44 (Differentiable transformation in Lebesgue integrals) *Let $T : \mathbb{R}^n \to \mathbb{R}^n$ be a continuously differentiable, one-to-one transformation with $\det(T'(x)) \neq 0$ for all x. Then for any Borel measurable function $g(x)$:*

$$\int_{\mathbb{R}^n} g(Tx) \left|\det(T'(x))\right| dm = \int_{\mathbb{R}^n} g(y) dm, \tag{4.43}$$

although both integrals may be infinite.

More generally, a Borel measurable function $g(x)$ is Lebesgue integrable if and only if $g(Tx) \left|\det(T'(x))\right|$ is Lebesgue integrable, and when integrable,

$$\int_{T^{-1}A} g(Tx) \left|\det(T'(x))\right| dm = \int_{A} g(y) dm, \tag{4.44}$$

for all $A \in \mathcal{B}(\mathbb{R}^n)$.

Proof.

1. **Reduction to** $\mu_T = m$: *The identity in (4.43) is (4.11) if* $\mu_T = m$ *as noted in (4.40), where* μ *is defined in (4.37). In detail, applying (4.38) to* $g(Tx)$, *then (4.11) yields:*

$$\int_{\mathbb{R}^n} g(Tx)\left|\det(T'(x))\right| dm = \int_{\mathbb{R}^n} g(Tx)d\mu = \int_{\mathbb{R}^n} g(y)d\mu_T,$$

which is (4.43) if $\mu_T = m$.

Similarly, (4.44) reflects (4.39) and then (4.12):

$$\int_{T^{-1}A} g(Tx)\left|\det(T'(x))\right| dm^n = \int_{T^{-1}A} g(Tx)d\mu = \int_A g(y)d\mu_T.$$

Turning to the proof of (4.40), continuity of T *obtains that* $T^{-1}(G)$ *is open for all open* G *by Proposition I.3.12. While this Book I result addressed continuous functions on* $\mathbb{R}^n$, *it is an exercise to verify that the identical proof works for transformations* $T : \mathbb{R}^n \to \mathbb{R}^n$ *with a change of notation. The inverse function theorem of Remark 4.36 obtains that* T^{-1} *is continuous and thus* $T(G)$ *is open for all open* G. *This then implies that* $\mathcal{B}(\mathbb{R}^n) = T\left[\mathcal{B}(\mathbb{R}^n)\right] = T^{-1}\left[\mathcal{B}(\mathbb{R}^n)\right]$. *Since* T *is assumed above to be one-to-one,* $\left(T^{-1}\right)^{-1} = T$, *and the definition in (4.9), that* $\mu_T(A) \equiv \mu\left[T^{-1}(A)\right]$ *for all* $A \in \mathcal{B}(\mathbb{R}^n)$, *is equivalent to* $\mu_T\left[T(A)\right] \equiv \mu\left[A\right]$ *for all* $A \in \mathcal{B}(\mathbb{R}^n)$.

2. **Reduction to** $\mu_T = m$ **on bounded** $\prod_{j=1}^n (a_j, b_j] \in \mathcal{A}'$: *To prove* $\mu_T = m$, *it is therefore sufficient to prove that for all* $A \in \mathcal{B}(\mathbb{R}^n)$:

$$m\left[T(A)\right] = \int_A |J(x)|\, dm, \tag{1}$$

where $J(x) \equiv \det(T'(x))$ *to simplify notation.*

We claim that to prove (1), *it is enough by the Book I extension theory summarized in Chapter 1 to restrict* A *to the semi-algebra* $\mathcal{A}' \equiv \left\{\prod_{j=1}^n (a_j, b_j]\right\}$ *of right semi-closed rectangles. To see this, define set functions on* $\mathcal{A}'$ *by:*

$$\mu_1(A) \equiv m\left[T(A)\right], \qquad \mu_2(A) \equiv \int_A |J(x)|\, dm,$$

and note that each is a pre-measure on $\mathcal{A}'$ *by Definition 1.12. Addressing countable additivity, let disjoint* $\{A_i\}_{i=1}^\infty \subset \mathcal{A}'$ *with* $A \equiv \bigcup_{i=1}^\infty A_i \in \mathcal{A}'$ *be given. Then* $\{T(A_i)\}_{i=1}^\infty \subset \mathcal{B}(\mathbb{R}^n)$ *as noted above, these sets are disjoint since* T *is one-to-one, and* $T(A) = \bigcup_{i=1}^\infty T(A_i)$. *Thus* μ_1 *is countably additive since* m *has this property, and this is true whether* $A \in \mathcal{A}'$ *or not. Countable additivity of* μ_2 *then follows from item 6 of Proposition III.2.49.*

Then by Proposition 1.20, both pre-measures can be extended additively to measures on the associated algebra $\mathcal{A}$. *This algebra is given by Exercise I.6.10 as the collection of all finite disjoint unions of* $\mathcal{A}'$*-sets, and thus this extension satisfies* $\mu_1(A) = \mu_2(A)$ *for all* $A \in \mathcal{A}$. *Since both measures are* σ*-finite, Proposition 1.22 applies. Thus* $\mu_1(A) = \mu_2(A)$ *for all* A *in* $\sigma(\mathcal{A})$, *the smallest sigma algebra that contains* $\mathcal{A}$, *which is* $\mathcal{B}(\mathbb{R}^n)$ *by Proposition I.8.1.*

Returning to (1), *it now follows by countable additivity on* $\mathcal{A}'$ *that we need to only prove* (1) *for bounded rectangles* $A \in \mathcal{A}'$, *since every* $A \in \mathcal{A}'$ *is an at most countable union of disjoint bounded rectangles by Proposition I.8.1.*

3. ***First bounds***: *So assume* $A = \prod_{j=1}^{n}(a_j, b_j]$ *is bounded. As* $J(x)$ *is continuous on* $\overline{A} = \prod_{j=1}^{n}[a_j, b_j]$, *it is uniformly continuous by Exercise III.1.14. So given* $\epsilon > 0$, *there is a* δ_1 *such that* $|J(x) - J(y)| < \epsilon$ *for all* $x, y \in \overline{A}$ *with* $|x - y| < \delta_1$. *Partition the intervals* $(a_j, b_j]$ *so that* $A = \bigcup_{k=1}^{N} R_k$ *is a finite disjoint union of right semi-closed rectangles with the property that if* $x, y \in R_k$, *then* $|x - y| < \delta_2 \leq \delta_1$, *where* δ_2 *is to be specified below.*

   Now if $x, y \in R_k$, *then* $|J(x) - J(y)| < \epsilon$ *implies that* $||J(x)| - |J(y)|| < \epsilon$, *and so:*

$$|J(y)| - \epsilon < |J(x)| < |J(y)| + \epsilon.$$

   Arbitrarily choose $x = x_k \in R_k$, *and using properties of Proposition III.2.49, integrate this inequality over* R_k *to obtain:*

$$\int_{R_k} |J(y)|\, dm - \epsilon m\left[R_k\right] < |J(x_k)|\, m\left[R_k\right] < \int_{R_k} |J(y)|\, dm + \epsilon m\left[R_k\right].$$

   Since $\{R_k\}_{k=1}^{N}$ *is a finite disjoint collection of rectangles, finite additivity of* m *and item 6 of Proposition III.2.49 obtain:*

$$\int_{A} |J(y)|\, dm - \epsilon m\left[A\right] < \sum\nolimits_{k=1}^{N} |J(x_k)|\, m\left[R_k\right] < \int_{A} |J(y)|\, dm + \epsilon m\left[A\right], \tag{2}$$

   where $x_k \in R_k$ *is arbitrary.*

4. ***Estimating*** $|J(x_k)|\, m\left[R_k\right]$: *The final challenge is to show that the middle terms in (2), each* $|J(x_k)|\, m\left[R_k\right]$, *can be made arbitrarily close to* $m\left[T(R_k)\right]$.

   To this end, let $x_k \in R_k$ *denote the "midpoint" of each rectangle, meaning that if* $R_k = \prod_{j=1}^{n}(c_j, d_j]$, *then the jth component of* x_k *satisfies* $x_{k_j} = (c_j + d_j)/2$. *For* $\epsilon > 0$ *above, where we can assume* $\epsilon < 1$, *define* $R_k^{+\epsilon}$ *and* $R_k^{-\epsilon}$ *to be rectangles centered on* x_k *but with respective sides of length* $(1 + \epsilon)(d_j - c_j)$ *and* $(1 + \epsilon)^{-1}(d_j - c_j)$. *In detail,* $R_k^{+\epsilon} = \prod_{j=1}^{n}(c_j - \alpha_j, d_j + \alpha_j]$ *where* $\alpha_j = \frac{\epsilon}{2}(d_j - c_j)$, *while* $R_k^{-\epsilon} = \prod_{j=1}^{n}(c_j + \alpha_j', d_j - \alpha_j']$ *where* $\alpha_j' = \frac{\epsilon'}{2}(d_j - c_j)$ *with* $(1 - \epsilon') = (1 + \epsilon)^{-1}$. *We show below that given such* ϵ, *there exists* $\delta_2 \leq \delta_1$ *so that with* $\{R_k\}_{k=1}^{N}$ *defined by such* δ_2:

$$T(x_k) + T'(x_k)\left[R_k^{-\epsilon} - x_k\right] \subset T\left(R_k\right) \subset T(x_k) + T'(x_k)\left[R_k^{+\epsilon} - x_k\right]. \tag{3}$$

   Here, $R_k^{\pm\epsilon} - x_k$ *denotes the respective rectangles translated to be centered at the origin, for example:*

$$R_k^{+\epsilon} - x_k = \prod\nolimits_{j=1}^{n}(c_j - \alpha_j - x_{k_j}, d_j + \alpha_j - x_{k_j}].$$

   Since Lebesgue measure is translation invariant, these inclusions obtain inequalities for $m\left[T(R_k)\right]$ *and* $m\left[A\right]$ *as follows.*

   Recalling (4.29) of Corollary 4.33 with $\det T'(x_k) \equiv J(x_k)$ *and:*

$$m\left[R_k^{\pm\epsilon}\right] = m\left[R_k^{\pm\epsilon} - x_k\right] = (1 + \epsilon)^{\pm n} m\left[R_k\right],$$

   the set inclusions in (3) *obtain by monotonicity of* m:

$$(1 + \epsilon)^{-n}\, |J(x_k)|\, m\left[R_k\right] \leq m\left[T(R_k)\right] \leq (1 + \epsilon)^{n}\, |J(x_k)|\, m\left[R_k\right].$$

   Since T *is one-to-one, it follows by summation:*

$$(1 + \epsilon)^{-n} \sum\nolimits_{k=1}^{N} |J(x_k)|\, m\left[R_k\right] \leq m\left[T(A)\right] \leq (1 + \epsilon)^{n} \sum\nolimits_{k=1}^{N} |J(x_k)|\, m\left[R_k\right].$$

Combining with the inequalities in (2) *produces:*

$$(1+\epsilon)^{-n}\left[\int_A |J(y)|\,dm - \epsilon m[A]\right] \leq m[T(A)] \leq (1+\epsilon)^n\left[\int_A |J(y)|\,dm + \epsilon m[A]\right],$$

and (1) *then follows since* $\epsilon > 0$ *is arbitrary.*

5. ***Left inclusion in*** (3)*: We next prove the existence of* $\{R_k\}_{k=1}^N$ *that obtains the inclusion on the left of* (3)*. Specifically, we show that for some* $\delta_1'' \leq \delta_1$*, with* δ_1 *defined above, and* R_k*-rectangles defined relative to* δ_1''*, that* $T(x_k) + T'(x_k)[y - x_k] \in T(R_k)$ *for* $y \in R_k^{-\epsilon}$*, and we do this by proving that* $T^{-1}[T(x_k) + T'(x_k)[y - x_k]] \in R_k$*.*

To this end, define $S : \overline{A} \times \overline{A} \to \mathbb{R}^n$ *by* $S(x,y) = T^{-1}[T(x) + T'(x)[y-x]]$*. As* T^{-1} *is continuously differentiable on an open set* V *that contains* $T(\overline{A})$ *by Proposition 4.38, the approximation in (4.31) applies. Then* $T^{-1}[T(x)] = x$ *since* T *is one-to-one, while the inverse function theorem obtains* $(T^{-1})'[T(x)] = [T'(x)]^{-1}$*. Combining:*

$$\begin{aligned} S(x,y) &= T^{-1}[T(x) + T'(x)[y-x]] \\ &= T^{-1}[T(x)] + (T^{-1})'[T(x)](T'(x)[y-x]) + o(|T'(x)[y-x]|) \\ &= y + o(|T'(x)[y-x]|). \end{aligned}$$

By the triangle inequality:

$$|S(x,y) - x| \leq |y - x| + o(|T'(x)[y-x]|),$$

and thus for $y \neq x$*:*

$$\frac{|S(x,y) - x|}{|y-x|} \leq 1 + \frac{o(|T'(x)[y-x]|)}{|y-x|}. \tag{4}$$

We claim that for all $x \in \overline{A}$*:*

$$\frac{o(|T'(x)[y-x]|)}{|y-x|} \to 0, \quad y \to x.$$

For this:

$$\frac{o(|T'(x)[y-x]|)}{|y-x|} = \frac{o(|T'(x)[y-x]|)}{|T'(x)[y-x]|}\frac{|T'(x)[y-x]|}{|y-x|},$$

and the first ratio on the right converges to 0 *by (4.33) as* $|T'(x)[y-x]| \to 0$*. Since* $\det T'(x) \neq 0$*,* $|T'(x)[y-x]| \to 0$ *if and only if* $|y-x| \to 0$*, meaning as* $y \to x$*. The second ratio is bounded by a constant* $k(x)$ *by (4.34), since* $T'(x)$ *is a linear transformation.*

Given $\epsilon > 0$ *above, it follows from* (4) *that for each* $x \in \overline{A}$*, there exists an open ball* $B(x, \delta_x) \equiv \{y \mid |y-x| < \delta_x\}$ *so that if* $y \in B(x,\delta_x)$*, then:*

$$\frac{|S(x,y) - x|}{|y-x|} \leq 1 + \epsilon. \tag{5}$$

We can, without loss of generality, require all $\delta_x \leq \delta_1$ *with* δ_1 *defined above, and by Remark 4.43, we assume that the norms in* (5) *are* l_∞*-norms. Since* $\{B(x,\delta_x)\}_{x\in\overline{A}}$ *is an open cover of compact* $\overline{A}$*, there exists a finite subcover* $\{B(x_k, \delta_{x_k})\}_{k=1}^N \equiv \{R_k\}_{k=1}^N$ *by the Heine-Borel theorem of Proposition I.2.27, from which we construct* $R_k^{-\epsilon} \subset R_k$ *for all* k*.*

Thus if $y \in R_k^{-\epsilon}$*, then by* (5) *and the definition of* $S(x,y)$ *:*

$$\left|T^{-1}[T(x_k) + T'(x_k)[y - x_k]] - x_k\right| \leq (1+\epsilon)|y - x_k|,$$

and so $T^{-1}[T(x_k) + T'(x_k)[y - x_k]] \in R_k$*, as was to be proved.*

6. ***Right inclusion in*** (3)*: To prove the right inclusion in* (3) *is to prove the existence of* $\{R_k\}_{k=1}^N$ *so that for* $y \in R_k$*:*

$$x_k + (T'(x_k))^{-1}[T(y) - T(x_k)] \subset R_k^{+\epsilon},$$

which as above is to prove that in the l_∞*-norm:*

$$\left|(T'(x_k))^{-1}[T(y) - T(x_k)]\right| \leq (1+\epsilon)|y - x_k|. \tag{6}$$

By the linear approximation in (4.31):

$$(T'(x))^{-1}[T(y) - T(x)] = y - x + (T'(x))^{-1}[o(|y-x|)],$$

where $(T'(x))^{-1}[o(|y-x|)]$ *denotes* $(T'(x))^{-1}$ *applied to the error function which is* $o(|y-x|)$*. By the triangle inequality, it follows that for* $y \neq x$*:*

$$\frac{\left|(T'(x))^{-1}[T(y) - T(x)]\right|}{|y-x|} \leq 1 + \frac{\left|(T'(x))^{-1}[o(|y-x|)]\right|}{|y-x|}.$$

An application of (4.34) obtains:

$$\frac{\left|(T'(x))^{-1}[T(y) - T(x)]\right|}{|y-x|} \leq 1 + k(x)\frac{|[o(|y-x|)]|}{|y-x|},$$

and the function on the right can for each x *be made arbitrarily small for* $|y-x|$ *small by (4.33).*

The remaining details are similar to part 5, using Heine-Borel, and left to the reader as an exercise.

■

Remark 4.45 *We note two observations on the above result:*

1. *As noted in part 1 of the above proof,* $\mathcal{B}(\mathbb{R}^n) = T^{-1}[\mathcal{B}(\mathbb{R}^n)]$*, and so (4.44) can be restated in the alternative version of (4.13), that for all* $A \in \mathcal{B}(\mathbb{R}^n)$*:*

$$\int_A g(Tx)\,|\det(T'(x))|\,dm = \int_{T(A)} g(y)dm. \tag{4.45}$$

2. *The above proposition also applies to* $T : U \to V$ *with open* $U, V \subset \mathbb{R}^n$ *and* $T(U) = V$ *when* T *is a continuously differentiable, one-to-one transformation with* $\det(T'(x)) \neq 0$ *for all* $x \in U$*.*

   To prove this, we need to apply the general results to measure spaces as follows:

$$(U, \mathcal{B}(U), m) \quad \xrightarrow{T} \quad (V, \mathcal{B}(V), m) \quad \xrightarrow{g} \quad (\mathbb{R}, \mathcal{B}(\mathbb{R}), m).$$

   The essential observation that makes the proof of this local result work is that the inverse mapping theorem is fundamentally a local result.

Generalizing Corollary 4.33, we have the following result.

Proposition 4.46 **($m[T(A)]$, $A \in \mathcal{B}(\mathbb{R}^n)$)** *Let $T : \mathbb{R}^n \to \mathbb{R}^n$ be a continuously differentiable, one-to-one transformation with $\det(T'(x)) \neq 0$. Then for any $A \in \mathcal{B}(\mathbb{R}^n)$:*

$$m[T(A)] = \int_A |\det(T'(x))| \, dm. \tag{4.46}$$

Proof. *This follows from (4.45) with $g(x) \equiv 1$.* ■

Example 4.47 (On Riemann integrals) *The reader may recall that the above proposition has been applied in Books II and IV in justifying that the normal density function integrates to 1. In the later book, this result was also needed in the derivation of the Box-Muller transform used for the simulation of normal variates, and the Bailey transform for the simulation of Student T variates.*

All such applications were made to Riemann integrals, to which the above results apply by Propositions III.2.18 and III.2.56. However, to apply the above transformations in a multivariate Riemann integral on $\mathbb{R}^n$ first required obtaining this integral from an iterated Riemann integral. This was justified by Corollary III.1.77 for Riemann integrals over bounded rectangles.

The general connections between iterated and multivariate Lebesgue (and other) integrals will be developed below in Fubini's theorem and Tonelli's theorem. These then also apply to certain Riemann integrals by the above noted Book III results.

5

Integrals in Product Spaces

The goal of this chapter is to develop **Fubini's theorem**, named for **Guido Fubini** (1879–1943), and **Tonelli's Theorem,** named for **Leonida Tonelli** (1885–1946). Both of these theorems provide results on the relationship between the value of a **product space integral**, by which is meant an integral defined relative to a product measure, and the value of the associated **iterated integrals**, whereby the integrand is in essence integrated one space variable at a time. This idea may sound familiar recalling the Riemann results of Section III.1.3.6, and in particular Corollary III.1.77.

We begin this discussion with a review of the construction and essential features of a product space measure developed in Chapter I.7 and summarized in Section 1.3.1, and compare that to a common alternative approach found in some references. While the resulting product space measures are quite similar, and indeed only differ on sets of measure 0, the statements of Fubini's and Tonelli's theorems reflect the approach taken and will be seen to differ in important ways which we attempt to elucidate.

5.1 Product Space Sigma Algebras

The approach to product space measures taken in Chapter I.7 is consistent with that of the **Royden** (1971), which also addresses the essential differences of that approach with a common alternative approach. The **Rudin** (1974) text develops the alternative approach and then discusses the approach taken in this book. The alternative approach is also seen in **Billingsley** (1995) and **Doob** (1994).

5.1.1 Sigma Algebra Constructions

In Chapter I.7, given measure spaces $\{(X_i, \sigma(X_i), \mu_i)\}_{i=1}^n$, the **product space** $X = \prod_{i=1}^n X_i$ is defined:

$$X = \{(x_1, x_2, ..., x_n)|x_i \in X_i\}.$$

A **measurable rectangle** in X is a set A so that with $A_i \in \sigma(X_i)$:

$$A = \prod\nolimits_{i=1}^{n} A_i = \{x \in X|x_i \in A_i\}.$$

Given a measurable rectangle $A = \prod_{i=1}^n A_i$, the **product set function** μ_0 is defined by:

$$\mu_0(A) = \prod\nolimits_{i=1}^{n} \mu_i(A_i). \tag{5.1}$$

The collection of measurable rectangles is a semi-algebra (Definition 1.14) denoted $\mathcal{A}'$, and the collection of all finite disjoint unions of $\mathcal{A}'$-sets is an algebra by Exercise I.6.10, denoted $\mathcal{A}$.

This is the starting point for both the approach taken in Book I as summarized in Chapter 1, and an alternative approach discussed below.

DOI: 10.1201/9781003264576-5

Remark 5.1 (On σ-finiteness) *We will assume throughout much of this chapter that the component measure spaces $\{(X_i, \sigma(X_i), \mu_i)\}_{i=1}^n$ are σ-finite. This is not the most general development possible, but is adequate for the applications of these volumes, covering all probability spaces, and all Borel measure spaces of interest.*

While the product space existence result of Proposition I.7.20 formally required σ-finiteness, this assumption was not strictly necessary as noted in Remark I.7.22. The Book I proof of countable additivity of the extension of μ_0 to the algebra $\mathcal{A}$ utilized a continuity from above argument of Proposition I.6.19, and this required σ-finiteness. But as seen in Section 3.2.4, this extension of μ_0 to $\mathcal{A}$ is countably additive without σ-finiteness, and thus so too is the ***existence result*** *of Proposition I.7.20.*

That said, the assumption of σ-finiteness of the resulting product space is, in any case, required to ensure the **uniqueness of extensions** *of Propositions I.6.14 and I.6.24. Such uniqueness is needed in the current context to employ the result of Proposition I.7.24, which proved that the product space of Proposition I.7.20 could be created in one step from the n component spaces, or in a sequence of steps, at each step creating product spaces of smaller dimensions. As will be seen in Proposition 5.18, this result will be needed to generalize the Fubini's and Tonelli's theorems from the case of $n = 2$, which we explicitly prove, to general n, when product spaces are constructed using the Book I approach.*

Another convenient implication of σ-finiteness is that it allows the application of Corollary 5.2 on the approximation of measurable sets by simpler sets.

The essential distinction between the Book I and alternative approach is in the construction of the sigma algebra on which the definition of μ_0 can be extended to the measure μ_X. In both constructions, the resulting sigma algebras contain $\mathcal{A}'$, the semi-algebra of measurable rectangles in X, and hence also $\mathcal{A}$, the associated algebra of all finite unions of disjoint measurable rectangles. Also, because μ_0 is defined as in (5.1) on $\mathcal{A}'$, the extension of μ_0 to $\mathcal{A}$ and denoted $\mu_{\mathcal{A}}$ is well defined by finite additivity. Hence, up to the algebra $\mathcal{A}$, the measurable sets and measures of these sets are identical between approaches.

The final step is the **construction of the sigma algebras.**

1. **Chapter I.7 Approach:** For the product measure space $(X, \sigma(X), \mu_X)$ of Proposition I.7.20, the sigma algebra $\sigma(X)$ is the collection of **Carathéodory measurable** sets defined in (1.13), using the outer measure $\mu_{\mathcal{A}}^*$ which is defined in (1.14). By the Hahn–Kolmogorov extension theorem of Proposition 1.18 and the first Carathéodory extension theorem of Proposition 1.17, $\sigma(X)$ is a complete sigma algebra, and μ_X is a measure on $\sigma(X)$ with μ_X defined as the restriction of $\mu_{\mathcal{A}}^*$ to $\sigma(X)$.

   It then follows from Corollary I.7.23 that such measurable sets can be approximated by simpler sets. We state this result for completeness, with the last statement from Proposition I.6.5.

Corollary 5.2 (Approximating measurable sets in product spaces) *Let $\mathcal{A}$ denote the algebra of finite disjoint unions of measurable rectangles on a product space X, $\mu_{\mathcal{A}}$ a σ-finite measure on $\mathcal{A}$, and $(X, \sigma(X), \mu_X)$ the complete σ-finite product measure space given in Proposition I.7.20.*

For $B \in \sigma(X)$ and $\epsilon > 0$:

(a) There is a set $A \in \mathcal{A}_\sigma$, the collection of countable unions of sets in the algebra $\mathcal{A}$, so that $B \subset A$ with:

$$\mu_X(A) \leq \mu_X(B) + \epsilon, \text{ and } \mu_X(A - B) < \epsilon. \tag{5.2}$$

(b) *There is a set* $C \in \mathcal{A}_\delta$*, the collection of countable intersections of sets in the algebra* $\mathcal{A}$*, so that* $C \subset B$ *with:*

$$\mu_X(B) \leq \mu_X(C) + \epsilon, \text{ and } \mu_X(B - C) < \epsilon. \tag{5.3}$$

(c) *There is a set* $A' \in \mathcal{A}_{\sigma\delta}$*, the collection of countable intersections of sets in* $\mathcal{A}_\sigma$*, and* $C' \in \mathcal{A}_{\delta\sigma}$*, the collection of countable unions of sets in* $\mathcal{A}_\delta$*, so that* $C' \subset B \subset A'$ *and:*

$$\mu_X(A' - B) = \mu_X(B - C') = 0. \tag{5.4}$$

If μ_X *is a general measure on* $\mathcal{A}$*, meaning not* σ*-finite, then the above results remain true for all* $B \in \sigma(X)$ *with* $\mu(B) < \infty$*.*

The key conclusion in part c is that every measurable set, meaning every $B \in \sigma(X)$, is within μ_X-measure 0 of subsets and supersets that can be constructed with countably many unions and intersections of sets from the algebra $\mathcal{A}$. Since the algebra $\mathcal{A}$ is the collection of finite unions of disjoint measurable rectangles in $\mathcal{A}'$, this conclusion can also be stated that every measurable set in $\sigma(X)$ is within μ_X-measure 0 of subsets and supersets that can be constructed with countably many unions and intersections of sets from the semi-algebra $\mathcal{A}'$ of measurable rectangles.

2. **Alternative Approach (For example, Section 18 of Billingsley (1995)):** For the product measure space $(X, \sigma'(X), \mu'_X)$, the sigma algebra $\sigma'(X)$ is defined as the smallest sigma algebra that contains the algebra $\mathcal{A}$, and so by definition:

$$\mathcal{A} \subset \sigma'(X) \subset \sigma(X).$$

Notation 5.3 *In earlier books, this smallest sigma algebra was often denoted* $\sigma(\mathcal{A})$*, but in the development below, the current notation is preferred.*

The conclusion from item 1.c. is that every $B \in \sigma(X)$ is within μ_X-measure 0 of subsets and supersets from $\sigma'(X)$. But then, how is it proved that $\mu_\mathcal{A}$, the μ_0 set function extended to $\mathcal{A}$, can be further extended to a measure μ'_X on $\sigma'(X)$? Of course, one could "simply" implement the above extension to $\sigma(X)$ and then define μ'_X as the restriction of μ_X to the sigma subalgebra $\sigma'(X)$.

Alternatively, the most common approach is to use results from integration theory. For example, assume that there are two component spaces, $(X_1, \sigma(X_1), \mu_1)$ and $(X_2, \sigma(X_2), \mu_2)$, to simplify notation. For a set $A' \in \sigma'(X)$, define the **projection sets** or **cross-sections** of A' :

$$\begin{aligned} A'(x_1) &= \{x_2 | (x_1, x_2) \in A'\}, \\ A'(x_2) &= \{x_1 | (x_1, x_2) \in A'\}. \end{aligned}$$

It is then shown that $A'(x_1) \in \sigma(X_2)$ **for all** x_1, and $A'(x_2) \in \sigma(X_1)$ **for all** x_2, and hence $\mu_2[A'(x_1)]$ and $\mu_1[A'(x_2)]$ are well-defined.

More importantly, it turns out that $\mu_2[A'(x_1)]$ is a measurable function on X_1, and $\mu_1[A'(x_2)]$ is a measurable function on X_2, relative to the respective sigma algebras, and hence one can define:

$$\int \mu_2[A'(x_1)]\, d\mu_1 \text{ and } \int \mu_1[A'(x_2)]\, d\mu_2.$$

The next step is to show that for a measurable rectangle $A' = A'_1 \times A'_2$, that these integrals have the desired common value $\mu_1(A'_1)\,\mu_2(A'_2) = \mu_0(A')$ as defined in (5.1), and further that these integrals agree for all $A' \in \sigma'(X)$. Finally, the common value of these integrals is then taken as the definition of $\mu'_X(A')$, and μ'_X is shown to be a measure using Lebesgue's monotone convergence theorem.

The use of such powerful integration results to prove a result on measures was exemplified in the relatively simple proof in Section 3.2.4 that the set function μ_0 was countably additive on the semi-algebra $\mathcal{A}'$.

3. **σ-Finite μ_X and Uniqueness:** Since for all $A \in \mathcal{A}'$:

$$\mu'_X(A') = \mu_X(A') = \mu_1(A'_1)\,\mu_2(A'_2),$$

the uniqueness result of Proposition 1.22 applies if μ_X is σ-finite. Namely, it follows from this proposition that for all $A \in \sigma'(X)$:

$$\mu'_X(A') = \mu_X(A'),$$

where $\sigma'(X)$ is defined as above as the smallest sigma algebra that contains the algebra $\mathcal{A}$.

Though not obvious, the sigma algebra $\sigma'(X)$ is not in general complete, and so:

$$\sigma'(X) \subsetneqq \sigma(X).$$

As the smallest sigma algebra that contains $\mathcal{A}$, $\sigma'(X)$ must contain any set that can be constructed from $\mathcal{A}$ based on countably many operations of unions or intersections, and thus:

$$\mathcal{A}, \mathcal{A}_\sigma, \mathcal{A}_\delta, \mathcal{A}_{\sigma\delta}, \mathcal{A}_{\delta\sigma} \ldots.. \subset \sigma'(X) \subsetneqq \sigma(X).$$

From Corollary 5.2, all $\sigma(X)$-sets are within μ_X-measure 0 of $\mathcal{A}_{\sigma\delta}$-sets and $\mathcal{A}_{\delta\sigma}$-sets, and we can conclude the following.

Conclusion 5.4 (On measure space constructions)

(1) For any $A \in \sigma(X)$, there is A'_1, $A'_2 \in \sigma'(X)$ so that $A'_1 \subset A \subset A'_2$ and:

$$\mu_X(A'_1) = \mu_X(A) = \mu_X(A'_2).$$

(2) If the original spaces $\{(X_i, \sigma(X_i), \mu_i)\}_{i=1}^n$ are σ-finite, then so too is $(X, \sigma(X), \mu_X)$, and thus $\mu'_X(A') = \mu_X(A')$ for all $A' \in \sigma'(X)$. Hence, from item (1):

$$\mu_X(A'_1) = \mu'_X(A'_1) = \mu_X(A) = \mu'_X(A'_2) = \mu_X(A'_2).$$

(3) When component measures spaces are σ-finite, $(X, \sigma(X), \mu_X)$ is the ***completion*** *of $(X, \sigma'(X), \mu'_X)$ by Proposition I.6.24.*

5.1.2 Implications for Chapter Results

The approach taken in constructing a product measure space has two important implications for the results below. The first relates to the measurability of cross-sections of measurable sets, and the second to the component-wise measurability of measurable functions.

1. **Measurability of Cross-Sections:**

   (a) **Alternative Approach:** Recalling the notion in item 2 of the previous section, **every cross-section of a measurable set** $A' \in \sigma'(X)$ **is measurable** in the respective sigma algebras of component spaces. In the notation above with two component spaces, $A'(x_1) \in \sigma(X_2)$ **for all** x_1, and $A'(x_2) \in \sigma(X_1)$ **for all** x_2.

   (b) **Chapter I.7 Approach:** If $A \in \sigma(X)$, then sets of μ_X-measure 0 can be added or removed from A by completeness and retain measurability. Thus we will only be able to assert that **cross-sections of measurable sets are measurable almost always.** In the notation above with two component spaces, $A'(x_1) \in \sigma(X_2)$ **for** μ_1**-almost all** x_1, and $A'(x_2) \in \sigma(X_1)$ **for** μ_2**-almost all** x_2.

Example 5.5 *We start with* $\{(\mathbb{R}_i, \sigma(\mathbb{R}_i), m)\}_{i=1}^2$*, two copies of the complete Lebesgue measure space, indexed for notational clarity.*

In the Lebesgue product measure space on $\mathbb{R}^2$*, the set* $A \equiv \{(1, y)|0 < y \leq 1\}$ *is* $\sigma'(\mathbb{R}^2)$*-measurable as the intersection of nested measurable rectangles,* $A_n \equiv (1-1/n, 1] \times (0, 1]$*. As these rectangles have product measure* $m(A_n) = 1/n$*, it follows from continuity from above that* $m(A) = 0$*. Then as a countable union of such sets:*

$$A' \equiv \{(q, y)|0 < y \leq 1, \; q \in \mathbb{Q} \bigcap [0, 1]\},$$

is again $\sigma'(\mathbb{R}^2)$*-measurable and* $m(A') = 0$*. All* x*- and* y*-cross-sections of* A' *are* $\sigma(\mathbb{R}_i)$*-measurable.*

Now let $A_0 \subset (0, 1]$ *be the nonmeasurable set constructed in step 2 of Proposition I.2.31 and define* $A'' \equiv \{(q, y)|y \in A_0\}$*. Then* $A'' \subset A'$*, and so by completeness* $A'' \in \sigma(\mathbb{R}^2)$*, and by monotonicity,* $m(A'') = 0$*. While all* y*-cross-sections are* $\sigma(\mathbb{R}_2)$*-measurable, since* $A''(y) \in \{\mathbb{Q} \bigcap [0, 1], \emptyset\}$*,* A'' *has* $\sigma(\mathbb{R}_2)$*-nonmeasurable* x*-cross-sections for all* $x = q$ *since* $A''(q) = A_0$*.*

By Proposition 5.15, it follows from the nonmeasurabilility of x*-cross-sections that* $A'' \notin \sigma'(\mathbb{R}^2)$*.*

2. **Component Measurability of Measurable Functions:**

   (a) **Alternative Approach:** To simplify notation, let $f(x, y)$ be a $\sigma'(\mathbb{R}^2)$-measurable function, meaning that $f^{-1}(B) \in \sigma'(\mathbb{R}^2)$ for all $B \in \mathcal{B}(\mathbb{R})$. Then for **all** x, the x-component function:

   $$f_x(y) \equiv f(x, y),$$

   will prove to be a measurable function of y. Similarly, for **all** y, the y-component function:

   $$f_y(x) \equiv f(x, y),$$

   will prove to be a measurable function of x.

   In other words, every **component function of a** $\sigma'(\mathbb{R}^2)$**-measurable function is** $\sigma(\mathbb{R}_i)$**-measurable.**

   (b) **Chapter I.7 Approach:** If $f(x, y)$ is a $\sigma(\mathbb{R}^2)$-measurable function, meaning that $f^{-1}(B) \in \sigma(\mathbb{R}^2)$ for any $B \in \mathcal{B}(\mathbb{R})$, then we will only prove that $f_x(y)$ is $\sigma(\mathbb{R}_2)$-measurable in y for **almost all** x, and similarly, $f_y(x)$ is $\sigma(\mathbb{R}_2)$-measurable in x for **almost all** y.

In other words, **component functions of a measurable function are almost always measurable**. This conclusion is a result of the fact that $\sigma(\mathbb{R}^2)$-measurability is a weaker condition than $\sigma'(\mathbb{R}^2)$-measurability, because the value of f can be arbitrarily changed on sets in $\mathbb{R}^2$ of μ-measure 0 and remain $\sigma(\mathbb{R}^2)$-measurable.

Example 5.6 *On the Example 5.5 Lebesgue product measure space on $\mathbb{R}^2$ define $f(x, y) = 1$. This is apparently a $\sigma'(\mathbb{R}^2)$-measurable function since $f^{-1}(B)$ is $\mathbb{R}^2$ or $\emptyset$ depending on whether $1 \in B$ or not.*

Now redefine $f(x, y)$ on a set of measure 0 to:

$$g(x,y) = \begin{cases} 2, & (x,y) \in A'', \\ 1, & (x,y) \notin A'', \end{cases}$$

where A'' is defined above. Then g is not a $\sigma'(\mathbb{R}^2)$-measurable function since $g^{-1}([2,3]) = A'' \notin \sigma'(\mathbb{R}^2)$.

But $g = f$, m-a.e., since $m(A'') = 0$, and thus g is a $\sigma(\mathbb{R}^2)$-measurable function by Proposition 2.14 and completeness of $\sigma(\mathbb{R}^2)$. For rational $q \in [0,1]$, $g_q(y)$ is never $\sigma(\mathbb{R}_2)$-measurable since $g_q(y) = 2$ for $y \in A_0$ and $g_q(y) = 1$ for $y \in \widetilde{A_0}$, and neither set is measurable.

Summary 5.7 *The Chapter I.7 approach to product spaces has the advantage that because the associated sigma algebras are complete, the product space associated with n, 1-dimensional Lebesgue measure spaces is an n-dimensional Lebesgue measure space. In contrast, the alternative approach requires the n-dimensional product space to be completed by Proposition I.6.14 to be equivalent to n-dimensional Lebesgue measure space.*

But completeness of $\sigma(X)$ imposes a weaker condition on a set to be $\sigma(X)$-measurable, and weaker conditions on a function to be $\sigma(X)$-measurable. Corresponding to these somewhat weaker conditions, one obtains somewhat weaker conclusions below. Many results will be stated to be true "except on a set of measure 0," while the alternative approach will obtain stronger conclusions that omit this qualifier.

More will be said on this below as the results emerge. We will derive results for the complete product measure space $(X, \sigma(X), \mu_X)$, and also for the product measure space $(X, \sigma'(X), \mu'_X)$ where $\sigma'(X)$ is the smallest σ-algebra that contains the algebra $\mathcal{A}$ of finite unions of disjoint measurable rectangles. In this latter case, we assume that the component spaces are σ-finite, so that as noted above, $\mu'_X = \mu_X$ on $\sigma'(X)$ by Proposition I.6.14.

Notation 5.8 *Given, for example, three σ-finite measure spaces $(X, \sigma(X), \mu)$, $(Y, \sigma(Y), \upsilon)$, and $(Z, \sigma(Z), \lambda)$, we denote by $\sigma(X \times Y \times Z)$ the complete sigma algebra of* ***Carathéodory-measurable*** *sets defined in terms of the set function $\mu \times \nu \times \lambda$ defined on semi-algebra $\mathcal{A}'$ of rectangles in $\sigma(X) \times \sigma(Y) \times \sigma(Z)$ and the associated outer measure $(\mu \times \nu \times \lambda)^*$. If needed for clarity, this could be denoted $\sigma(\sigma(X) \times \sigma(Y) \times \sigma(Z))$.*

Similarly, $\sigma'(X \times Y \times Z)$ will denote the smallest sigma algebra that contains the semi-algebra $\mathcal{A}'$ of rectangles from $\sigma(X) \times \sigma(Y) \times \sigma(Z)$. This smallest sigma algebra was sometimes denoted $\sigma_0(X \times Y \times Z)$ in Book I, for example, in Section I.7.5. If needed for clarity, this latter sigma algebra can also be denoted $\sigma'(\sigma(X) \times \sigma(Y) \times \sigma(Z))$.

5.2 Preliminary Results

5.2.1 Introduction to Fubini/Tonelli Theorems

We first develop results in a "2-dimensional" product space to simplify notation, and generalize below. But note that here, "dimension" is merely a notational device in that there is nothing in the following development that precludes these component spaces from being product spaces of other component spaces.

Assume that the component spaces $(X, \sigma(X), \mu)$ and $(Y, \sigma(Y), v)$ are σ-finite, and that the complete product space $(X \times Y, \sigma(X \times Y), \mu \times v)$ is constructed as in Chapter I.7 as summarized above. As in that development, $\mathcal{A}'$ denotes the semi-algebra of measurable rectangles in $X \times Y$, $\mathcal{A}$ the associated algebra of finite disjoint unions of $\mathcal{A}'$-sets, and $\mathcal{A}_\sigma, \mathcal{A}_\delta, \mathcal{A}_{\sigma\delta}$, and $\mathcal{A}_{\delta\sigma}$ are as defined in Corollary 5.2.

Given a function $f(x, y)$ **which is integrable on** $X \times Y$, meaning

$$\int_{X \times Y} |f(x, y)| \, d(\mu \times v) < \infty,$$

Fubini's theorem will state that:

$$\int_X \left[\int_Y f(x, y) dv \right] d\mu = \int_{X \times Y} f(x, y) d(\mu \times v) = \int_Y \left[\int_X f(x, y) d\mu \right] dv.$$

Before addressing the equality of the values of the integrals, this theorem also has to justify that this statement makes sense, meaning that it must also address:

1. The integrability of the component functions, $f_x(y)$ and $f_y(x)$, so that $\int_Y f_x(y) dv$ and $\int_X f_y(x) d\mu$ are well-defined.
2. The integrability of the **iterated integrals**, $\int_Y f_x(y) dv$ and $\int_X f_y(x) d\mu$, so that $\int_X \left[\int_Y f_x(y) dv\right] d\mu$ and $\int_Y \left[\int_X f_y(x) d\mu\right] dv$ are well-defined.

Tonelli's theorem states the same conclusion on the equality of these integrals without assuming integrability of $f(x, y)$ on $X \times Y$. Instead, this result assumes that $f(x, y)$ **is nonnegative and measurable** and allows the possibility that all of these integrals are infinite. The point is, they will be infinite or finite together, and when finite, the values will agree.

The approach to the proof of either result is to first confirm that these results are true for characteristic functions of measurable sets of finite measure, and then to generalize using the integration tools developed in Chapter 3.

To address characteristic functions, we begin with sets $A \in \mathcal{A}_{\sigma\delta} \subset \sigma'(X \times Y)$ defined above. Then we will show that all cross-sections are measurable, and the measure of these cross-sections can be expressed as measurable functions of the defining variables.

We then generalize to characteristic functions of sets $A \in \sigma(X \times Y)$, recalling the above approximations that such sets are within $\mu \times v$-measure 0 of $\mathcal{A}_{\sigma\delta}$-sets. In this case we will see that cross-section measurability is now only true almost everywhere, and the measure of these cross-sections can be expressed as a measurable function that is defined almost everywhere. Results for characteristic functions of sets $A \in \sigma'(X \times Y)$ are then addressed.

5.2.2 Integrals of Characteristic Functions

We begin by formalizing definitions and notations introduced above and then show that Fubini's theorem holds for characteristic functions of sets $A \in \sigma(X \times Y)$, where $(X, \sigma(X), \mu)$

and $(Y, \sigma(Y), \upsilon)$ are complete measure spaces, and $(X \times Y, \sigma(X \times Y), \mu \times \upsilon)$ is the complete product space of Proposition I.7.20. For this section's results, we do not require σ-finiteness.

The final result considers characteristic functions of sets $A \in \sigma'(X \times Y)$.

Definition 5.9 (Cross-sections of sets; Component functions) *Given a set $A \subset X \times Y$, the **cross-sections of** A are defined by:*

$$\begin{aligned} &x\text{-cross-section:} && A_x = \{y|(x,y) \in A\}, \\ &y\text{-cross-section:} && A_y = \{x|(x,y) \in A\}. \end{aligned} \tag{5.5}$$

Thus $A_x \subset Y$ for all x, and $A_y \subset X$ for all y.

*If $f(x,y)$ is a function defined on $X \times Y$, the **component functions of** $f(x,y)$ are defined by:*

$$\begin{aligned} &x\text{-component function} &:& \quad f_x(y) = f(x,y), \\ &y\text{-component function} &:& \quad f_y(x) = f(x,y). \end{aligned} \tag{5.6}$$

As above, $\mathcal{A}' \subset \sigma(X \times Y)$ denotes the semi-algebra of measurable rectangles:

$$\mathcal{A}' = \{A \times B | A \in \sigma(X),\ B \in \sigma(Y)\},$$

$\mathcal{A}$ the associated algebra of finite disjoint unions, and related collections such as $\mathcal{A}_{\sigma\delta}$ are defined as in Corollary 5.2.

The first result states that for sets $A \in \mathcal{A}_{\sigma\delta}$, all cross-sections are measurable, and when $\mu \times \upsilon(A) < \infty$, the measures of these cross-sections are integrable functions of the defining variable.

Remark 5.10 (Fubini for characteristic functions) *The next proposition is in fact Fubini's theorem restricted to characteristic functions $f(x,y) = \chi_A(x,y)$ for $A \in \mathcal{A}_{\sigma\delta}$ with $\mu \times \upsilon(A) < \infty$.*

This follows because for all x, y:

$$\chi_A(x,y) = \chi_{A_x}(y) = \chi_{A_y}(x),$$

and thus if A_x and A_y are appropriately measurable, then by (3.1):

$$\upsilon(A_x) = \int \chi_A(x,y) d\upsilon(y), \qquad \mu(A_y) = \int \chi_A(x,y) d\mu(x).$$

Similarly:

$$\int \chi_A(x,y) d(\mu \times \upsilon) \equiv \mu \times \upsilon(A).$$

Thus the result in (5.8) can be expressed:

$$\int \left[\int \chi_A(x,y) d\upsilon \right] d\mu = \int \chi_A(x,y) d(\mu \times \upsilon) = \int \left[\int \chi_A(x,y) d\mu \right] d\upsilon, \tag{5.7}$$

which is Fubini's theorem applied to the function $f(x,y) = \chi_A(x,y)$ for $A \in \mathcal{A}_{\sigma\delta}$ with $\mu \times \upsilon(A) < \infty$.

Since we do not assume σ-finiteness of measure spaces for this next result, we require that $\mu \times \upsilon(A) < \infty$ in part 2 to utilize Corollary 5.2.

Proposition 5.11 (Fubini's theorem: $\chi_A(x,y)$ for $A \in \mathcal{A}_{\sigma\delta}$) *Let $(X, \sigma(X), \mu)$ and $(Y, \sigma(Y), \upsilon)$ be complete measure spaces and $A \in \mathcal{A}_{\sigma\delta}$. Then:*

1. *$A_x \in \sigma(Y)$ for all x, and $A_y \in \sigma(X)$ for all y.*

2. *If $\mu \times \upsilon(A) < \infty$, then $g(x) \equiv \upsilon(A_x)$ is $\sigma(X)$-measurable, $h(y) \equiv \upsilon(A_y)$ is $\sigma(Y)$-measurable, and:*

$$\int g(x)d\mu = \mu \times \upsilon(A) = \int h(y)d\upsilon. \tag{5.8}$$

Proof.

1. *By symmetry, only the statement regarding A_x needs proof.*
   If $A = B \times C \in \mathcal{A}'$, a measurable rectangle, then by definition $A_x \in \{C, \emptyset\} \subset \sigma(Y)$. If $A \in \mathcal{A}_\sigma$, say $A = \bigcup_{j=1}^{\infty} A_j$ with $A_j \in \mathcal{A}$, then as each A_j is a finite disjoint union of $\mathcal{A}'$-sets, this characterization of A can be modified by relabelling to assume that $A_j \in \mathcal{A}'$. Then $A_x = \bigcup_{j=1}^{\infty} (A_j)_x \in \sigma(Y)$. Finally, if $A \in \mathcal{A}_{\sigma\delta}$, say $A = \bigcap_{j=1}^{\infty} A_j$ with $A_j \in \mathcal{A}_\sigma$, then $A_x = \bigcap_{j=1}^{\infty} (A_j)_x$, and the result follows.

2. *Again by symmetry, we only prove the result for $g(x) \equiv \upsilon(A_x)$, and use the same sequential approach as part 1.*

   (a) *If $A = B \times C \in \mathcal{A}'$ with $\mu \times \upsilon(A) < \infty$, then from part 1, $g(x) = \chi_B(x)\upsilon(C)$, which is μ-measurable since $\upsilon(C) < \infty$.*
   Then by (3.1) and (1.18):

   $$\int g(x)d\mu = \upsilon(C) \int \chi_B(x)\, d\mu = \upsilon(C)\,\mu(B) = \mu \times \upsilon(A).$$

   (b) *Assume $A \in \mathcal{A}_\sigma$ with $\mu \times \upsilon(A) < \infty$, say $A = \bigcup_{j=1}^{\infty} A_j$ with $A_j \in \mathcal{A}'$ as noted in part 1. Defining $A^{(n)} = \bigcup_{j=1}^{n} A_j$, then $\{A^{(n)}\}_{n=1}^{\infty}$ is an increasing nested sequence of measurable sets, $A^{(n)} \subset A^{(n+1)}$.*
   By Proposition I.7.13, each $A^{(n)}$ can be expressed as a finite union of disjoint measurable rectangles.
   Specifically, if $A_j = B_j \times C_j$, then $\bigcup_{j=1}^{n} B_j$ and $\bigcup_{j=1}^{n} C_j$ can each be expressed as unions of no more than $M = 2^n - 1$ disjoint sets. If $\{B_j'\}_{j=1}^{M}$ and $\{C_j'\}_{j=1}^{M}$ are the resulting disjoint collections, including empty sets for notational simplicity, then $\{B_j' \times C_k'\}_{j,k=1}^{M}$ are M^2 disjoint measurable rectangles. For every j, k, either $B_j' \times C_k' \subset A^{(n)}$ or $B_j' \times C_k' \bigcap A^{(n)} = \emptyset$, and we choose the collections of subset rectangles and note by construction that they are disjoint and with union $A^{(n)}$.
   Relabel these sets so $A^{(n)} = \bigcup_{j=1}^{N(n)} A_j^{(n)}$ with $\{A_j^{(n)}\}_{j=1}^{N(n)}$ disjoint measurable rectangles. Defining $g_j^{(n)}(x) \equiv \upsilon\left[\left(A_j^{(n)}\right)_x\right]$, then $g_j^{(n)}(x)$ is nonnegative and $\sigma(X)$-measurable by part 2.a. since $A_j^{(n)} \in \mathcal{A}'$, and thus $g^{(n)}(x) \equiv \sum_{j=1}^{N(n)} g_j^{(n)}(x)$ is $\sigma(X)$-measurable and increasing in n.
   Further, disjointedness of $\left\{\left(A_j^{(n)}\right)_x\right\}_{j=1}^{N(n)}$ and finite additivity of υ obtain:

   $$g^{(n)}(x) \equiv \sum_{j=1}^{N(n)} \upsilon\left[\left(A_j^{(n)}\right)_x\right] = \upsilon\left[A_x^{(n)}\right].$$

But then by continuity from below of υ:

$$g(x) \equiv \upsilon\left[A_x\right] = \lim_{n\to\infty} g^{(n)}(x),$$

and so $g(x)$ *is* $\sigma(X)$*-measurable by Proposition 2.19.*
Linearity of the integral, part 2.a., and finite additivity of $\mu\times\upsilon$ *then yield:*

$$\begin{aligned}\int g^{(n)}(x)d\mu &= \sum\nolimits_{j=1}^{N(n)} \int g_j^{(n)}(x)d\mu \\ &= \sum\nolimits_{j=1}^{N(n)} \mu\times\upsilon\left(A_j^{(n)}\right) \\ &= \mu\times\upsilon(A^{(n)}).\end{aligned}$$

Finally, Lebesgue's monotone convergence theorem of Proposition 3.23 and continuity from below of $\mu\times\upsilon$ *obtain:*

$$\begin{aligned}\int g(x)d\mu &= \lim_{n\to\infty}\int g^{(n)}(x)d\mu \\ &= \lim_{n\to\infty}\mu\times\upsilon(A^{(n)}) \\ &= \mu\times\upsilon\,(A)\,.\end{aligned}$$

Hence, part 2 of the proposition is proved for $A\in\mathcal{A}_\sigma$.

(c) *Assume* $A\in\mathcal{A}_{\sigma\delta}$ *with* $\mu\times\upsilon\,(A)<\infty$, *where* $A=\bigcap_{j=1}^\infty A_j$ *with* $A_j\in\mathcal{A}_\sigma$. *We can assume that* $\{A_j\}_{j=1}^\infty$ *are nested with* $A_{j+1}\subset A_j$ *since given a general representation* $A=\bigcap_{i=1}^\infty B_i$ *with* $B_i\in\mathcal{A}_\sigma$, *define* $A_j=\bigcup_{i=j}^\infty B_i$, *noting that* $A_j\in\mathcal{A}_\sigma$ *by definition.*
The assumption that $\mu\times\upsilon\,(A)<\infty$ *assures by Corollary 5.2 that* A *has a superset* $D\in\mathcal{A}_\sigma$, *so* $A\subset D$, *and:*

$$\mu\times\upsilon\,(D)\leq\mu\times\upsilon\,(A)+\epsilon.$$

By redefining the A_j*-sets by* $A_j\bigcap D\in\mathcal{A}_\sigma$, *we can assume that* $\mu\times\upsilon\,(A_1)<\infty$. *If* $g_j(x)\equiv\upsilon\left[(A_j)_x\right]$, *then* $g_j(x)$ *is nonnegative and* $\sigma(X)$*-measurable by part 2.b.,*

$$\int g_j(x)d\mu=\mu\times\upsilon\,(A_j)<\infty,$$

and so in particular, $g_1(x)<\infty$, μ*-a.e.*
Hence, for any x *with* $g_1(x)<\infty$, $\{(A_j)_x\}_{j=1}^\infty$ *is a decreasing nested collection of sets of finite measure with intersection* A_x. *By continuity from above of* υ:

$$\upsilon\,(A_x)\equiv g(x)=\lim_j g_j(x)\ \mu\text{-a.e.},$$

and hence $g(x)$ *is measurable by completeness and Proposition 2.14.*
Finally, since $0\leq g_j(x)\leq g_1(x)$ *and* $g_1(x)$ *is integrable, we can apply Lebesgue's dominated convergence theorem and continuity from above of* $\mu\times\upsilon$, *justified by* $\mu\times\upsilon\,(A_1)<\infty$, *to conclude:*

$$\begin{aligned}\int g(x)d\mu &= \lim_{j\to\infty}\int g_j(x)d\mu \\ &= \lim_{j\to\infty}\mu\times\upsilon\,(A_j) \\ &= \mu\times\upsilon\,(A)\,.\end{aligned}$$

■

Remark 5.12 (On cross-sections) *One conclusion that can be drawn from the above proposition is that if $A \in \mathcal{A}_{\sigma\delta}$, then $\mu \times \upsilon(A) = 0$ if and only if both $\upsilon(A_x) = 0$ for (μ-)almost all x, and, $\mu(A_y) = 0$ for (υ-)almost all y.*

This follows because $\mu \times \upsilon(A) = 0$ if and only if:

$$\int g(x)d\mu = \int h(y)d\upsilon = 0.$$

Since $g(x)$ and $h(y)$ are nonnegative, this can be true if and only if $g(x) = 0$ μ-a.e., and $h(y) = 0$ υ-a.e., by Proposition 3.29.

Put another way, if $\mu \times \upsilon(A) < \infty$ and $\mu \times \upsilon(A) \neq 0$, then both $\upsilon(A_x)$ and $\mu(A_y)$ must be non-zero on sets of positive μ/υ-measure, respectively.

An important extension of this observation is that for any set $A \in \sigma(X \times Y)$ with $\mu \times \upsilon(A) = 0$, ***almost all*** *cross-sections have measure 0 relative to the component measures. This transition from $\mathcal{A}_{\sigma\delta}$ to $\sigma(X \times Y)$ is based on another application of the approximation results of Corollary 5.2.*

Corollary 5.13 (Cross-sections of $A \in \sigma(X \times Y)$ with $\mu \times \upsilon(A) = 0$) *Let $(X, \sigma(X), \mu)$ and $(Y, \sigma(Y), \upsilon)$ be complete measure spaces. If $A \in \sigma(X \times Y)$ and $\mu \times \upsilon(A) = 0$, then:*

1. *$A_x \in \sigma(Y)$ for μ-a.e. x, and $A_y \in \sigma(X)$ for υ-a.e. y.*

2. *$\upsilon(A_x) = 0$ for μ-a.e. x, and $\mu(A_y) = 0$ for υ-a.e. y.*

Proof. *By Corollary 5.2, there is a superset $A' \in \mathcal{A}_{\sigma\delta}$ so that $A \subset A'$ and $\mu \times \upsilon(A') = 0$. By Proposition 5.11, $A'_x \in \sigma(Y)$ for all x, and $A'_y \in \sigma(X)$ for all y. Also, $\upsilon(A'_x) = 0$ for μ-a.e. x, and $\mu(A'_y) = 0$ for υ-a.e. y by Remark 5.12.*

Then $A_x \subset A'_x$ and $A_y \subset A'_y$ and the completeness of the measure spaces obtains both results. ■

The next result generalizes Proposition 5.11 to $A \in \sigma(X \times Y)$. It should be noted that for measurable sets outside $\mathcal{A}_{\sigma\delta}$, the conclusions in general switch from "all x" to "almost all x", and similarly for the y-statements. However, this proposition is one step closer to Fubini's theorem in that by Remark 5.10, (5.7) now applies to characteristic functions $f(x, y) = \chi_A(x, y)$ for general $A \in \sigma(X \times Y)$ with $\mu \times \upsilon(A) < \infty$.

Proposition 5.14 (Fubini's theorem: $\chi_A(x, y)$ for $A \in \sigma(X \times Y)$) *Let $(X, \sigma(X), \mu)$ and $(Y, \sigma(Y), \upsilon)$ be complete measure spaces. If $A \in \sigma(X \times Y)$ with $\mu \times \upsilon(A) < \infty$:*

1. *$A_x \in \sigma(Y)$ for μ-a.e. x, and $A_y \in \sigma(X)$ for υ-a.e. y.*

2. *$g(x) \equiv \upsilon(A_x)$ is defined for μ-a.e. x and is $\sigma(X)$-measurable, $h(y) \equiv \upsilon(A_y)$ is defined for υ-a.e. y and is $\sigma(Y)$-measurable, and:*

$$\int g(x)d\mu = \mu \times \upsilon(A) = \int h(y)d\upsilon. \tag{5.9}$$

Proof. *As in the proof of Corollary 5.13, given such A, there is a superset $A' \in \mathcal{A}_{\sigma\delta}$ so that $A \subset A'$ and $\mu \times \upsilon(A') = \mu \times \upsilon(A)$. By Proposition 5.11, $A'_x \in \sigma(Y)$ for all x, and $A'_y \in \sigma(X)$ for all y.*

If $B \equiv A' - A$, then $B \in \sigma(X \times Y)$ and finite additivity applied to $A' = A \bigcup B$ obtains $\mu \times \upsilon(B) = 0$. By Corollary 5.13, $B_x \in \sigma(Y)$ for μ-a.e. x, and $B_y \in \sigma(X)$ for υ-a.e. y. Then $A_z = A'_z - B_z$ for $z = x, y$ obtains part 1.

Also by Corollary 5.13, $\upsilon(B_x) = 0$ *for* μ*-a.e.* x*, and* $\mu(B_y) = 0$ *for* υ*-a.e.* y*. Hence,* $g(x) \equiv \upsilon(A_x) = \upsilon(A'_x)$ μ*-a.e., and* $h(y) \equiv \upsilon(A_y) = \upsilon(A'_y)$ υ*-a.e. But* $\upsilon(A'_x)$ *is* $\sigma(X)$*-measurable by Proposition 5.11, and so* $g(x)$ *is* $\sigma(X)$*-measurable by completeness of* $(X, \sigma(X), \mu)$ *and Proposition 2.14. The same argument applies to the* $\sigma(Y)$*-measurability of* $h(y)$*.*

Then Proposition 3.29 applied to $g(x) = \upsilon(A'_x)$*,* μ*-a.e, and (5.8), obtains:*

$$\int g(x)d\mu = \int \upsilon(A'_x)\, d\mu = \mu \times \upsilon(A') = \mu \times \upsilon(A).$$

The same derivation applies to $\int h(y)d\upsilon$*, which proves (5.9).* ■

We end this section with a parallel result for the sigma algebra $\sigma'(X \times Y)$.

Proposition 5.15 (Fubini's theorem: $\chi_A(x,y)$ for $A \in \sigma'(X \times Y)$) *Let* $(X, \sigma(X), \mu)$ *and* $(Y, \sigma(Y), \upsilon)$ *be measure spaces and* $\sigma'(X \times Y)$ *the smallest* σ*-algebra that contains the algebra* $\mathcal{A}$ *of finite disjoint unions of* $\mathcal{A}'$*-sets, with* $\mathcal{A}'$ *the semi-algebra of measurable rectangles. If* $A \in \sigma'(X \times Y)$ *with* $\mu \times \upsilon(A) < \infty$*:*

1. $A_x \in \sigma(Y)$ *for all* x*, and* $A_y \in \sigma(X)$ *for all* y*.*

2. $g(x) \equiv \upsilon(A_x)$ *is well-defined for all* x *and* $\sigma(X)$*-measurable,* $h(y) \equiv \upsilon(A_y)$ *is well-defined for all* y *and* $\sigma(Y)$*-measurable, and (5.9) is satisfied, though all integrals may be infinite.*

Proof. *By symmetry, only the first statement of part 1 requires proof.*

Given $x \in X$*, define a mapping* $f_x : Y \to X \times Y$ *by* $f_x(y) = (x, y)$*. If* $A \in \mathcal{A}' \subset \sigma'(X \times Y)$ *is a measurable rectangle, say* $A = E \times F$*, then* $f_x^{-1}(A) \in \sigma(Y)$ *since* $f_x^{-1}(A) = F$ *if* $x \in E$ *and* $f_x^{-1}(A) = \emptyset$ *if* $x \notin E$*.*

Similarly, if $A \in \mathcal{A}$ *is a finite union of measurable rectangles,* $A = \bigcup_{j=1}^n E_j \times F_j$*, then* $f_x^{-1}(A) \in \sigma(Y)$ *as a finite union of the* F_j*-sets for which* $x \in E_j$*. For such* $A \in \mathcal{A}$*, it follows that:*

$$f_x^{-1}(\widetilde{A}) = Y - f_x^{-1}(A) \in \sigma(Y),$$

while if $\{A_j\}_{j=1}^\infty \subset \mathcal{A}$*, then:*

$$f_x^{-1}(\bigcup_{j=1}^\infty A_j) = \bigcup_{j=1}^\infty f_x^{-1}(A_j) \in \sigma(Y).$$

Since the algebra $\mathcal{A}$ *generates* $\sigma'(X \times Y)$*, it follows that* $f_x^{-1}[\sigma'(X \times Y)] \subset \sigma(Y)$ *for all* x*. As* $f_x^{-1}(A) = A_x$*, this proves that* $A_x \in \sigma(Y)$ *for all* x*.*

Hence, $g(x) \equiv \upsilon(A_x)$ *is well-defined for all* x*, and analogously* $h(y) \equiv \upsilon(A_y)$ *is well-defined for all* y*.*

For the rest of the proof, we defer to the result of Proposition 5.20, which proves (5.7) for integrable functions $f(x,y)$ *on a finite measure space, and apply this to* $\chi_A(x,y)$ *for* $A \in \sigma'(X \times Y)$ *with* $\mu \times \upsilon(A) < \infty$*.* ■

5.3 Fubini's Theorem

Fubini's theorem is named for **Guido Fubini** (1879–1943). By Remark 5.10 and Proposition 5.14, this result is already proved for $f(x,y) = \chi_A(x,y)$ for $A \in \sigma(X \times Y)$ with

$\mu \times \upsilon(A) < \infty$. Such functions are automatically $\mu \times \upsilon$-integrable by (3.1). The final result below states that for any $\mu \times \upsilon$-integrable function, the properties proved above for characteristic functions generalize.

The transition from characteristic functions to general measurable functions will require the earlier results of Corollary 2.32 on approximating measurable functions with simple functions.

See Section 5.3.2 for Fubini's theorem in the measure space $(X \times Y, \sigma'(X \times Y), \mu \times \upsilon)$ discussed above. In this case, the "almost all" statements of 1, 1′, 2, and 2′ are changed to "all."

Proposition 5.16 (Fubini's theorem on $(X \times Y, \sigma(X \times Y), \mu \times \upsilon)$**)** *Let* $(X, \sigma(X), \mu)$ *and* $(Y, \sigma(Y), \upsilon)$ *be complete, σ-finite measure spaces, and* $f(x,y)$ *an integrable function on* $(X \times Y, \sigma(X \times Y), \mu \times \upsilon)$*:*

$$\int_{X \times Y} |f(x,y)|\, d(\mu \times \upsilon) < \infty. \tag{5.10}$$

Then:

1. *For* μ*-a.e.* x, $f_x(y)$ *is* $\sigma(Y)$*-measurable, and,*

1'. *For* υ*-a.e.* y, $f_y(x)$ *is* $\sigma(X)$*-measurable.*

2. $\int_Y f_x(y) d\upsilon \equiv \int_Y f(x,y) d\upsilon$ *is defined for* μ*-a.e.* x *and is* $\sigma(X)$*-measurable, and,*

2'. $\int_X f_y(x) d\mu \equiv \int_X f(x,y) d\mu$ *is defined for* υ*-a.e.* y *and is* $\sigma(Y)$*-measurable.*

3. *The* $\mu \times \upsilon$*-integral of* $f(x,y)$ *can be evaluated as an iterated integral:*

$$\int_X \left[\int_Y f(x,y) d\upsilon\right] d\mu = \int_{X \times Y} f(x,y) d(\mu \times \upsilon) = \int_Y \left[\int_X f(x,y) d\mu\right] d\upsilon. \tag{5.11}$$

Proof. *By symmetry, we only prove parts 1, 2, and the first half of 3. In addition, we can assume that* $f(x,y)$ *is nonnegative, since by linearity of integrals from Proposition 3.42 and the assumption that* $f(x,y)$ *is integrable, this result can then be applied to the components of general* $f \equiv f^+ - f^-$ *to obtain the final result.*

By Remark 5.10, Proposition 5.14 proves these results for $f(x,y) = \chi_A(x,y)$ *for* $A \in \sigma(X \times Y)$ *with* $\mu \times \upsilon(A) < \infty$*. Hence, by linearity of integrals, these results are true for simple functions that are zero outside sets of finite measure. Given nonnegative integrable* $f(x,y)$*, Corollary 2.32 obtains the existence of an increasing sequence of nonnegative simple functions* $\{\varphi_n(x,y)\}_{n=1}^{\infty}$*, each equal to zero outside a set of finite measure, and such that* $\varphi_n(x,y) \to f(x,y)$ *for all* (x,y).

Hence for each x, $(\varphi_n)_x(y) \to f_x(y)$ *for all* y*. Since* $(\varphi_n)_x(y)$ *is* $\sigma(Y)$*-measurable for almost all* x, $f_x(y)$ *is* $\sigma(Y)$*-measurable by Proposition 2.19 for all* x *for which all* $(\varphi_n)_x(y)$ *are measurable, which is almost all* x*. Lebesgue's monotone convergence theorem then obtains for almost all* x*:*

$$\int_Y f_x(y) d\upsilon = \lim_{n \to \infty} \int_Y \varphi_n(x,y) d\upsilon.$$

Then $\sigma(X)$*-measurability of* $\int_Y \varphi_n(x,y) d\upsilon$ *and* μ*-a.e. convergence obtains* $\sigma(X)$*-measurability of* $\int_Y f_x(y) d\upsilon$ *by completeness and Corollary 2.22.*

Applying Lebesgue's theorem again, and Proposition 5.14:

$$\begin{aligned}\int_X\left[\int_Y f(x,y)dv\right]d\mu &= \lim_{n\to\infty}\int_X\left[\int_Y \varphi_n(x,y)dv\right]d\mu \\ &= \lim_{n\to\infty}\int_{X\times Y}\varphi_n(x,y)d(\mu\times v) \\ &= \int_{X\times Y} f(x,y)d(\mu\times v),\end{aligned}$$

which proves (5.11). ■

Remark 5.17 (On integrability of $f(x,y)$) *It should be noted that the assumption of integrability of $f(x,y)$ allowed the proof to focus on nonnegative $f(x,y)$ and then generalize by decomposing $f\equiv f^+-f^-$. If we assumed here that $f(x,y)$ was nonnegative, the same proof would apply without the assumption of integrability.*

The application of Corollary 2.32 would then be justified by σ-finiteness of the product space, but in the applications of Lebesgue's monotone convergence theorem, there would be no assurance that the resulting limit integrals would be finite. Thus in this case, all integrals in (5.11) could be infinite, or all finite, depending on the integrability of $f(x,y)$.

See Tonelli's theorem below, which addresses nonnegative measurable functions.

5.3.1 Generalizing Fubini's Theorem

The above theorem and Tonelli's theorem below are explicitly stated and proved in terms of a bivariate function $f(x,y)$. As noted in the introduction, while these results appear to apply only in a "2-dimensional" product space, "dimension" here is merely a notational device and there was nothing in the above development that precluded these component spaces from also being product spaces. More importantly, these results are more generally applicable to higher "dimensions" as is verified by an iterative application of the 2-dimensional result.

For example, assume that $(X,\sigma(X),\mu)$, $(Y,\sigma(Y),v)$, and $(Z,\sigma(Z),\lambda)$ are complete, σ-finite measure spaces and $f(x,y,z)$ is an integrable function on $(X\times Y\times Z,\sigma(X\times Y\times Z),\mu\times v\times\lambda)$:

$$\int_{X\times Y\times Z}|f(x,y,z)|\,d(\mu\times v\times\lambda)<\infty.$$

The key step in iterating the above result is the identification of $(X\times Y\times Z,\sigma(X\times Y\times Z),\mu\times v\times\lambda)$ and $(X\times(Y\times Z),\sigma(X\times(Y\times Z)),\mu\times(v\times\lambda))$ as measure spaces. The notation implies that the triple product measure space is identical with the product measure space created with $(X,\sigma(X),\mu)$ and $(Y\times Z,\sigma(Y\times Z),v\times\lambda)$. This identification was proved in Proposition I.7.24 using the uniqueness result of that book's using the uniqueness result of Proposition I.6.14, which in turn required that the measure spaces be σ-finite.

We can conclude from this result the following re-interpretation of Proposition 5.16. We leave further generalizations to the reader.

Proposition 5.18 (Fubini's theorem on $(X\times Y\times Z,\sigma(X\times Y\times Z),\mu\times v\times\lambda)$) *Assume that $(X,\sigma(X),\mu)$, $(Y,\sigma(Y),v)$, and $(Z,\sigma(Z),\lambda)$ are complete, σ-finite measure spaces and that $f(x,y,z)$ is an integrable function on $(X\times Y\times Z,\sigma(X\times Y\times Z),\mu\times v\times\lambda)$. Then:*

1. *For μ-a.e. x, $f_x(y,z)=f(x,y,z)$ is $\sigma(Y\times Z)$-measurable, and,*

1'. *For $v\times\lambda$-a.e. (y,z), $f_{(y,z)}(x)=f(x,y,z)$ is $\sigma(X)$-measurable.*

2. $\int_{Y\times Z} f(x,y,z)d(\upsilon\times\lambda)$ *is* $\sigma(X)$*-measurable, and,*

2'. $\int_X f(x,y,z)d\mu$ *is* $\sigma(Y\times Z)$*-measurable.*

3. The $\mu\times\upsilon\times\lambda$*-integral of* $f(x,y,z)$ *can be evaluated as an iterated integral:*

$$\int_X\left[\int_{Y\times Z} f(x,y,z)d(\upsilon\times\lambda)\right]d\mu = \int_{X\times Y\times Z} f(x,y,z)d(\mu\times\upsilon\times\lambda) = \int_{Y\times Z}\left[\int_X f(x,y,z)d\mu\right]d(\upsilon\times\lambda).$$

Remark 5.19 (Further iterations) *The above result is not the final statement of Fubini's theorem in three variables but provides the iterative step. By items 2' and 3,* $\int_X f(x,y,z)d\mu$ *is integrable and we can apply Fubini's theorem to the* $\upsilon\times\lambda$*-integral on the right in 3 to conclude:*

$$\int_Y\left[\int_Z\left[\int_X f(x,y,z)d\mu\right]d\lambda\right]d\upsilon = \int_{Y\times Z}\left[\int_X f(x,y,z)d\mu\right]d(\upsilon\times\lambda) = \int_Z\left[\int_Y\left[\int_X f(x,y,z)d\mu\right]d\upsilon\right]d\lambda.$$

Similarly, by 2 and 3 we can apply Fubini's theorem to the $\upsilon\times\lambda$*-integral on the left of 3 to conclude:*

$$\int_X\left[\int_Z\left[\int_Y f(x,y,z)d\upsilon\right]d\lambda\right]d\mu = \int_X\left[\int_{Y\times Z} f(x,y,z)d(\upsilon\times\lambda)\right]d\mu = \int_X\left[\int_Y\left[\int_Z f(x,y,z)d\lambda\right]d\upsilon\right]d\mu.$$

Hence, under the assumption of integrability under the product measure, the value of the integral under this measure equals the value of the iterated integrals. Two such iterated integral decompositions are:

$$\begin{aligned}&\int_X\left[\int_Y\left[\int_Z f(x,y,z)d\lambda\right]d\upsilon\right]d\mu\\ =& \int_{X\times Y\times Z} f(x,y,z)d(\mu\times\upsilon\times\lambda)\\ =& \int_Z\left[\int_Y\left[\int_X f(x,y,z)d\mu\right]d\upsilon\right]d\lambda.\end{aligned} \tag{5.12}$$

5.3.2 Fubini's Theorem on $\sigma'(X\times Y)$

In the above development of Fubini's theorem, the first step was the verification that this theorem was true for characteristic functions of measurable sets of finite measure. The first and easiest step in this verification was for characteristic functions of $A\in\mathcal{A}'$, the semi-algebra of measurable rectangles, and for this step, completeness of the component spaces was not needed.

Recalling the **functional monotone class theorem** of Proposition 2.41, this step also provides one of the key properties to verify to prove that the same Fubini result is true for all bounded measurable functions. There, measurability is defined relative to $\sigma(\mathcal{A}')$, the smallest sigma algebra generated by $\mathcal{A}'$. It is thus natural to investigate the potential for this earlier result to prove this version of Fubini's theorem.

For this inquiry, we will therefore work with the sigma algebra $\sigma'(X \times Y)$, defined as the smallest sigma algebra that contains the algebra $\mathcal{A}$, which in turn is generated by the semi-algebra of measurable rectangles $\mathcal{A}'$. So in the notation of the functional monotone class theorem:

$$\sigma(\mathcal{A}') \equiv \sigma'(X \times Y) \subset \sigma(X \times Y).$$

To ensure the uniqueness of the extension from $\mathcal{A}'$ to $\sigma(\mathcal{A}')$ of the product set function in (5.1), we will require that $(X, \sigma(X), \mu)$ and $(Y, \sigma(Y), \upsilon)$ be σ-finite measure spaces as noted in Section 5.1.1 on sigma algebra constructions.

With this introduction, let $(X \times Y, \sigma'(X \times Y), \mu \times \upsilon)$ denote the product space referenced in this chapter's introduction under the **Alternative Approach,** whereby $\sigma'(X \times Y)$ is defined as above, and $\mu \times \upsilon$ is uniquely determined on $\sigma'(X \times Y)$ as the extension of the measure of measurable rectangles $A \in \mathcal{A}'$ as defined in (5.1). In the notation of the functional monotone class theorem, let $\mathcal{L}$ denote a class of functions $f(x, y)$ on $X \times Y$, which satisfy a given statement of Fubini's theorem. To demonstrate that $\mathcal{L}$ contains all bounded $\sigma'(X \times Y)$-measurable functions on $X \times Y$, it must be shown that:

1. $\chi_A \in \mathcal{L}$ for all $A \in \mathcal{A}'$, and $\chi_X \in \mathcal{L}$.

2. $\mathcal{L}$ is a vector space: If $f, g \in \mathcal{L}$ then $af + bg \in \mathcal{L}$ for all $a, b \in \mathbb{R}$.

3'. If $f : X \to \mathbb{R}^+$ is bounded and nonnegative and the pointwise limit of increasing $\{f_n\}_{n=1}^{\infty} \subset \mathcal{L}$, then $f \in \mathcal{L}$.

Note that condition 3 of Proposition 2.41 is replaced by $3'$ as proved in Exercise 2.43.

The next result is Fubini's theorem applied to $(X \times Y, \sigma'(X \times Y), \mu \times \upsilon)$ in the case of finite component measure spaces, such as probability spaces, without the assumption on completeness. See Exercise 5.22 for the generalization to σ-finite spaces.

Proposition 5.20 (Fubini's theorem on finite $(X \times Y, \sigma'(X \times Y), \mu \times \upsilon)$**)**
Let $(X, \sigma(X), \mu)$ and $(Y, \sigma(Y), \upsilon)$ be finite measure spaces, and $f(x, y)$ an integrable function on the finite measure space $(X \times Y, \sigma'(X \times Y), \mu \times \upsilon)$. Then:

1. *For all x, $f_x(y)$ is $\sigma(Y)$-measurable, and,*

1'. *For all y, $f_y(x)$ is $\sigma(X)$-measurable.*

2. *$\int_Y f_x(y) d\upsilon \equiv \int_Y f(x, y) d\upsilon$ is $\sigma(X)$-measurable, and,*

2'. *$\int_X f_y(x) d\mu \equiv \int_X f(x, y) d\mu$ is $\sigma(Y)$-measurable.*

3. *The $\mu \times \upsilon$-integral of $f(x, y)$ can be evaluated as an iterated integral:*

$$\int_X \left[\int_Y f(x, y) d\upsilon\right] d\mu = \int_{X \times Y} f(x, y) d(\mu \times \upsilon) = \int_Y \left[\int_X f(x, y) d\mu\right] d\upsilon. \tag{5.13}$$

Proof. *Let $\mathcal{L}$ denote the class of integrable functions $f(x, y)$, which satisfy all five statements above. We prove this result in two steps, first proving that $\mathcal{L}$ contains all bounded, measurable functions, which are of necessity integrable, and then extending to all integrable functions.*

Checking the criteria for the functional monotone class theorem, Proposition 5.11 proved that $\chi_A(x, y) \in \mathcal{L}$ for all $A \in \mathcal{A}'$ with $\mu \times \upsilon(A) < \infty$. Since $\mu \times \upsilon$ is a finite measure, this result applies to all such A, including $X \times Y$. Also, $\mathcal{L}$ is a vector space over $\mathbb{R}$ by properties of measurable functions from Proposition 2.13 and linearity of integrals from Proposition 3.42.

Now if $f : X \times Y \to \mathbb{R}^+$ is bounded and nonnegative and the pointwise limit of increasing $\{f_n\}_{n=1}^{\infty} \subset \mathcal{L}$, then f satisfies 1 and 1′ by Proposition 2.19, while 2 and 2′ are satisfied by Lebesgue's monotone convergence theorem and Proposition 2.19.

For 3, by definition of $\mathcal{L}$:

$$\int_X \left[\int_Y f_n(x,y) dv \right] d\mu = \int_{X \times Y} f_n(x,y) d(\mu \times v) = \int_Y \left[\int_X f_n(x,y) d\mu \right] dv,$$

for all n. Lebesgue's monotone convergence theorem then obtains:

$$\int_{X \times Y} f_n(x,y) d(\mu \times v) \to \int_{X \times Y} f(x,y) d(\mu \times v). \tag{1}$$

Similarly, for all x :

$$\int_Y f_n(x,y) dv \to \int_Y f(x,y) dv, \tag{2}$$

while another application of monotone convergence yields the final result.

Thus by the functional monotone class theorem, $\mathcal{L}$ contains all bounded measurable functions.

To prove that the class $\mathcal{L}$ contains all nonnegative integrable functions, given such $f(x,y)$ define $f_n(x,y) = \max[f(x,y), n]$. Then $f_n(x,y) \to f(x,y)$ pointwise, and $f_n(x,y) \in \mathcal{L}$ since bounded. Thus (5.13) obtains:

$$\int_X \left[\int_Y f_n(x,y) dv \right] d\mu = \int_{X \times Y} f_n(x,y) d(\mu \times v) = \int_Y \left[\int_X f_n(x,y) d\mu \right] dv. \tag{3}$$

Since $\{f_n(x,y)\}_{n=1}^{\infty}$ is increasing, Lebesgue's monotone convergence theorem yields (1), and (2) for all x. Letting $g_n(x) \equiv \int_Y f_n(x,y) dv$, then $g_n(x)$ is $\sigma(X)$-measurable by the above result, and increasing with n. Again by Lebesgue's monotone convergence theorem,

$$\int_X \left[\int_Y f_n(x,y) dv \right] d\mu \to \int_X \left[\int_Y f(x,y) dv \right] d\mu.$$

By monotone convergence and (3), this limit function is finite and equals $\int_{X \times Y} f(x,y) d(\mu \times v)$.

Thus:

$$\int_X \left[\int_Y f(x,y) dv \right] d\mu = \int_{X \times Y} f(x,y) d(\mu \times v),$$

and a similar argument applies to the second equality in (5.13).

The result now applies to general integrable f, using the decomposition into nonnegative functions: $f = f^+ - f^-$, and linearity of integrals. ■

Remark 5.21 (On integrable $f(x,y)$) *Since all measure spaces are finite, if $f(x,y)$ is bounded, then $\int_Y f(x,y) dv$ is finite* ***for all*** *x and $\int_X f(x,y) d\mu$ is finite* ***for all*** *y by item 5 of Proposition 3.42. If $f(x,y)$ is integrable but not bounded, then while $\int_Y f_n(x,y) dv \to \int_Y f(x,y) dv$ for all x, this limit need not be finite for all x. But as $\int_X \left[\int_Y f(x,y) dv\right] d\mu$ is finite, this assures that this limit is finite for μ-a.e. x by Proposition 3.40.*

Exercise 5.22 (Fubini's theorem on σ-finite $(X \times Y, \sigma'(X \times Y), \mu \times v)$) *Generalize Proposition 5.20 to σ-finite measure spaces. Hint: By Definition 1.21, there exists a countable collections $\{X_j\}_{j=1}^{\infty} \subset \sigma(X)$ with $\mu(X_j) < \infty$ for all j and $\{Y_j\}_{j=1}^{\infty} \subset \sigma(Y)$ with $v(Y_j) < \infty$ for all j, and $X \times Y = \bigcup_{j=1}^{\infty} X_j \times \bigcup_{j=1}^{\infty} Y_j$. Let $B^{(N)} \equiv \bigcup_{j=1}^{N} X_j \times \bigcup_{j=1}^{N} Y_j$, so*

$\mu \times \upsilon\left(B^{(N)}\right) < \infty$ for all N, and define $g_N(x,y) = \chi_{B^{(N)}}(x,y)$. Given integrable $f(x,y)$ on $(X \times Y, \sigma'(X \times Y), \mu \times \upsilon)$, define $f_N(x,y) = g_N(x,y)f(x,y)$. Then recalling Exercise II.1.29, $f_N(x,y)$ is an integrable function on the finite measure space $(B^{(N)}, \sigma'(X \times Y) \bigcap B^{(N)}, \mu \times \upsilon)$, and thus Proposition 5.20 applies for each N:

$$\int_X \left[\int_Y f_N(x,y)d\upsilon\right] d\mu = \int_{X \times Y} f_N(x,y)d(\mu \times \upsilon) = \int_Y \left[\int_X f_N(x,y)d\mu\right] d\upsilon.$$

Note that by definition of $f_N(x,y)$ and (3.5) of Definition 3.7, that we can fix the domain spaces as noted. Justify letting $N \to \infty$, addressing the necessary measurability requirements.

5.4 Tonelli's Theorem

Tonelli's theorem is named for **Leonida Tonelli** (1885–1946) and addresses the same question as does Fubini's theorem. Specifically, it addresses the relationship between product space and iterated integrals but circumvents Fubini's sometimes difficult-to-establish assumption on the integrability of $f(x,y)$. Tonelli's theorem requires only that the function be nonnegative and measurable.

Because integrability of the function $f(x,y)$ is not assumed, the identity in (5.11), as repeated below in (5.14), must now allow for the case where all integrals are infinite. Recalling Remark 5.17, measurability of $f(x,y)$ and σ-finiteness of the component spaces does not assure the integrability of $f(x,y)$. But these conditions do assure that when integrable, the product space and iterated integrals agree, and when not integrable, all integrals are infinite.

The proof is nearly identical with that of Fubini's result and is included for completeness. As noted in Section 5.3.1 on generalizing Fubini's theorem, this next result applies more generally to integrals in product spaces of more than two factors. Details are left as an exercise.

Proposition 5.23 (Tonelli's theorem on $(X \times Y, \sigma(X \times Y), \mu \times \upsilon)$) *Let $(X, \sigma(X), \mu)$ and $(Y, \sigma(Y), \upsilon)$ be complete, σ-finite measure spaces, and $f(x,y)$ a nonnegative measurable function on $(X \times Y, \sigma(X \times Y), \mu \times \upsilon)$, then:*

1. *For μ-a.e. x, $f_x(y) = f(x,y)$ is $\sigma(Y)$-measurable, and,*

1'. *For υ-a.e. y, $f_y(x) = f(x,y)$ is $\sigma(X)$-measurable.*

2. *$\int_Y f(x,y)d\upsilon$ is $\sigma(X)$-measurable, and,*

2'. *$\int_X f(x,y)d\mu$ is $\sigma(Y)$-measurable.*

3. *The $\mu \times \upsilon$-integral of $f(x,y)$ can be evaluated as an iterated integral:*

$$\int_X \left[\int_Y f(x,y)d\upsilon\right] d\mu = \int_{X \times Y} f(x,y)d(\mu \times \upsilon) = \int_Y \left[\int_X f(x,y)d\mu\right] d\upsilon, \tag{5.14}$$

though such integrals need not be finite.

Proof. *By symmetry, only half of the results need be proved, and specifically we address parts 1, 2 and the first equality of 3.*

Proposition 5.14 provides the desired measurability results and (5.14) for $f(x,y) = \chi_A(x,y)$ *for* $A \in \sigma(X \times Y)$ *with* $\mu \times \upsilon(A) < \infty$, *and hence these results are true for simple functions that are zero outside sets of finite measure. By Corollary 2.32, given nonnegative* $f(x,y)$ *on a* σ*-finite measure space, there is an increasing sequence* $\{\varphi_n(x,y)\}_{n=1}^{\infty}$ *of nonnegative simple functions, each equal to zero outside a set of finite measure, and with* $\varphi_n(x,y) \to f(x,y)$ *pointwise. Hence* $(\varphi_n)_x(y) \to f_x(y)$, *and so* $f_x(y)$ *is* $\sigma(Y)$*-measurable for all* x *for which all* $(\varphi_n)_x(y)$ *are measurable, which is* μ*-a.e.* x.

By Lebesgue's monotone convergence theorem:

$$\int_Y f(x,y)d\upsilon = \lim_{n\to\infty} \int_Y \varphi_n(x,y)d\upsilon,$$

and the μ*-measurability of* $\int_Y \varphi_n(x,y)d\upsilon$ *proves the same for* $\int_Y f(x,y)d\upsilon$ *by Proposition 2.19. Applying Lebesgue's theorem again, and Proposition 5.14:*

$$\begin{aligned}\int_X \left[\int_Y f(x,y)d\upsilon\right] d\mu &= \lim_{n\to\infty} \int_X \left[\int_Y \varphi_n(x,y)d\upsilon\right] d\mu \\ &= \lim_{n\to\infty} \int_{X\times Y} \varphi_n(x,y)d(\mu\times\upsilon) \\ &= \int_{X\times Y} f(x,y)d(\mu\times\upsilon),\end{aligned}$$

which proves (5.14). ■

5.4.1 Tonelli's Theorem on $\sigma'(X \times Y)$

As was the case for Fubini's theorem, Tonelli's result on $\sigma'(X \times Y)$ omits the almost everywhere qualifiers.

Proposition 5.24 (Tonelli's theorem on σ-finite $(X \times Y, \sigma'(X \times Y), \mu \times \upsilon)$**)** *Let* $(X, \sigma(X), \mu)$ *and* $(Y, \sigma(Y), \upsilon)$ *be* σ*-finite measure spaces, and* $f(x,y)$ *a nonnegative measurable function on the* σ*-finite measure space* $(X \times Y, \sigma'(X \times Y), \mu \times \upsilon)$. *Then:*

1. *For all* x, $f_x(y)$ *is* $\sigma(Y)$*-measurable, and,*

1'. *For all* y, $f_y(x)$ *is* $\sigma(X)$*-measurable.*

2. $\int_Y f_x(y)d\upsilon \equiv \int_Y f(x,y)d\upsilon$ *is* $\sigma(X)$*-measurable, and,*

2'. $\int_X f_y(x)d\mu \equiv \int_X f(x,y)d\mu$ *is* $\sigma(Y)$*-measurable.*

3. *The* $\mu \times \upsilon$*-integral of* $f(x,y)$ *can be evaluated as an iterated integral:*

$$\int_X \left[\int_Y f(x,y)d\upsilon\right] d\mu = \int_{X\times Y} f(x,y)d(\mu\times\upsilon) = \int_Y \left[\int_X f(x,y)d\mu\right] d\upsilon,$$

though such integrals need not be finite.

Proof. *This result is a special case of Proposition 5.20 and Exercise 5.22, restricted to nonnegative measurable* $f(x,y)$.

As in Exercise 5.22, there exists $\{X_j\}_{j=1}^{\infty} \subset \sigma(X)$ *with* $\mu(X_j) < \infty$ *for all* j *and* $\{Y_j\}_{j=1}^{\infty} \subset \sigma(Y)$ *with* $\upsilon(Y_j) < \infty$ *for all* j, *so that* $X \times Y = \bigcup_{j=1}^{\infty} X_j \times \bigcup_{j=1}^{\infty} Y_j$. *Let*

$B^{(N)} \equiv \bigcup_{j=1}^{N} X_j \times \bigcup_{j=1}^{N} Y_j$, *so* $\mu \times \upsilon\left(B^{(N)}\right) < \infty$ *for all* N. *Given nonnegative* $f(x, y)$ *on* $(X \times Y, \sigma'(X \times Y), \mu \times \upsilon)$, *define:*

$$f_N(x,y) = f(x,y)\chi_{B^{(N)}}(x,y)\chi_{\{f \leq N\}}(x,y).$$

Then $f_N(x, y)$ *is integrable and Proposition 5.20 applies for all* N.

But note that in the final application of Lebesgue's monotone convergence theorem as $N \to \infty$, *it cannot be asserted that the resulting integrals are finite but only that they will be finite or infinite together.* ■

Remark 5.25 (Fubini vs. Tonelli) *From a theoretical point of view, Fubini's theorem appears more satisfying because it assures that under the assumption of the integrability of* $f(x, y)$ *and the iterated integrals also exist and provide the same numerical value. Thus integrability of* $f(x, y)$ *assures that the component functions are integrable almost everywhere and that the iterated integrals of these component functions are also integrable. So the step-by-step evaluation of the iterated integral makes sense at each step.*

On the other hand, from a practical point of view Tonelli's theorem is more useful than is Fubini's because we do not need to ascertain integrability of $f(x, y)$ *in advance. The Tonelli result states that we can, in fact, establish integrability of nonnegative measurable* $f(x, y)$ *by demonstrating the finiteness of the iterated integrals, and this can sometimes be easier than establishing the finiteness of a product integral. In addition, for a general measurable function* $f(x, y)$, *Tonelli's theorem can be applied to* $|f(x, y)|$ *to determine if* $f(x, y)$ *is integrable by verifying the finiteness of the iterated integrals of* $|f(x, y)|$. *If integrable, we can conclude by Fubini's theorem that* $\int_{X \times Y} f(x, y) d(\mu \times \upsilon)$ *can also be evaluated as an iterated integral.*

5.5 Examples

Fubini's and Tonelli's theorems have appeared in prior books in various developments that were otherwise independent of these deep results. Put another way, these theorems are pervasive in even more elementary developments that require either evaluations, or various manipulations, of multivariate integrals. These earlier appearances were sometimes in proofs, and sometimes in applications.

In this section, we revisit the integral of the normal density function and develop a new example that will be useful later in this book. The application to the normal density is quite similar to that needed in the development of the Box-Muller and Bailey transforms of Book IV. Other applications of these results were seen on transformed random variables in Chapter IV.2, as well as on large deviation theory in Section IV.7.1.

The reader may be interested in revisiting some of these applications.

Example 5.26 (The normal density function) *A common example of the application of Fubini's or Tonelli's theorem is, perhaps surprisingly, in the evaluation of the one-variable integral of the* ***normal probability density function*** *investigated in Books II and IV.*

Recall that this density function is nonnegative, defined on $(-\infty, \infty)$, *depends on a location parameter* $\mu \in \mathbb{R}$ *and a scale parameter* $\sigma > 0$, *and is defined by:*

$$f_N(x) = \frac{1}{\sigma\sqrt{2\pi}} \exp\left(-(x-\mu)^2 / \left(2\sigma^2\right)\right). \tag{5.15}$$

By a change of variables, $y = (x - \mu)/\sigma$*, we obtain that:*

$$\int_{-\infty}^{\infty} f_N(x)dx = \int_{-\infty}^{\infty} \phi(x)dx,$$

where $\phi(x)$ *denotes the* ***unit normal probability density:***

$$\phi(x) = \frac{1}{\sqrt{2\pi}} \exp\left(-x^2/2\right). \tag{5.16}$$

Because for any N*:*

$$\exp\left(-x^2/2\right) < |x|^{-N} \quad as \quad |x| \to \infty,$$

it follows that $\int_{-\infty}^{\infty} \phi(x)dx < \infty$ *as an improper Riemann integral. By Proposition III.2.56, this integral therefore equals the corresponding Lebesgue integral.*

To use this section's results, we consider the square of the integral of $\phi(x)$*:*

$$\begin{aligned}\left[\int_{-\infty}^{\infty} \phi(x)dx\right]^2 &= \frac{1}{2\pi}\int_{-\infty}^{\infty}\int_{-\infty}^{\infty} \exp\left[-\left(x^2+y^2\right)/2\right] dxdy \\ &= \frac{1}{2\pi}\int\int_{\mathbb{R}^2} \exp\left[-\left(x^2+y^2\right)/2\right] dm \\ &= \frac{2}{\pi}\int\int_{Q} \exp\left[-\left(x^2+y^2\right)/2\right] dm.\end{aligned}$$

While the first step is notational, the transition from iterated integrals to product space integral is justified by Tonelli's theorem since $\exp\left[-\left(x^2+y^2\right)/2\right]$ *is continuous and hence measurable, or by Fubini's theorem, noting that this function is integrable because as noted above,* $\exp\left[-\left(x^2+y^2\right)/2\right] < \left[x^2+y^2\right]^{-N}$ *for all* N *as* $x^2+y^2 \to \infty$*. Then by symmetry this integral over* $\mathbb{R}^2$ *equals four times the integral over the first quadrant,* $Q = \{(x,y)|x > 0, y > 0\}$*.*

The final step is to use ***polar coordinates*** *to evaluate this joint integral, using the measurable transformation* $T : (0,\infty) \times (0,\pi/2) \to Q$ *defined by:*

$$\begin{gathered} T : (r,\theta) \to (x,y), \\ x = r\cos\theta, \qquad y = r\sin\theta. \end{gathered}$$

This transformation is continuously differentiable with Jacobian matrix:

$$T'(r,\theta) \equiv \begin{pmatrix} \cos\theta & -r\sin\theta \\ \sin\theta & r\cos\theta \end{pmatrix},$$

and Jacobian determinant $\det\left(T'(r,\theta)\right) = r$*. The inverse transformation is given:*

$$\begin{gathered} T^{-1} : (x,y) \to (r,\theta), \\ r = \sqrt{x^2+y^2}, \qquad \theta = \arctan(y/x). \end{gathered}$$

From (4.43) of Proposition 4.44 and item 2 of Remark 4.45, it follows that with $T : (0,\infty) \times (0,\pi/2) \to Q$ *and* $g(x,y) = \exp\left[-\left(x^2+y^2\right)/2\right]$*:*

$$\begin{aligned}\frac{2}{\pi}\int_{Q} \exp\left[-\left(x^2+y^2\right)/2\right] dm &= \frac{2}{\pi}\int_{T^{-1}Q} g(T(r,\theta))\left|\det(T'(r,\theta))\right| dm \\ &= \frac{2}{\pi}\int_{T^{-1}Q} r\exp\left(-r^2/2\right) dm.\end{aligned}$$

Another application of Fubini's theorem to the last integral, justified since $r\exp\left(-r^2/2\right)$ *is integrable over* $T^{-1}Q = (0,\infty)\times(0,\pi/2)$, *produces the iterated integral:*

$$\frac{2}{\pi}\int_Q \exp\left[-\left(x^2+y^2\right)/2\right]dm = \frac{2}{\pi}\int_0^{\pi/2}\int_0^{\infty} r\exp\left(-r^2/2\right)drd\theta = 1.$$

As the integral on the left equals $\left[\int_{-\infty}^{\infty}\phi(y)dy\right]^2$, *this completes the proof that the normal density integrates to* 1.

The following result is another example of the application of Fubini's theorem to the evaluation of a one-variable integral, and one which is needed in the section below on the Fourier transform. The trick is to embed the one-variable function that we seek to integrate into a two-variable function for which the multiple integral is well-defined, and for which the integral relative to the new variable is easy. Then by Fubini's theorem, we can reverse the orders of integration to put the easy integral first.

Proposition 5.27 *The function* $f(x) = \frac{\sin x}{x}$ *is Riemann and Lebesgue integrable on* $[0,\infty)$, *and:*

$$\lim_{t\to\infty}\int_0^t \frac{\sin x}{x}dx = \frac{\pi}{2}. \tag{5.17}$$

Proof. *For* $x > 0$:

$$f(x) = \int_0^{\infty} e^{-ux}\sin x du,$$

exists as a Riemann integral, and also as a Lebesgue integral by Proposition III.2.56, since $|e^{-ux}\sin x| \leq e^{-ux}$ *is Riemann integrable.*

Further, the integrand $e^{-ux}\sin x$ *is continuous and hence measurable on* $A_t \equiv \{(x,u)|0 < x \leq t,\ 0 < u < \infty\}$. *In fact, this function is integrable over* A_t *by Tonelli's theorem:*

$$\begin{aligned}\int\int_{A_t}\left|e^{-ux}\sin x\right|dm &= \int_0^t\left[\int_0^{\infty}\left|e^{-ux}\sin x\right|du\right]dx \\ &= \int_0^t \frac{|\sin x|}{x}dx \\ &< t.\end{aligned}$$

The last inequality follows because $|\sin x| \leq x$ *for* $x \geq 0$.

As noted in Remark 5.25, Fubini's theorem now applies and allows the interchange in the order of integration:

$$\begin{aligned}\int_0^t \frac{\sin x}{x}dx &= \int_0^{\infty}\int_0^t e^{-ux}\sin x dx du \\ &= \int_0^{\infty}\left[\frac{1}{u^2+1}\left(1-e^{-ut}\left[u\sin t+\cos t\right]\right)\right]du \\ &= \int_0^{\infty}\frac{1}{u^2+1}du - \int_0^{\infty}\frac{e^{-ut}\left[u\sin t+\cos t\right]}{u^2+1}du,\end{aligned}$$

where the middle equation follows by two applications of integration by parts.

The final step is to note that by a substitution of $u = \tan x$, *the first integral equals* $\pi/2$, *and by the substitution* $y = ut$, *the second integral converges to* 0 *as* $t\to\infty$. *In detail:*

$$\int_0^{\infty}\frac{1}{u^2+1}du = \int_0^{\pi/2}\frac{\sec^2 x}{1+\tan^2 x}dx = \frac{\pi}{2},$$

while:

$$\begin{aligned}\left|\int_0^\infty \frac{e^{-ut}\left[u\sin t+\cos t\right]}{u^2+1}du\right| &\leq \int_0^\infty \frac{e^{-y}}{y^2+t^2}\left|y\sin t+t\cos t\right|dy \\ &\leq \int_0^\infty \frac{y+t}{y^2+t^2}e^{-y}dy \\ &\leq \frac{c}{t}.\end{aligned}$$

The last inequality follows from a standard calculus maximization of $f(y) = (y+t)/(y^2+t^2)$ *for* $y \geq 0$ *to show that* $f(y) \leq c/t$ *for* $t > 0$. ■

6

Two Applications of Fubini/Tonelli

As noted in the prior chapter, applications of the Fubini and Tonelli theorems are found throughout mathematical analysis and its applications, and Books II and IV have required several applications of these results. In this chapter, we can by no means summarize the extent of such applications but will instead apply the results of the Fubini/Tonelli theorems in two very different contexts:

1. **Lebesgue-Stieltjes Integration by Parts**

   The first section will develop results for Lebesgue-Stieltjes integrals that extend the corresponding Riemann result of III.(1.30), the Lebesgue result of Proposition III.3.64, and the Riemann-Stieltjes result of Proposition III.4.14 as supplemented by Remark III.4.29.

2. **Convolution of Integrable Functions**

   Convolutions of functions were informally introduced in Notation IV.2.13, related to the distribution function of the sum of independent random variables. This notion will be formally defined in the second section and the all-important question of integrability of convolutions studied. The Book IV application of convolutions to the distribution and density functions of sums of independent random variables will be continued in Book VI.

Chapter 7 will then develop another application of the Fubini/Tonelli results, which is the development of Fourier transform theory. This theory will be critical in Book VI where the **characteristic function** of a distribution function will be introduced and applied. Characteristic functions will be seen to provide a far more robust tool for studying properties of distribution functions than do moment generating functions, since as will be proved, they always exist. With this tool, we will prove more general central limit theorems than was possible in Book IV, and derive important results on sums of random variables.

6.1 Lebesgue-Stieltjes Integration by Parts

In this section, we present an extension of the Lebesgue and Riemann-Stieltjes integration by parts results developed in Book III, which in turn extended the familiar Riemann integration by parts formula. Recall that the Riemann result of III.(1.30) and the Lebesgue result of Proposition III.3.63 looked identical:

$$\int_a^b f'(x)g(x)dx = f(b)g(b) - f(a)g(a) - \int_a^b f(x)g'(x)dx.$$

In other words, both results required at least some level of differentiability of the given functions, as well as other assumptions that would ensure that both integrals would exist as Riemann integrals, or both as Lebesgue integrals.

DOI: 10.1201/9781003264576-6

- For the Riemann result, the simplest statement required that $f(x)$ and $g(x)$ be continuously differentiable. Then both $f'(x)g(x)$ and $f(x)g'(x)$ are continuous and the Riemann integrals exist by Proposition III.1.15. But by Lebesgue's existence theorem of Proposition III.1.22, it is enough to have $f(x)$ and $g(x)$ continuously differentiable almost everywhere if $f'(x)g(x)$ and $f(x)g'(x)$ are bounded.

- The Lebesgue version required both $f(x)$ and $g(x)$ to be absolutely continuous, and thus $f(x)g(x)$ is absolutely continuous by Exercise III.3.63, and differentiable almost everywhere by Proposition III.3.59. Lebesgue integrability of $f'(x)g(x)$ and $f(x)g'(x)$ is then derived in the proof, as is the above identity.

Beyond the Riemann and Lebesgue contexts, integration by parts results tend to exchange the roles of integrand functions and integrator "measures" rather than exchange differentiation from one integrand factor to the other. This was first seen in the Riemann-Stieltjes context of Proposition III.4.14:

$$\int_a^b g(x)df = f(b)g(b) - f(a)g(a) - \int_a^b f(x)dg.$$

Here it was assumed that $g(x)$ was bounded and that the integral on the left existed, and from this the existence of the integral on the right was derived.

Given bounded $g(x)$, this result did not identify conditions on $f(x)$ that would assure existence of the integral on the left. If $g(x)$ is continuous and thus automatically bounded, then Proposition III.4.17 assures existence for increasing $f(x)$, and, if $g(x)$ is increasing and continuous, then this integral exists if $f(x)$ is of bounded variation.

In this latter case of bounded variation $f(x)$, Proposition III.4.23 obtains that this df integral can be evaluated as a difference of integrals with increasing integrators. This result followed from Proposition III.3.29 that derived that $f(x)$ is of bounded variation if and only if $f(x) = I_1(x) - I_2(x)$ for increasing functions $I_j(x)$, and then by Proposition III.4.23:

$$\int_a^b g(x)df = \int_a^b g(x)dI_1 - \int_a^b g(x)dI_2.$$

For Lebesgue-Stieltjes integration by parts, we follow a similar approach. The integral $\int_{(a,b]} gd\mu_f$ will be transformed to $\int_{(a,b]} fd\mu_g$ plus some constants, and this raises existence questions. As summarized in Section 1.1.3, Book I investigated the creation of a Borel measure μ_f from an increasing, right continuous function $f(x)$. Thus with the current level of theory, it is feasible to connect these integrals for increasing, right continuous $f(x)$ and $g(x)$.

But to have results that apply to more general functions, we again consider the class of functions of bounded variation, the topic of the next section. We will then see that by Proposition III.3.29, a "signed" set function μ_f can be derived for such functions, which is formally related to the notion of a **signed measure,** and that this set function can be characterized as a difference of Borel measures. Signed measures will be studied again in Section 8.2.1.

After developing properties of functions of bounded variation and associated signed set functions, we return to the main topic of integration by parts.

6.1.1 Functions of Bounded Variation

Recall the notion of a function of bounded variation from Definition III.3.23.

Definition 6.1 (Function of bounded variation on $[a,b]$) *Given a real-valued function $f(x)$ on an interval $[a,b]$, we say that $f(x)$* ***is of bounded variation on*** *$[a,b]$, and abbreviated $f \in B.V.$ or $f \in B.V.[a,b]$, if the* ***total variation*** *$T_{[a,b]} < \infty$, where:*

$$T_{[a,b]} \equiv \sup_{\Pi} \sum\nolimits_{i=1}^{n} |f(x_i) - f(x_{i-1})| .$$

Here the supremum is taken over all partitions $\Pi = \{x_0, x_1, ..., x_n\}$ with:

$$a = x_0 < x_1 ... < x_n = b.$$

As noted above, it was proved in Proposition III.3.29 that a function $f(x)$ defined on $[a,b]$ is of bounded variation if and only if:

$$f(x) = I_1(x) - I_2(x),$$

where $I_1(x)$ and $I_2(x)$ are increasing real-valued functions. This decomposition cannot be unique since $I_j(x) + k(x)$ also works for all increasing $k(x)$. Increasing functions are Borel (and hence Lebesgue)-measurable since inverse images of intervals are intervals, and differentiable m-a.e., meaning relative to Lebesgue measure, by Proposition III.3.12. Thus as a difference of increasing functions, functions of bounded variation are measurable and differentiable m-a.e.

For the current application, we would like to generalize Chapter 3 to be able define integrals relative to certain "signed" measures that will be defined relative to functions of bounded variation. To this end, we first generalize the bounded variation definition from $[a,b]$ to $\mathbb{R}$ and then identify a useful subset of this class of functions in the following definition. The notion of a signed measure will be studied in more detail in Section 8.2.1.

Definition 6.2 (Function of bounded variation on $\mathbb{R}$) *A real-valued function $f(x)$ defined on $\mathbb{R}$* ***is of bounded variation on*** *$\mathbb{R}$, abbreviated $f \in B.V.[\mathbb{R}]$, if the* ***total variation*** *$T < \infty$ where:*

$$T \equiv \sup_{[a,b]} T_{[a,b]},$$

where the supremum is taken over all such intervals.

Such a function $f(x)$ ***is of normalized bounded variation on*** *$\mathbb{R}$, abbreviated $f \in N.B.V.[\mathbb{R}]$, if $f \in B.V.[\mathbb{R}]$, f is right continuous, and $f(-\infty) = 0$.*

Remark 6.3 (On $B.V.[\mathbb{R}]$ and $N.B.V.[\mathbb{R}]$) *We identify a number of results below, leaving some of the details as exercises.*

1. *If $f \in B.V.[\mathbb{R}]$, then by definition $f \in B.V.[a,b]$ for all $[a,b]$. Hence by Proposition III.3.29, for any such interval:*

$$f(x) = I_1^{[a,b]}(x) - I_2^{[a,b]}(x),$$

for increasing real-valued functions $I_1^{[a,b]}(x)$ and $I_2^{[a,b]}(x)$. This implies that there exist increasing real-valued functions $I_1(x)$ and $I_2(x)$, defined on $\mathbb{R}$, so that $f(x)$ has the representation:

$$f(x) = I_1(x) - I_2(x). \tag{1}$$

Hint: On arbitrary $[c,d]$, define $I_j(x) = I_j^{[c,d]}(x)$, so (1) is satisfied on this interval. If $[c,d] \subset [a,b]$, extend the definition of $I_j(x)$ by:

$$I_j(x) = \begin{cases} I_j(c) + I_j^{[a,b]}(x) - I_j^{[a,b]}(c), & x \in [a,c], \\ I_j(x), & x \in [c,d], \\ I_j(d) + I_j^{[a,b]}(x) - I_j^{[a,b]}(d), & x \in [d,b]. \end{cases}$$

Check that $I_j(x)$ is increasing and (1) is satisfied on $[a,b]$. Now iterate.

2. *If $f \in B.V.[\mathbb{R}]$, then by item 1 and Proposition III.3.12, f is Borel measurable, and differentiable m-a.e.*

3. *If $f \in B.V.[\mathbb{R}]$, then f is bounded since $|f(b) - f(a)| \leq T_{[a,b]} \leq T$ for all intervals, and so $|f(x)| \leq M$ for some M. This does not imply that $I_1(x)$ and $I_2(x)$ are bounded, but since:*
$$I_2(x) - M \leq I_1(x) \leq I_2(x) + M,$$
either both $I_1(x)$ and $I_2(x)$ are bounded or neither is bounded.

4. *If $f \in B.V.[\mathbb{R}]$, then:*
$$\lim_{x \to \pm\infty} f(x) \text{ exist and both limits are finite.}$$
Hint: If these limits exist, they are bounded by item 3. Using proof by contradiction, recall Exercise 2.18 and assume that:
$$\lim_{x \to \infty} [\inf\{f(y)|y \geq x\}] = l < L = \lim_{x \to \infty} [\sup\{f(y)|y \geq x\}],$$
Then by Proposition I.3.45, for any $\epsilon > 0$ there exist $x_n, x'_n \to \infty$ so that $l \leq f(x_n) \leq l + \epsilon$ and $L - \epsilon \leq f(x'_n) \leq L$. Hence if $\epsilon < (L - l)/2$, any collection of disjoint intervals of the form $[x_n, x'_m]$ or $[x'_n, x_m]$ would provide a contradiction to the assumption that $f \in B.V.[\mathbb{R}]$.

5. *If $f \in B.V.[\mathbb{R}]$, then f is continuous m-a.e. by item 2. Indeed, f is continuous with at most countably many exceptions by the decomposition in item 1 and Proposition I.5.8. Then by this last result, $g(x) \equiv f(x^+)$ is right continuous, $g(x) = f(x)$, m-a.e., and $g \in B.V.[\mathbb{R}]$, where $f(x^+) \equiv \lim_{y \to x^+} f(y)$ (Definition III.3.1). Hint: Apply Proposition I.5.8 to the component increasing functions to confirm the existence of $f(x^+)$, then derive:*
$$g(x) = I_1(x^+) - I_2(x^+).$$
Thus $g(x) \in B.V.[a,b]$ for all $[a,b]$ by Proposition III.3.29, so left to prove that T in Definition 6.2 is finite.

6. *If $f \in B.V.[\mathbb{R}]$ is right continuous, while there exists increasing functions with $f(x) = I_1(x) - I_2(x)$, right continuity of f does not assure right continuity $I_1(x)$ and $I_2(x)$. However, $f(x) = f(x^+)$ assures that $I_1(x) - I_2(x) = I_1(x^+) - I_2(x^+)$, and thus such $f(x)$ can be equivalently decomposed with right continuous increasing, functions:*
$$f(x) \equiv I_1(x^+) - I_2(x^+).$$

7. *By items 4 and 5, if $f \in B.V.[\mathbb{R}]$, then $h(x) \equiv f(x^+) - f(-\infty)$ is right continuous, $h \in B.V.[\mathbb{R}]$, $h(-\infty) = 0$, and so $h \in N.B.V.[\mathbb{R}]$. Thus every $f \in B.V.[\mathbb{R}]$ obtains such $h \in N.B.V.[\mathbb{R}]$, where outside an at most countable set, $h(x) = f(x) + k$.*

8. *If $f \in N.B.V.[\mathbb{R}]$ is defined by right continuous, increasing $I_1(x)$ and $I_2(x)$ by item 6, then by item 4:*
$$\lim_{x \to -\infty} [I_1(x) - I_2(x)] = 0, \quad \lim_{x \to \infty} [I_1(x) - I_2(x)] = c.$$
But as noted in item 3, the separate I_j-limits need not be finite.

Example 6.4 (Borel measures) *If μ is a finite Borel measure, then $F_\mu(x) \equiv \mu(-\infty, x] \in N.B.V.[\mathbb{R}]$ since it is increasing, right continuous, $F(-\infty) = 0$, and $F(\infty) < \infty$, by Proposition I.5.7. By this same result, if μ is a general Borel measure, then $F_\mu(x) \in B.V.[a,b]$ for all $[a,b]$, with:*

$$F_\mu(x) = \begin{cases} \mu((0,x]), & x > 0, \\ 0, & x = 0, \\ -\mu((x,0]), & x < 0. \end{cases}$$

Such $F_\mu(y)$ is again right continuous reflecting continuity from above of μ.

Conversely, if $F(x)$ is right continuous and increasing, then $F \in B.V.[a,b]$ for all $[a,b]$, and by Proposition I.5.23, F induces a measure μ_F on the Borel sigma algebra $\mathcal{B}(\mathbb{R})$. If F is bounded, then $F \in B.V.[\mathbb{R}]$ and μ_F is a finite measure. In this case, μ_F is identical with the measure μ_G where $G(x) \equiv F(x) - F(-\infty) \in N.B.V.[\mathbb{R}]$.

This identification between right continuous increasing functions and Borel measures is generalized in the next result that identifies right continuous $B.V.[\mathbb{R}]$ with countably additive set functions definable on bounded $A \in \mathcal{B}(\mathbb{R})$. This set function is "nearly" a measure as is discussed below, and also "nearly" a signed measure as defined in Section 8.2.1.

Proposition 6.5 (μ_f for $f \in B.V.[\mathbb{R}]$) *Let $f \in B.V.[\mathbb{R}]$ be right continuous. With $\mathcal{A}'$ the semi-algebra of right semi-closed intervals, define the set function μ_f on bounded $(a,b] \in \mathcal{A}'$:*

$$\mu_f\left[(a,b]\right] = f(b) - f(a).$$

Then there exist Borel measures μ_1 and μ_2 so that on bounded intervals in $\mathcal{A}'$:

$$\mu_f = \mu_1 - \mu_2.$$

In addition, μ_f has a well-defined extension to bounded $A \in \mathcal{B}(\mathbb{R})$.

Proof. *By 6 of Remark 6.3, $f(x) = I_1(x) - I_2(x)$ where $I_1(x)$ and $I_2(x)$ are right continuous, increasing functions. Consequently, by the development of Chapter I.5 and summarized in Section 1.1.3, there exist Borel measures μ_1 and μ_2 so that for all right semi-closed intervals $(a,b]$:*

$$\mu_j\left[(a,b]\right] = I_j(b) - I_j(a).$$

Thus for all bounded $(a,b] \in \mathcal{A}'$:

$$\mu_f\left[(a,b]\right] = \mu_1\left[(a,b]\right] - \mu_2\left[(a,b]\right],$$

and the set function μ_f has an extension to all bounded Borel sets $A \in \mathcal{B}(\mathbb{R})$:

$$\mu_f[A] \equiv \mu_1[A] - \mu_2[A]. \tag{6.1}$$

To see that this extension of μ_f is well-defined, assume that also $f(x) = I_1'(x) - I_2'(x)$, where $I_1'(x)$ and $I_2'(x)$ are right continuous, increasing functions. Then since $\mu_f\left[(a,b]\right] = f(b) - f(a)$, it follows by the same argument that for all bounded $(a,b]$:

$$\mu_1\left[(a,b]\right] - \mu_2\left[(a,b]\right] = \mu_1'\left[(a,b]\right] - \mu_2'\left[(a,b]\right].$$

Rewriting:

$$\mu_1\left[(a,b]\right] + \mu_2'\left[(a,b]\right] = \mu_1'\left[(a,b]\right] + \mu_2\left[(a,b]\right]. \tag{1}$$

On $\mathcal{A}'$, define the measures:

$$v_1 = \mu_1 + \mu_2', \quad v_2 = \mu_1' + \mu_2.$$

Since $\upsilon_1 = \upsilon_2$ on $\mathcal{A}'$, and $\mathcal{B}(\mathbb{R})$ is σ-finite, the uniqueness result of Proposition 1.22 obtains that $\upsilon_1 = \upsilon_2$ on $\mathcal{B}(\mathbb{R})$, noting that $\mathcal{B}(\mathbb{R})$ is the smallest sigma algebra that contains $\mathcal{A}'$ by Proposition I.8.1. Thus $\mu_1 + \mu_2' = \mu_1' + \mu_2$ on $\mathcal{B}(\mathbb{R})$.

Now Borel measures are finite on compact sets by Definition 1.8, and thus if $A \in \mathcal{B}(\mathbb{R})$ is bounded it is contained in a compact set. Since then all the measures of A are finite, we can subtract to obtain:

$$\mu_1[A] - \mu_2[A] = \mu_1'[A] - \mu_2'[A].$$

Hence, μ_f is well-defined by (6.1) for bounded $A \in \mathcal{B}(\mathbb{R})$. ■

Remark 6.6 (On μ_f) *In general, μ_f is not a measure since by construction, it need not be the case that $\mu_f[A] \geq 0$. Further, this proposition falls short of declaring that μ_f is a signed measure, meaning by Definition 8.5 that it extends to a countably additive set function on the Borel sigma algebra $\mathcal{B}(\mathbb{R})$.*

The problem for countable additivity and this extension to $\mathcal{B}(\mathbb{R})$ is that as noted in item 3 of Remark 6.3, either both $I_1(x)$ and $I_2(x)$ are bounded functions or neither is bounded. In the later case, if A is an unbounded interval, then formally, $\mu_f[A] = \infty - \infty$, which is not well defined. Similarly, such μ_f is not countably additive on disjoint $\{A_i\}_{i=1}^{\infty}$ with $\bigcup_{i=1}^{\infty} A_i$ unbounded.

However, μ_f is perfectly well-defined on bounded Borel measurable sets and countably additive when both $\mu_j[\bigcup A_j] < \infty$, which is enough for the development below.

Exercise 6.7 ($I_j(x)$ bounded) *Prove that if $f(x) = I_1(x) - I_2(x)$ where $I_1(x)$ and $I_2(x)$ are right continuous, increasing and bounded functions, then μ_f is a countably additive set function on $\mathcal{B}(\mathbb{R})$. This is then an example of a* ***signed measure,*** *to be discussed in Section 8.2.1.*

6.1.2 Lebesgue-Stieltjes integration by parts

In this section, we first investigate the extension of the Lebesgue-Stieltjes integration theory of Chapter 3 to integrals with respect to the signed set function μ_f of the previous section. Once defined, a result on integration by parts will be derived.

Proposition 6.8 ($\int_A g d\mu_f$, $f \in B.V.[\mathbb{R}]$) *If $f \in B.V.[\mathbb{R}]$ is right continuous, then for all bounded Borel measurable functions g and bounded $A \in \mathcal{B}(\mathbb{R})$, the* ***Lebesgue-Stieltjes integral:***

$$\int_A g d\mu_f,$$

is well-defined by:

$$\int_A g d\mu_f = \int_A g d\mu_1 - \int_A g d\mu_2, \tag{6.2}$$

where μ_1 and μ_2 are any Borel measures as defined in Proposition 6.5.

Proof. *Given a decomposition $f(x) = I_1(x) - I_2(x)$, where $I_1(x)$ and $I_2(x)$ are right continuous, increasing functions, the integrals on the right in (6.2) are well-defined by item 5 of Proposition 3.42. This follows since $|g(x)| \leq M$, and $\mu_j(A) < \infty$ by Definition 1.8 since A is bounded and thus contained in a compact interval.*

For well-definedness of (6.2), assume that also $f(x) = I_1'(x) - I_2'(x)$, where $I_1'(x)$ and $I_2'(x)$ are right continuous, increasing functions. Then for all bounded $A \in \mathcal{B}(\mathbb{R})$, $\mu_1 - \mu_2 = \mu_1' - \mu_2'$ by Proposition 6.5, and hence by Definition 3.2, for all simple functions φ with bounded domains:

$$\int \varphi d\mu_1 - \int \varphi d\mu_2 = \int \varphi d\mu_1' - \int \varphi d\mu_2'. \tag{1}$$

If $g(x)$ is nonnegative, bounded and measurable, and A is bounded, then by Corollary 2.32 there exists an increasing sequence $\{\varphi_n\}_{n=1}^{\infty}$ of simple functions so that $\varphi_n \to g$ pointwise on A. Lebesgue's monotone convergence theorem of Proposition 3.23 then obtains:

$$\int_A g d\mu = \lim_{n\to\infty} \int_A \varphi_n d\mu,$$

where μ here denotes any of the four measures in (1). As each integral in (1) is finite by the assumptions on g and A, $\int_A g d\mu_f$ is well-defined by (6.2) for nonnegative $g(x)$.

For bounded and measurable $g(x)$, this nonnegative result applies to g^+ and g^- of Definition 3.37, and the proof is complete. ■

With the aid of Fubini's theorem, we are now ready to state and prove an integration by parts result for Lebesgue-Stieltjes integrals. The first result compares most closely with the corresponding result on Riemann-Stieltjes integrals of Proposition III.4.14.

Remark 6.9 (On $f, g \in B.V.[\mathbb{R}]$) *To simplify the statements, the following proposition and corollary assume that $f, g \in B.V.[\mathbb{R}]$ and then present results that are true for any bounded interval $(a, b]$. But given fixed $(a, b]$, a local condition on f, g, such as $f, g \in B.V.[[a - \epsilon, b]]$, is enough to assure the same conclusions on this interval. This follows since such functions can be extended to $B.V.[\mathbb{R}]$ functions by defining, for example, $f(x) = f(a-\epsilon)$ for $x < a - \epsilon$ and $f(x) = f(b)$ for $x > b$.*

Proposition 6.10 (Lebesgue-Stieltjes integration by parts on $(a, b]$)
If $f, g \in B.V.[\mathbb{R}]$ are right continuous and at least one continuous, then for any bounded interval $(a, b]$:

$$\int_{(a,b]} g d\mu_f = f(b)g(b) - f(a)g(a) - \int_{(a,b]} f d\mu_g. \tag{6.3}$$

Proof. *Both f and g are bounded and measurable since $f, g \in B.V.[\mathbb{R}]$, and so both integrals are well-defined by item 5 of Proposition 3.42. Define $E \subset \mathbb{R}^2$ by:*

$$E = \{(x, y) | a < x \leq y \leq b\}.$$

Then $E \in \mathcal{B}(\mathbb{R}^2)$ as the intersection of open and closed sets:

$$E = \{x \leq y\} \bigcap \{y \leq b\} \bigcap \{x > a\}.$$

Since $E \subset A$ with $A = (a, b] \times (a, b]$, and:

$$(\mu_f \times \mu_g) [A] \equiv (f(b) - f(a)) (g(b) - g(a)) < \infty,$$

it follows from Definition 3.7 and monotonicity that:

$$\int_A \chi_E(x, y) d (\mu_f \times \mu_g) = (\mu_f \times \mu_g) (E) < \infty. \tag{1}$$

Thus Fubini's theorem applies and the integral in (1) can be evaluated as an iterated integral. Without loss of generality, we can assign either measure to either variate. Temporarily denoting these functions h and k, Fubini's theorem obtains:

$$(\mu_f \times \mu_g) (E) = \int_{(a,b]} \left[\int_{(a,y]} d\mu_h(x) \right] d\mu_k(y) = \int_{(a,b]} \left[\int_{[x,b]} d\mu_k(y) \right] d\mu_h(x). \tag{2}$$

For any Borel measure:

$$\int_{(a,y]} d\mu_h(x) = h(y) - h(a),$$

but the integral $\int_{[x,b]} d\mu_k(y)$ *is more subtle. If the B.V function* k *underlying* μ_k *is continuous, then* $\mu_k\,[[x,b]] = \mu_k\,[(x,b]]$ *by Exercise I.5.22, and then* $\int_{[x,b]} d\mu_k(y) = k(b) - k(x)$*. So whichever of* f *and* g *is continuous is assigned the role of* k*, notationally.*

To be specific, assume that g *is the continuous function. Then from (2):*

$$\begin{aligned}(\mu_f \times \mu_g)(E) &= \int_{(a,b]} \left[\int_{(a,y]} d\mu_f(x)\right] d\mu_g(y) \\ &= \int_{(a,b]} [f(y) - f(a)]\, d\mu_g(y) \\ &= \int_{(a,b]} f d\mu_g - f(a)\,[g(b) - g(a)]\,.\end{aligned}$$

Similarly,

$$\begin{aligned}(\mu_f \times \mu_g)(E) &= \int_{(a,b]} \left[\int_{[x,b]} d\mu_g(y)\right] d\mu_f(x) \\ &= \int_{(a,b]} [g(b) - g(x)]\, d\mu_f(x) \\ &= g(b)[f(b) - f(a)] - \int_{(a,b]} g d\mu_f,\end{aligned}$$

The proof is now complete by simplifying the resulting expressions. ■

For the following corollary, note that at least one of $f(a^-)$ or $g(a^-)$ equals the value of the respective function at a by the continuity requirement.

Corollary 6.11 (Lebesgue-Stieltjes integration by parts on $[a,b]$**)** *If* $f, g \in B.V.[\mathbb{R}]$ *are right continuous and at least one continuous, then for any bounded interval* $[a,b]$ *:*

$$\int_{[a,b]} g d\mu_f = f(b)g(b) - f(a^-)g(a^-) - \int_{[a,b]} f d\mu_g, \tag{6.4}$$

where $f(a^-)$ *and* $g(a^-)$ *denote the lefts limits at* a*.*

Thus if both f *and* g *are continuous:*

$$\int_{[a,b]} g d\mu_f = f(b)g(b) - f(a)g(a) - \int_{[a,b]} f d\mu_g. \tag{6.5}$$

Proof. *Define* $E' = \{(x,y) | a \le x \le y \le b\}$*, then* $(\mu_f \times \mu_g)(E') < \infty$ *and again Fubini's theorem applies. The details are left as an exercise.* ■

6.2 Convolution of Integrable Functions

If $f(x)$ and $g(x)$ are μ-integrable functions, then the product $f(x)g(x)$ needs not be μ-integrable. For example on $[0,1]$, a function $f(x) = x^a$ is Lebesgue integrable if and only if

$a > -1$, so the product of such integrable functions need not be integrable. However, there is a special type of product, called a **convolution product** or just a **convolution**, for which integrability is preserved. Introduced in Section IV.2.2.1 and to be further developed in Book VI, convolutions play an important role in probability theory, and in mathematical analysis generally.

Notation 6.12 (Lebesgue integrals) *To simplify notation, we will denote Lebesgue integrals by $\int f(x)dm$ rather than $(\mathcal{L})\int f(x)dx$. In integrals involving more than one variable, the notation such as $\int f(x,y)dm(y)$ will be used for clarity.*

Definition 6.13 (Convolution of functions) *Let $f(x)$ and $g(x)$ be measurable functions on the Lebesgue measure space $(\mathbb{R}, \mathcal{M}_L(\mathbb{R}), m)$. The* **convolution of** f **and** g*, denoted $f*g$, is defined pointwise by:*

$$f * g(x) = \int f(x-y)g(y)dm(y), \tag{6.6}$$

when this integral exists.

Exercise 6.14 ($f * g = g * f$) *Using the Proposition 4.24 result on change of variables, prove that for any x for which the integral in (6.6) exists, that $f * g(x) = g * f(x)$. In other words:*

$$\int f(x-y)g(y)dm(y) = \int g(x-y)f(y)dm(y). \tag{6.7}$$

The goal of this section is to prove that if f and g are integrable, then $f(x-y)g(y)$ is y-integrable for almost all x, and $f * g(x)$ is, in fact, integrable. Moreover, this last result will also yield an upper bound for the value of the integral $\int f * g(x)dm$.

Lebesgue product spaces are σ-finite (Definition 1.21), as is proved by considering the collection of compact rectangles defined by $A_M = \{(x_1, x_2, ..., x_n) | -M \leq x_j \leq M$ for all $j\}$, and noting that A_M has Lebesgue measure $(2M)^n$. Thus to justify the use of Fubini's theorem, we must prove "only" that $f(x-y)g(y)$ is a Lebesgue integrable function on the product space $(\mathbb{R}^2, \mathcal{M}_L(\mathbb{R}^2), m)$, constructed with two copies of $(\mathbb{R}, \mathcal{M}_L(\mathbb{R}), m)$ as in Section I.7.6, and summarized in Section 1.3.2.

Regarding Lebesgue measurability of $h(x,y) \equiv f(x-y)g(y)$, meaning measurability of:

$$h : (\mathbb{R}^2, \mathcal{M}_L(\mathbb{R}^2), m) \to (\mathbb{R}, \mathcal{B}(\mathbb{R}), m),$$

it is enough to establish measurability of the component functions by Proposition 2.13. While the measurability of $g(y)$ is readily established, perhaps surprisingly, the Lebesgue measurability of $f(x-y)$ on $\mathbb{R}^2$ is not obvious and in fact somewhat challenging to prove.

One reference that explicitly addresses this question is **Hewitt and Stromberg** (1965), whose approach we follow. Given Lebesgue measurable $f : \mathbb{R} \to \mathbb{R}$, their idea is to identify a property of $\varphi : \mathbb{R}^2 \to \mathbb{R}$ that assures the Lebesgue measurability of $f(\varphi(x,y))$ and then to prove that $\varphi(x,y) \equiv x - y$ has this property.

Before proceeding, the reader may want to recall Remark I.3.31 and Proposition I.3.33, which addressed Lebesgue measurability of composite functions.

Proposition 6.15 (Lebesgue measurability of $f(\varphi) : \mathbb{R}^2 \to \mathbb{R}$) *Let $\varphi : \mathbb{R}^2 \to \mathbb{R}$ be Borel measurable and with the additional property that if $N \in \mathcal{M}_L(\mathbb{R})$ with $m(N) = 0$, then $\varphi^{-1}(N) \in \mathcal{M}_L(\mathbb{R}^2)$ and $m\left[\varphi^{-1}(N)\right] = 0$.*

Then $f(\varphi) : \mathbb{R}^2 \to \mathbb{R}$ is Lebesgue measurable for every Lebesgue measurable function f.

Proof. *Given a set $A \in \mathcal{M}_L(\mathbb{R})$, let $f(x) \equiv \chi_A(x)$, the characteristic function of A, defined to equal 1 when $x \in A$ and equal 0 otherwise. By (2.20) of Proposition I.2.42, there exists disjoint $G \in \mathcal{B}(\mathbb{R})$ and $N \in \mathcal{M}_L(\mathbb{R})$ so that $A = G \bigcup N$ and $m(N) = 0$.*

Then $\chi_A(\varphi)(x,y) \equiv \chi_{\varphi^{-1}(A)}(x,y)$ *and:*

$$\chi_{\varphi^{-1}(A)}(x,y) = \chi_{\varphi^{-1}(G)}(x,y) + \chi_{\varphi^{-1}(N)}(x,y).$$

Since $\varphi^{-1}(G) \in \mathcal{B}(\mathbb{R}^2)$ *by Borel measurability of* φ, *and* $\varphi^{-1}(N) \in \mathcal{M}_L$ *by hypothesis,* $\chi_A(\varphi)$ *is Lebesgue measurable for all* $A \in \mathcal{M}_L$.

It then follows from Proposition 2.13 that $f(\varphi)$ *is Lebesgue measurable for all simple functions* $f = \sum_{j=1}^{n} a_j \chi_{A_j}$, *since then* $f(\varphi) = \sum_{j=1}^{n} a_j \chi_{A_j}(\varphi)$.

Given a Lebesgue measurable function $f(x)$, *Definition 3.37 and Proposition 2.28 assure the existence of a sequence of simple functions* $\psi_k(x)$ *with* $\psi_k(x) \to f(x)$ *for all* x. *Thus* $\psi_k(\varphi)(x,y) \to f(\varphi)(x,y)$ *for all* (x,y), *and measurability of* $f(\varphi)$ *follows from Proposition 2.19.* ■

We next prove that $\varphi(x-y) = x-y$ is an example of a function that satisfies the requirements of Proposition 6.15.

Proposition 6.16 ($\varphi(x-y) = x-y$) *The function* $\varphi(x-y) = x-y$ *satisfies the requirements of Proposition 6.15.*

Proof. *First,* $\varphi : \mathbb{R}^2 \to \mathbb{R}$ *is Borel measurable by continuity and Proposition III.3.13. If* $N \in \mathcal{M}_L(\mathbb{R})$ *with* $m(N) = 0$, *then* $\varphi^{-1}(N) = \bigcup_{n=1}^{\infty} P_n$ *with:*

$$P_n \equiv \{(x,y) | x-y \in N \text{ and } |y| \leq n\}.$$

By countable subadditivity of m, *the desired result that* $m\left[\varphi^{-1}(N)\right] = 0$ *will follow from* $P_n \in \mathcal{M}_L$ *and* $m\left[P_n\right] = 0$ *for all* n.

To this end, m *is outer regular by Proposition I.2.43, and so there exists open sets* $\{G'_k\}_{k=1}^{\infty}$ *with* $N \subset G'_k$ *for all* k, *and* $\inf_k m\left[G'_k\right] = m(N) = 0$. *Given* $\epsilon > 0$, *such sets can be chosen so that* $m[G'_k] < \epsilon$ *for all* k *by Proposition I.2.42. Defining* $G_k = \bigcap_{j \leq k} G'_j$, *it follows that* $\{G_k\}_{k=1}^{\infty}$ *are open and nested, with* $G_{k+1} \subset G_k$ *and* $m[G_1] < \epsilon$, *and* $\lim_{k\to\infty} m\left[G_k\right] = m(N) = 0$.

Fixing n, *define* $\{H_k\}_{k=1}^{\infty}$ *by:*

$$H_k = \{(x,y) | x-y \in G_k\} \bigcap \{|y| \leq n\}. \tag{1}$$

Now $\{(x,y) | x-y \in G_k\} = \varphi^{-1}(G_k)$ *is open by Proposition I.3.12 as the pre-image of an open set under a continuous function, and thus* $H_k \in \mathcal{B}(\mathbb{R}^2)$ *as the intersection of an open and closed set. In addition,* $H_{k+1} \subset H_k$ *for all* k, *and* $P_n \subset \bigcap_{k=1}^{\infty} H_k$.

To prove that $P_n \in \mathcal{M}_L$ *and* $m\left[P_n\right] = 0$, *it is enough by completeness of* m *to prove that* $m\left[\bigcap_{k=1}^{\infty} H_k\right] = 0$, *and for this, continuity from above of* m *by Proposition I.2.44 is needed.*

Since $\{H_k\}_{k=1}^{\infty}$ *are nested with* $H_{k+1} \subset H_k$, *if* $m[H_1] < \infty$, *it will follow from continuity from above of* m *that:*

$$m\left[\bigcap_{k=1}^{\infty} H_k\right] = \lim_{k\to\infty} m\left[H_k\right]. \tag{2}$$

For $m[H_1]$, *recall that* $G_1 = G'_1$, *an open set in* $\mathbb{R}$ *with* $m[G'_1] < \epsilon$. *By Proposition I.2.12,* $G'_1 = \bigcup_{j=1}^{\infty} I_j$ *where* $\{I_j\}_{j=1}^{\infty}$ *is a collection of disjoint open intervals, and thus:*

$$H_1 = \bigcup_{j=1}^{\infty} \left[\{(x,y) | x-y \in I_j\} \bigcap \{|y| \leq n\}\right],$$

a disjoint union. Finally, $m\left[\{(x,y) | x-y \in I_j\} \bigcap \{|y| \leq n\}\right] = 2nm\left(I_j\right)$ *since this set is a parallelogram, and so* $m[H_1] < 2n\epsilon$ *by countable additivity and (2) is justified.*

Now with χ_{H_k} denoting the characteristic function of the set H_k, it follows from either Fubini's or Tonelli's theorem that:

$$\begin{aligned} m\left[H_k\right] &\equiv \int \chi_{H_k}(x,y)dm \\ &= \int_{-\infty}^{\infty}\left[\int_{-\infty}^{\infty}\chi_{H_k}(x,y)dm(x)\right]dm(y) \\ &= \int_{-n}^{n}\left[\int_{-\infty}^{\infty}\chi_{G_k}(x-y)dm(x)\right]dm(y). \end{aligned}$$

By a change of variable with Proposition 4.24:

$$\int_{-\infty}^{\infty}\chi_{G_k}(x-y)dm(x) = \int_{-\infty}^{\infty}\chi_{G_k}(x)dm(x) = m(G_k),$$

and so:

$$m\left[H_k\right] = 2nm(G_k). \tag{3}$$

Since $\lim_{k\to\infty} m\left[G_k\right] = m(N) = 0$, it follows from (2) that $m\left[\bigcap_{k=1}^{\infty} H_k\right] = 0$.

Thus $m\left[P_n\right] = 0$ for all n, which obtains $P_n \in \mathcal{M}_L$ and $m\left[\varphi^{-1}(N)\right] = 0$, as was to be proved. ■

With the background work done, we now prove that $f*g(x)$ is Lebesgue integrable when $f(x)$ and $g(x)$ are Lebesgue integrable functions.

Proposition 6.17 (Lebesgue integrability of $f*g(x)$) *Let $f(x)$ and $g(x)$ be Lebesgue measurable functions, integrable on $\mathbb{R}$.*

Then the function $f(x-y)g(y)$ is Lebesgue integrable in y for almost all x. That is,

$$\int |f(x-y)g(y)|\,dm(y) < \infty, \ m\text{-a.e.}$$

Defining $f*g(x)$ as in (6.6), then $f*g(x)$ is Lebesgue integrable:

$$\int |f*g(x)|\,dm \le \int |f(x)|\,dm \int |g(y)|\,dm, \tag{6.8}$$

and:

$$\int f*g(x)dm = \int f(x)dm \int g(y)dm. \tag{6.9}$$

Proof. *By Propositions 6.15 and 6.16, if $f(x)$ is Lebesgue measurable on $\mathbb{R}$, then $f(x-y)$ is Lebesgue measurable on $\mathbb{R}^2$. That $g(y)$ is Lebesgue measurable on $\mathbb{R}^2$ follows by noting that for $A \in \mathcal{B}(\mathbb{R})$:*

$$g^{-1}(A) = \mathbb{R}\times g_0^{-1}(A),$$

where $g_0^{-1}(A)$ denotes the pre-image of A under the Lebesgue measurable function $g:\mathbb{R}\to\mathbb{R}$. Thus $F(x,y)\equiv f(x-y)g(y)$ is Lebesgue measurable on $\mathbb{R}^2$ by Proposition 2.13.

To prove integrability of $|F(x,y)|$, an application of Tonelli's theorem obtains:

$$\int\int |f(x-y)g(y)|\,dm(x,y) = \int\left(\int |f(x-y)|\,dm(x)\right)|g(y)|\,dm(y).$$

A change of variable in the x-integral by Proposition 4.24 obtains:

$$\int |f(x-y)|\,dm(x) = \int |f(x)|\,dm(x), \text{ for all } y,$$

and hence:

$$\int\int |f(x-y)g(y)|\, dm(x,y) = \int |f(x)|\, dm(x) \int |g(y)|\, dm(y).$$

Consequently, integrability of $f(x)$ and $g(x)$ assures the integrability of $|f(x-y)g(y)|$ in $(\mathbb{R}^2, \mathcal{M}_L(\mathbb{R}^2), m)$. Fubini's theorem now obtains that $f(x-y)g(y)$ is Lebesgue integrable in y for almost all x. By this result, the triangle inequality, and a change of variable:

$$\begin{aligned}\int |f*g(x)|\, dm(x) &\leq \int\int |f(x-y)g(y)|\, dm(y)dm(x)\\ &= \int |f(x)|\, dm(x) \int |g(y)|\, dm(y).\end{aligned}$$

This proves the Lebesgue integrability of $|f*g(x)|$ and (6.8).

Applying Fubini's theorem to integrable $f(x-y)g(y)$ obtains:

$$\begin{aligned}\int f*g(x)dm(x) &= \int\int f(x-y)g(y)dm(y)dm(x)\\ &= \int\int f(x-y)dm(x)g(y)dm(y)\\ &= \int f(x)dm(x)\int g(y)dm(y),\end{aligned}$$

which is (6.9). ■

Example 6.18 (Convolution of unit normal densities) *Let $f(x)$ and $g(y)$ be the density functions for unit normal random variables X and Y as in (5.16), for example:*

$$f(x) = \frac{1}{\sqrt{2\pi}}\exp\left(-x^2/2\right).$$

Then:

$$\begin{aligned}f*g(x) &= \frac{1}{2\pi}\int \exp\left[-(x-y)^2/2\right]\exp\left(-y^2/2\right)dy\\ &= \frac{1}{2\pi}\exp\left(-x^2/4\right)\int \exp\left[-(y-x/2)^2\right]dy.\end{aligned}$$

This second integral equals $\sqrt{\pi}$ by substitution, since the normal density integrates to one, and hence:

$$f*g(x) = \frac{1}{\sqrt{4\pi}}\exp\left(-x^2/4\right).$$

Comparing this with (5.15) obtains that $f*g(x)$ is the density of a normal variate with $\mu = 0$ and $\sigma^2 = 2$.

By Proposition IV.2.17, $f*g(x)$ is the density function for the random variable $Z = X+Y$, assuming that X and Y are independent random variables.

With more algebra this conclusion generalizes.

Exercise 6.19 (Convolution of normal densities) *Prove that if $f(x)$ and $g(y)$ are density functions for normal variates as in (5.15) with parameters (μ_1, σ_1^2) and (μ_2, σ_2^2), then $f*g(x)$ is the density of a normal variate with $\mu = \mu_1 + \mu_2$ and $\sigma^2 = \sigma_1^2 + \sigma_2^2$. Generalize this to n such density functions by induction.*

7

The Fourier Transform

In this chapter, we introduce the Fourier transform of an integrable function, and then set out to develop a number of its properties that will be instrumental in applications to come. By "transform" is meant that an integrable function will be transformed to another function, much like the Laplace transform identified in Remark IV.4.15 transformed certain density functions to moment generating functions. Fourier-Stieltjes transforms will also be investigated, whereby a finite measure is transformed to a function. In both cases, these resultant functions will be complex-valued functions of a real variable, meaning functions $\mathbb{R} \to \mathbb{C}$.

Fourier transform theory will be applied in Book VI, where the **characteristic function** of a density function will be introduced and its properties developed. Characteristic functions provide a far more robust tool for studying properties of distribution functions than do the moment generating functions of Book IV, since they always exist. With this new tool, more general versions of the central limit theorem are possible than what was proved in Book IV, and several important results on sums of random variables will be derived.

This chapter will focus on the 1-dimensional theory of Fourier transforms, as this is sufficient for both a 1-dimensional and n-dimensional theory of characteristic functions in Book VI. For a more advanced development, see **Stein and Weiss** (1971).

7.1 Integration of Complex-Valued Functions

The study of Fourier transforms and Fourier-Stieltjes transforms requires an integration theory applicable to **complex-valued functions** of a real variable. That is, functions $f(x)$ with $f : \mathbb{R} \to \mathbb{C}$. As it turns out, the earlier theory of Chapter 3 readily extends. However, for the record, it should be noted that a significant amount of additional work would be needed for an integration theory for complex-valued functions of a complex variable, $f : \mathbb{C} \to \mathbb{C}$, which is not a theory needed for our purposes.

Given a function $f(x)$ with $f : \mathbb{R} \to \mathbb{C}$, it follows by definition that $f(x) = u(x) + iv(x)$ where $u, v : \mathbb{R} \to \mathbb{R}$, and i is standard notation for the "imaginary" unit, $i \equiv \sqrt{-1}$.

Definition 7.1 (Measurability, integrability of $f : \mathbb{R} \to \mathbb{C}$) *Given a real measure space $(\mathbb{R}, \sigma(\mathbb{R}), \mu)$, a function $f(x) : \mathbb{R} \to \mathbb{C}$ is defined to be **measurable,** and sometimes $\sigma(\mathbb{R})$-**measurable,** if $f(x) = u(x) + iv(x)$ and both $u(x)$ and $v(x)$ are $\sigma(\mathbb{R})$-measurable. That is, $f(x)$ is measurable if $u^{-1}(A) \in \sigma(\mathbb{R})$ and $v^{-1}(A) \in \sigma(\mathbb{R})$ for all Borel sets $A \in \mathcal{B}(\mathbb{R})$.*

*More generally, given a measure space $(X, \sigma(X), \mu)$, a function $f : X \to \mathbb{C}$ is defined to be **measurable** and sometimes $\sigma(X)$-**measurable** if $u(x)$ and $v(x)$ are $\sigma(X)$-measurable.*

*In either case, if $u(x)$ and $v(x)$ are μ-integrable, we define μ-**integral of** $f(x)$ by:*

$$\int f(x)d\mu \equiv \int u(x)d\mu + i\int v(x)d\mu,$$

DOI: 10.1201/9781003264576-7

and recalling Definition 3.7, define $\int_A f(x)d\mu$ *for* $A \in \sigma(\mathbb{R})$ *(respectively,* $A \in \sigma(X)$*) by:*

$$\int_A f(x)d\mu \equiv \int \chi_A(x)f(x)d\mu = \int \chi_A(x)u(x)d\mu + i\int \chi_A(x)v(x)d\mu.$$

Remark 7.2 (Integrability of $|f(x)|$) *Recall Definition 3.38, that an integral such as* $\int f(x)d\mu$ *is defined for* $f : \mathbb{R} \to \mathbb{R}$ *only when* $\int |f(x)|\, d\mu$ *is finite. While it may not be apparent given the above definition, we show that this definition assures integrability of* $|f(x)|$.

If $f(x) = u(x) + iv(x)$*, then* $|f(x)| = \sqrt{u^2(x) + v^2(x)}$ *by Definition 9.4 and thus:*

$$\max[|u(x)|, |v(x)|] \leq |f(x)| \leq |u(x)| + |v(x)|.$$

The lower bound is true by definition, while the upper bound follows by squaring. Hence, $|f(x)|$ *is* μ*-integrable if and only if* $u(x)$ *and* $v(x)$ *are* μ*-integrable.*

In summary, if $f(x)$ *is integrable by the above definition, then so too is* $|f(x)|$.

With the exception of the triangle inequality, all of the properties of the integral summarized in Proposition 3.42 are readily seen to be satisfied. Moreover, this integral is linear over complex scalars. If $\alpha \in \mathbb{C}$ and $f : X \to \mathbb{C}$ is integrable, then αf is integrable, with:

$$\int \alpha f(x)d\mu = \alpha \int f(x)d\mu. \tag{7.1}$$

To justify a triangle inequality requires a clever trick, because the naive approach using the triangle inequality for real functions does not produce the desired result. Specifically, using the triangle inequality of Proposition 3.42 and Definition 9.4:

$$\begin{aligned} \left|\int f(x)d\mu\right|^2 &\equiv \left|\int u(x)d\mu\right|^2 + \left|\int v(x)d\mu\right|^2 \\ &\leq \left[\int |u(x)|\, d\mu\right]^2 + \left[\int |v(x)|\, d\mu\right]^2 \\ &\leq \left[\int |u(x)|\, d\mu + \int |v(x)|\, d\mu\right]^2. \end{aligned}$$

Hence:

$$\left|\int f(x)d\mu\right| \leq \int [|u(x)| + |v(x)|]\, d\mu.$$

By Remark 7.2, this upper bound exceeds the desired result of $\int |f(x)|\, d\mu = \int \sqrt{u^2(x) + v^2(x)}d\mu$.

But note that if $z \in \mathbb{C}$ with $z \neq 0$, then there exists $\alpha \in \mathbb{C}$ with $|\alpha| = 1$ and $\alpha z = |z|$. Indeed, let $\alpha = \bar{z}/|z|$, where $\bar{z} = a - bi$ is the **complex conjugate** of $z = a + bi$ (Definition 9.4).

Proposition 7.3 (Triangle inequality) *If* $f : X \to \mathbb{C}$ *is* μ*-integrable, then for all measurable* $A \in \sigma(X)$*:*

$$\left|\int_A f(x)d\mu\right| \leq \int_A |f(x)|\, d\mu. \tag{7.2}$$

Proof. *By incorporating* $\chi_A(x)$ *into the definition of* $f(x)$*, it is enough to prove (7.2) for* $A = X$.

Since $\int f(x)d\mu \in \mathbb{C}$, there exists $\alpha \in \mathbb{C}$ with $|\alpha| = 1$ and:

$$\left|\int f(x)d\mu\right| = \alpha \int f(x)d\mu.$$

Then by (7.1):

$$\left|\int f(x)d\mu\right| = \int \alpha f(x)d\mu = \mathrm{Re}\left[\int \alpha f(x)d\mu\right], \tag{1}$$

where $\mathrm{Re}(z) = a$ *if* $z = a + bi$. *The last equality follows since this integral is real.*

By Definition 7.1, and noting that $\mathrm{Re}(z) \leq |z|$:

$$\begin{aligned} \mathrm{Re}\left[\int \alpha f(x)d\mu\right] &= \int \mathrm{Re}\left[\alpha f(x)\right] d\mu \\ &\leq \int |\alpha f(x)|\, d\mu \qquad (2) \\ &= \int |f(x)|\, d\mu. \end{aligned}$$

This last equality follows from $|\alpha\beta| = |\alpha|\,|\beta|$ *for all* $\alpha, \beta \in \mathbb{C}$, *as an be verified.*

Combining (1) *and* (2) *completes the proof.* ■

Finally, **Lebesgue's dominated convergence theorem** and corollaries again apply in this context. We prove the main theorem and leave it to the reader to verify the corollaries.

Proposition 7.4 (Lebesgue's dominated convergence theorem) *Let* $\{f_n(x)\}_{n=1}^{\infty} = \{u_n(x) + iv_n(x)\}_{n=1}^{\infty}$ *be a sequence of complex-valued,* $\sigma(X)$*-measurable functions on a measure space* $(X, \sigma(X), \mu)$ *that converge pointwise on a measurable set* E *to* $f(x) = u(x) + iv(x)$.

If there exists a function $g(x)$, μ*-integrable on* E, *so that for all* n:

$$|f_n(x)| \leq g(x),$$

then $f(x)$ *is integrable on* E *and:*

$$\int_E f(x)d\mu = \lim_{n\to\infty} \int_E f_n(x)d\mu.$$

Further:

$$\lim_{n\to\infty} \int_E |f_n(x) - f(x)|\, d\mu \to 0.$$

Proof. *As noted in Remark 7.2:*

$$\max[|u(x) - u_n(x)|, |v(x) - v_n(x)|] \leq |f(x) - f_n(x)|,$$

so $\lim_{n\to\infty} f_n(x) = f(x)$ *obtains both* $\lim_{n\to\infty} u_n(x) = u(x)$ *and* $\lim_{n\to\infty} v_n(x) = v(x)$. *Similarly,* $|f_n(x)| \leq g(x)$ *obtains both* $|u_n(x)| \leq g(x)$ *and* $|v_n(x)| \leq g(x)$. *Then by Lebesgue's dominated convergence theorem of Proposition 3.45, both* $u(x)$ *and* $v(x)$ *are* μ*-integrable on* E, *and:*

$$\int_E u(x)d\mu = \lim_{n\to\infty} \int_E u_n(x)d\mu, \quad \int_E v(x)d\mu = \lim_{n\to\infty} \int_E v_n(x)d\mu.$$

Hence, $f(x)$ *is* μ*-integrable on* E *and:*

$$\int_E f(x)d\mu = \lim_{n\to\infty} \int_E f_n(x)d\mu.$$

Further, since $|a+bi| \leq |a|+|b|$, *another application of Proposition 3.45 obtains as* $n \to \infty$:

$$\begin{aligned}\int_E |f_n(x) - f(x)|\, d\mu &\leq \int_E |u_n(x) - u(x)|\, d\mu + \int_E |v_n(x) - v(x)|\, d\mu \\ &\to 0.\end{aligned}$$

■

7.2 Fourier Transforms

Fourier transforms are named for **Jean Baptiste Joseph Fourier** (1768–1830), who introduced the underlying mathematical tools in the context of **Fourier series,** which he developed to solve certain partial differential equations. An example of a Fourier series will be seen in (9.19) of Chapter 9. As will be seen in Book VI, Fourier transforms also have important applications to probability theory, where this transform will be recast as the **characteristic function** of a density function.

With $i \equiv \sqrt{-1}$, recall that as is verified by Taylor series analysis, e^{ix} is well defined for $x \in \mathbb{R}$ by the power series:

$$e^{ix} \equiv \sum\nolimits_{j=0}^{\infty} \frac{(ix)^j}{j!}. \tag{7.3}$$

This series is absolutely convergent for all $x \in \mathbb{R}$, and so the summation can be split into even and odd terms:

$$\sum\nolimits_{j=0}^{\infty} \frac{(ix)^j}{j!} = \sum\nolimits_{j=0}^{\infty} \frac{x^{2j}}{(2j)!} + i \sum\nolimits_{j=0}^{\infty} \frac{x^{2j+1}}{(2j+1)j!}.$$

Recognizing the two Taylor series on the right, this obtains **Euler's formula:**

$$e^{ix} = \cos x + i \sin x, \tag{7.4}$$

named for **Leonhard Euler** (1707–1783). The formula in (7.3) generalizes to e^z for $z = a + bi$, $a, b \in \mathbb{R}$. Justified by absolute convergence, it follows that $e^z = e^a e^{ib}$ and thus:

$$e^z \equiv \sum\nolimits_{k=0}^{\infty} \frac{z^k}{k!} = e^a (\cos b + i \sin b). \tag{7.5}$$

A simple consequence of Euler's formula is that

$$\left| e^{ix} \right| = 1, \text{ all } x \in \mathbb{R}, \tag{7.6}$$

since $\cos^2 x + \sin^2 x = 1$ for all such x. This formula also contains the beautiful result known as **Euler's identity:**

$$e^{i\pi} = -1. \tag{7.7}$$

Finally, considering $e^{\pm ix}$ yields:

$$\cos x = \frac{e^{ix} + e^{-ix}}{2}, \qquad \sin x = \frac{e^{ix} - e^{-ix}}{2i}. \tag{7.8}$$

The error in the partial summations in (7.3) is given by the following proposition and will be useful in the development that follows.

Proposition 7.5 (Bound on $\left|e^{ix} - \sum_{j=0}^{n} (ix)^j / j!\right|$) *For every n:*

$$\left| e^{ix} - \sum\nolimits_{j=0}^{n} \frac{(ix)^j}{j!} \right| \leq \min \left[\frac{|x|^{n+1}}{(n+1)!}, \frac{2|x|^n}{n!} \right]. \tag{7.9}$$

Proof. *Since $e^{ix} = 1 + i\int_0^x e^{it}dt$, an integration by parts yields:*

$$\int_0^x (x-t)^n e^{it}dt = \frac{x^{n+1}}{n+1} + \frac{i}{n+1} \int_0^x (x-t)^{n+1} e^{it}dt. \tag{1}$$

Starting with $n = 0$ and applying this formula iteratively:

$$e^{ix} = \sum\nolimits_{j=0}^{n} \frac{(ix)^j}{j!} + \frac{i^{n+1}}{n!} \int_0^x (x-t)^n e^{it}dt. \tag{2}$$

By the triangle inequality and (7.6), and considering $x < 0$ and $x > 0$ separately, produces:

$$\left| e^{ix} - \sum\nolimits_{j=0}^{n} \frac{(ix)^j}{j!} \right| \leq \frac{|x|^{n+1}}{(n+1)!}. \tag{3}$$

Rewriting the equality in (1)*:*

$$\begin{aligned} \frac{i}{n} \int_0^x (x-t)^n e^{it}dt &= \int_0^x (x-t)^{n-1} e^{it}dt - \frac{x^n}{n} \\ &= \int_0^x (x-t)^{n-1} \left(e^{it} - 1\right) dt, \end{aligned}$$

and so:

$$\frac{i^{n+1}}{n!} \int_0^x (x-t)^n e^{it}dt = \frac{i^n}{(n-1)!} \int_0^x (x-t)^{n-1} \left(e^{it} - 1\right) dt.$$

Since $\left|e^{it} - 1\right| \leq 2$ by the triangle inequality and (7.6), we again estimate the error term in (2) *with the above expression to obtain:*

$$\left| e^{ix} - \sum\nolimits_{j=0}^{n} \frac{(ix)^j}{j!} \right| \leq \frac{2|x|^n}{n!}. \tag{4}$$

The proof is complete with (3) *and* (4). ■

With the above preliminary results, we now turn to the main topic of this section.

Definition 7.6 (Fourier transform, Fourier-Stieltjes transform) *If $f(x)$ is a Lebesgue integrable function, $f : \mathbb{R} \to \mathbb{C}$, the* ***Fourier transform of*** *$f(x)$, denoted $\widehat{f}(t)$, is defined by:*

$$\widehat{f}(t) = \int_{-\infty}^{\infty} f(x)e^{itx}dm, \tag{7.10}$$

where dm denotes Lebesgue measure.

If μ is a finite Borel measure on $\mathbb{R}$, the ***Fourier-Stieltjes transform of*** *μ, denoted $\phi_\mu(t)$, is defined by:*

$$\phi_\mu(t) = \int_{-\infty}^{\infty} e^{itx}d\mu. \tag{7.11}$$

Notation 7.7 **($\phi_\mu(t) \equiv \phi_F(t)$)** *As derived in Chapter I.5 and summarized in Section 1.1.3, every finite Borel measure μ is induced by an increasing, bounded and right continuous function $F(x)$, in the sense that for all $(a,b]$:*

$$\mu[(a,b]] \equiv F(b) - F(a).$$

Thus it is common to denote $\phi_\mu(t)$ in (7.11) as $\phi_F(t)$, and then it is also common to express this integral as a Riemann-Stieltjes integral by Proposition 3.63:

$$\phi_F(t) = \int_{-\infty}^{\infty} e^{itx} dF.$$

Remark 7.8 *Several observations are warranted on the above definition:*

1. *Remark 7.32 summarizes other common conventions for the definition of $\widehat{f}(t)$. The particular formulation here was chosen to be consistent with the notion of a characteristic function studied in Book VI.*

2. *By the above discussion on integration of complex valued functions, both $\widehat{f}(t)$ and $\phi_\mu(t)$ are well-defined for all t for the integrable functions and finite measures considered.*

   By (7.6) and the triangle inequality:

$$\left|\widehat{f}(t)\right| \leq \int_{-\infty}^{\infty} |f(x)|\, dm, \qquad |\phi_\mu(t)| \leq \int_{-\infty}^{\infty} d\mu,$$

   and by definition:

$$\widehat{f}(0) = \int_{-\infty}^{\infty} f(x) dm, \qquad \phi_\mu(0) = \int_{-\infty}^{\infty} d\mu.$$

3. *When the distribution function F induced by μ in Notation 7.7 is given by a nonnegative, Lebesgue integrable density function $f(x)$:*

$$F(x) = \int_{-\infty}^{x} f(y) dm,$$

   then by an application of the change of variables formula in (4.3):

$$\phi_F(t) = \widehat{f}(t),$$

   and hence these definitions agree.

   That is, the Fourier-Stieltjes transform of the finite measure μ equals the Fourier transform of the underlying "density" function f when this function exists. Consequently, results below for $\phi_F(t)$ are automatically true for $\widehat{f}(t)$ when $F(x)$ is defined by an integrable function $f(x)$. By Proposition III.3.62, $F(x)$ is defined by such $f(x)$ if and only if $F(x)$ is absolutely continuous, and then $f(x) = F'(x)$ m-a.e. This proposition applies to the integral starting at $-\infty$ since $F(x) \to 0$ as $x \to -\infty$.

   Conversely, given a nonnegative integrable function $f(x)$, we can always create the associated finite Borel measure μ by the above steps, and thus $\widehat{f}(t) = \phi_F(t)$ again. Consequently, results below for $\widehat{f}(t)$ are automatically true for $\phi_F(t)$ for F defined by f.

4. *By (7.4):*

$$\widehat{f}(t) = \int_{-\infty}^{\infty} f(x) \cos x dm + i \int_{-\infty}^{\infty} f(x) \sin x dm. \tag{7.12}$$

Hence:

(a) If $f(x)$ is an ***even function,*** *meaning $f(-x) = f(x)$, then $\int_{-\infty}^{\infty} f(x)\sin xdm = 0$ and $\widehat{f}(t)$ is real for all t.*

(b) If $f(x)$ is an ***odd function,*** *meaning $f(-x) = -f(x)$, then $\int_{-\infty}^{\infty} f(x)\cos xdm = 0$ and $\widehat{f}(t)$ is purely "imaginary" for all t.*

7.3 Properties of Fourier Transforms

Beginning with an exercise, some of the important properties of $\widehat{f}(t)$ and $\phi_F(t)$ are then developed in the following propositions.

Exercise 7.9 (Properties of Fourier transform) *Derive the following properties of the Fourier transform assuming that $f(x)$ and $g(x)$ are integrable. Justify the integrability of $h(x)$ in each case.*

1. *If $h(x) = f(x)e^{iax}$ for $a \in \mathbb{R}$, then $\widehat{h}(t) = \widehat{f}(t+a)$.*
2. *If $h(x) = f(x-a)$ for $a \in \mathbb{R}$, then $\widehat{h}(t) = \widehat{f}(t)e^{iat}$.*
3. *If $h(x) = f(x/a)$ for real $a \neq 0$, then $\widehat{h}(t) = |a|\,\widehat{f}(at)$.*
4. *If $h(x) = af(x) + bg(x)$ for $a, b \in \mathbb{R}$, then $\widehat{h}(t) = a\widehat{f}(t) + b\widehat{g}(t)$.*
5. *If $h(x) = f(ax+b)$ for $a \neq 0, b \in \mathbb{R}$, then $\widehat{h}(t) = \frac{1}{|a|}e^{-ibt/a}\widehat{f}(t/a)$.*

The first proposition states that the smoothness of the Fourier-Stieltjes transform $\phi_F(t)$ is determined by the rate of decay of the μ_F-measure of sets as these sets move toward $\pm\infty$. In a probability context, this decay rate is captured in the notion of **fat tails** for probability measures that decay slowly, and **thin** or **skinny tails** for probability measures that decay rapidly.

More formally, part 1 of the next proposition states that given a finite measure μ_F:

$$\int_M^N d\mu_F \leq C_0 < \infty, \text{ all } N, M,$$

then $\phi_F(t)$ will be uniformly continuous. However, if:

$$\int_M^N |x|\, d\mu_F \leq C_1 < \infty, \text{ all } N, M,$$

then $\phi_F(t)$ will be once continuously differentiable, while if:

$$\int_M^N |x|^n\, d\mu_F \leq C_n < \infty, \text{ all } N, M,$$

$\phi_F(t)$ will be n-times continuously differentiable. Infinite differentiability is also characterized.

The connection between these integral constraints and the μ_F-measures of sets "near" $\pm\infty$ can be appreciated by an application of Corollary 3.49:

$$C_n \geq \int_{-\infty}^{\infty} |x|^n \, d\mu_F = \sum\nolimits_{m=-\infty}^{\infty} \int_m^{m+1} |x|^n \, d\mu_F \geq \sum\nolimits_{m=-\infty}^{\infty} |m|^n \, \mu_F\left[(m, m+1]\right].$$

Thus this integral constraint requires $\mu_F\left[(m, m+1]\right] \to 0$ fast enough as $|m| \to \infty$ to ensure convergence of this summation. For item 3 below, this summation must converge for all n.

In the case where μ_F is given by a distribution function $F(x)$ with associated integrable density function $f(x)$, this general result can be restated with (4.3) as follows. If:

$$\int_M^N |x|^n \, f(x) dm \leq C_n < \infty, \text{ all } N, M,$$

then $\widehat{f}(t)$ will be n-times continuously differentiable. Thus the decay rate of the density function $f(x)$ at $\pm\infty$ determines the degree of smoothness of $\widehat{f}(t)$. Specifically, if $f(x)$ is integrable, and as $|x| \to \infty$:

$$|f(x)| \leq C \, |x|^{-n-1-\epsilon},$$

then $\widehat{f}(t)$ will be n-times continuously differentiable.

When μ_F is a **probability measure**, finiteness of the above integrals can be stated in terms of the existence of moments by Chapter IV.4. More will be made of this connection in Book VI.

Proposition 7.10 (On smoothness of $\phi_F(t)$) *Given a finite measure μ_F:*

1. *$\phi_F(t)$ is uniformly continuous on $\mathbb{R}$.*
2. *If $\int_{-\infty}^{\infty} |x|^n \, d\mu_F < \infty$ for a positive integer n, then $\phi_F(t)$ is differentiable up to order n,*

$$\phi_F^{(k)}(t) = \int_{-\infty}^{\infty} (ix)^k \, e^{itx} d\mu_F, \qquad \text{for } 1 \leq k \leq n, \tag{7.13}$$

   and $\phi_F^{(k)}(t)$ is uniformly continuous for $1 \leq k \leq n$.

3. *If $\int_{-\infty}^{\infty} e^{|sx|} d\mu_F < \infty$ for any $s \neq 0$, then $\phi_F(t)$ is infinitely differentiable, and for $|t| \leq |s|$:*

$$\phi_F(t) = \sum\nolimits_{j=0}^{\infty} \frac{(it)^j}{j!} \int_{-\infty}^{\infty} x^k d\mu_F. \tag{7.14}$$

Proof. *We prove these statements in turn.*

1. *By (7.6),*

$$\left| e^{i(t+h)x} - e^{itx} \right| = \left| e^{ihx} - 1 \right|,$$

   and so by the triangle inequality:

$$\begin{aligned} |\phi_F(t+h) - \phi_F(t)| &\leq \int_{-\infty}^{\infty} \left| e^{ihx} - 1 \right| d\mu_F \\ &\leq 2 \int_{-\infty}^{\infty} d\mu_F. \end{aligned}$$

   By continuity, $e^{ihx} - 1 \to 0$ for all x as $h \to 0$, and Lebesgue's dominated convergence theorem of Proposition 7.4 obtains as $h \to 0$:

$$|\phi_F(t+h) - \phi_F(t)| \to 0.$$

   As this convergence is independent of t, continuity of $\phi_F(t)$ is uniform.

2. *We prove this result by induction. If* $n = 1$, *then by (7.9):*

$$\begin{aligned}\left|\frac{\phi_F(t+h)-\phi_F(t)}{h}\right| &\leq \int_{-\infty}^{\infty}\left|\frac{e^{ihx}-1}{h}\right| d\mu_F \\ &\leq \int_{-\infty}^{\infty}|x|\, d\mu_F < \infty.\end{aligned}$$

Since $\frac{e^{ihx}-1}{h} \to ixe^{itx}$, *Lebesgue's dominated convergence obtains:*

$$\phi_F'(t) \equiv \lim_{h\to 0}\int_{-\infty}^{\infty}\left[\frac{e^{i(t+h)x}-e^{itx}}{h}\right] d\mu_F = \int_{-\infty}^{\infty} ixe^{itx}\, d\mu_F.$$

Moreover, by the argument in part 1, $\phi_F'(t)$ *is again seen to be uniformly continuous. For the induction step, assume that (7.13) is true for* $k-1$. *Then:*

$$\begin{aligned}\left|\frac{\phi_F^{(k)}(t+h)-\phi_F^{(k)}(t)}{h}\right| &\leq \int_{-\infty}^{\infty}|ix|^{k-1}\left|\frac{e^{ihx}-1}{h}\right| d\mu_F \\ &\leq \int_{-\infty}^{\infty}|x|^k\, d\mu_F < \infty,\end{aligned}$$

and by Lebesgue's dominated convergence:

$$\phi_F^{(k)}(t) \equiv \lim_{h\to 0}\int_{-\infty}^{\infty}[ix]^{k-1}\left[\frac{e^{i(t+h)x}-e^{itx}}{h}\right] d\mu_F = \int_{-\infty}^{\infty}[ix]^k\, e^{itx}\, d\mu_F.$$

Uniform continuity of $\phi_F^{(k)}(t)$ *follows as above.*

3. *For any* n, $|x|^n \leq e^{|sx|}$ *for* $|x|/\ln|x| \geq n/|s|$. *Since* $|x|/\ln|x|$ *is increasing and unbounded, it follows that* $|x|^n \leq e^{|sx|}$ *for* $|x| \geq c_n$, *so the assumed integrability assures the requirement of part 2 is satisfied for all* n. *In detail:*

$$\int_{-\infty}^{\infty}|x|^n\, d\mu_F \leq \int_{-c_n}^{c_n}|x|^n\, d\mu_F + \int_{|x|\geq c_n} e^{|sx|} d\mu_F,$$

and this is finite since μ_F *is a finite measure. Thus* $\phi_F(t)$ *is infinitely differentiable. For the expansion in (7.19), (7.9) provides:*

$$\left|\phi_F(t) - \sum\nolimits_{j=0}^{n}\frac{(it)^j}{j!}\int_{-\infty}^{\infty} x^k d\mu_F\right| \leq \frac{|t|^{n+1}}{(n+1)!}\int_{-\infty}^{\infty}|x|^{n+1}\, d\mu_F, \tag{1}$$

and the proof will be complete by showing that if $|t| \leq |s|$, *this upper bound can be made small uniformly in* n.

To this end, the μ_F*-integrability of* $e^{|sx|}$ *justifies the application of Corollary 3.49 to Lebesgue's dominated convergence theorem to conclude that:*

$$\int_{-\infty}^{\infty} e^{|sx|} d\mu_F = \sum\nolimits_{j=0}^{\infty}\frac{(|s|)^j}{j!}\int_{-\infty}^{\infty}|x|^j\, d\mu_F.$$

Thus for any $\epsilon > 0$, *there is an* N *so that for* $n \geq N$:

$$\sum\nolimits_{j=n}^{\infty}\frac{(|s|)^j}{j!}\int_{-\infty}^{\infty}|x|^j\, d\mu_F < \epsilon.$$

If $|t| \leq |s|$, this can be applied in (1) to obtain that for any $\epsilon > 0$, there exists $N = N(\epsilon)$ so that for $n \geq N(\epsilon)$:

$$\left| \phi_F(t) - \sum\nolimits_{j=0}^{n} \frac{(it)^j}{j!} \int_{-\infty}^{\infty} x^k d\mu_F \right| < \epsilon,$$

which is (7.14).

■

As noted above, when the finite measure μ_F has an associated and thus integrable density function $f(x)$, this proposition states that the **rate of decay of** $f(x)$ **at** $\pm\infty$ determines the **degree of smoothness of** $\widehat{f}(t)$, defined in terms of the existence of continuous derivatives. This connection also applies to Lebesgue integrable $f(x)$.

Proposition 7.11 (On smoothness of $\widehat{f}(t)$) *Given a Lebesgue integrable function $f(x)$:*

1. *$\widehat{f}(t)$ is uniformly continuous on $\mathbb{R}$.*
2. *If $\int_{-\infty}^{\infty} |x|^n |f(x)| \, dm < \infty$ for a positive integer n, then $\widehat{f}(t)$ is differentiable up to order n,*
$$\widehat{f}^{(k)}(t) = \int_{-\infty}^{\infty} (ix)^k e^{itx} f(x) dm, \qquad \text{for } 1 \leq k \leq n, \tag{7.15}$$
and $\widehat{f}^{(k)}(t)$ is uniformly continuous for $1 \leq k \leq n$.
3. *If $\int_{-\infty}^{\infty} e^{|sx|} |f(x)| \, dm < \infty$ for any $s \neq 0$, then $\widehat{f}(t)$ is infinitely differentiable, and for $|t| \leq |s|$:*
$$\widehat{f}(t) = \sum\nolimits_{j=0}^{\infty} \frac{(it)^j}{j!} \int_{-\infty}^{\infty} x^k f(x) dm. \tag{7.16}$$

Proof. *Let $f(x) = f^+(x) - f^-(x)$ as in Definition 3.37, where $f^{\pm}(x)$ are nonnegative and Lebesgue integrable. By item 4 of Exercise 7.9, and then (4.3):*

$$\begin{aligned} \widehat{f}(t) &= \widehat{f^+}(t) - \widehat{f^-}(t) \\ &= \phi_{F^+}(t) - \phi_{F^-}(t), \end{aligned}$$

where $\mu_{F^\pm}$ are the Borel measures induced by the distribution functions $F^\pm$ associated with the density functions $f^\pm$. Thus item 1 follows from item 1 of Proposition 7.10.

For item 2:

$$\infty > \int_{-\infty}^{\infty} |x|^n |f(x)| \, dm = \int_{-\infty}^{\infty} |x|^n \, d\mu_{F^+} + \int_{-\infty}^{\infty} |x|^n \, d\mu_{F^-},$$

and so $\widehat{f}^{(k)}(t)$ exists and is uniformly continuous since $\phi_{F^\pm}^{(k)}(t)$ have these properties by (7.13). Further, reversing the above steps:

$$\begin{aligned} \widehat{f}^{(k)}(t) &= \int_{-\infty}^{\infty} (ix)^k e^{itx} d\mu_{F^+} - \int_{-\infty}^{\infty} (ix)^k e^{itx} d\mu_{F^-} \\ &= \int_{-\infty}^{\infty} (ix)^k e^{itx} f(x) dm, \end{aligned}$$

which is (7.15).

Item 3 follows similarly and is left as an exercise. ■

As noted above, if $f(x)$ is integrable and $|f(x)| \leq C|x|^{-n-1-\epsilon}$ as $|x| \to \infty$, then the integrability condition for item 2 is assured. But the converse is not true.

Exercise 7.12 (Integrable $|x|^n f(x) \nRightarrow |f(x)| \leq C|x|^{-n-1-\epsilon}$) *Develop an example for which $|x|^n f(x)$ is integrable for arbitrary n, but for no m is $|f(x)| \leq C|x|^{-m}$ as $|x| \to \infty$. Hint: $f(x)$ can be unbounded, but only on sets of small measure.*

The next proposition provides for $f(x)$ and $\widehat{f}(t)$ another connection between decay rate and smoothness. But here it is the **smoothness of** $f(x)$, defined in terms of the integrability of $f(x)$ and its derivatives, that determines the **decay rate of** $\widehat{f}(t)$ as $t \to \pm\infty$.

Part 1 of this result is the **Riemann-Lebesgue lemma,** named for **Bernhard Riemann** (1826–1866) and **Henri Lebesgue** (1875–1941). It states that if $f(x)$ is integrable, then $\left|\widehat{f}(t)\right| \to 0$ as $t \to \pm\infty$.

Part 2 generalizes this result. If $f'(x)$ exists and is integrable, then $\left|\widehat{f}(t)\right| = o(t^{-1})$ as $t \to \pm\infty$, and if all derivatives up to the nth derivative $f^{(n)}(x)$ exist and are integrable, then $\left|\widehat{f}(t)\right| = o(t^{-n})$ as $t \to \pm\infty$.

Recall the following "big-O" and "little-o" terminology.

Definition 7.13 (big O, little o) *Let $f(x)$ and $g(x) > 0$ be two functions defined on $\mathbb{R}$. The expression:*

$$f(x) = O(g(x)) \text{ as } x \to \infty,$$

*or in words, "$f(x)$ **is big-O of** $g(x)$ **as** $x \to \infty$," means that there exists x_0 and a positive constant M so that:*

$$|f(x)| \leq Mg(x) \text{ for all } x \geq x_0.$$

The expression:

$$f(x) = o(g(x)) \text{ as } x \to \infty,$$

*or in words, "$f(x)$ **is little-o of** $g(x)$ **as** $x \to \infty$," if:*

$$\frac{|f(x)|}{g(x)} \to 0 \text{ as } x \to \infty.$$

When $|f(x)| \to 0$ as $x \to \infty$, it is common to express this as $f(x) = o(1)$ as $x \to \infty$.

Item 1 of the following result can be seen as a special case of item 2 since by conventional, $f^{(0)}(x) \equiv f(x)$, and (7.17) is equivalent to $\left|\widehat{f}(t)\right| = o(1)$ as $t \to \pm\infty$.

Remark 7.14 (On $|\phi_F(t)|$ as $t \to \pm\infty$) *The next result addresses $\left|\widehat{f}(t)\right|$ as $t \to \pm\infty$ for Lebesgue integrable $f(x)$. By item 3 of Remark 7.8, this result also provides the same estimates for $|\phi_F(t)|$ as $t \to \pm\infty$ when $F(x)$ has a density function $f(x)$ that satisfies the necessary criteria.*

By Proposition III.3.62, which is applicable since $F(x) \to 0$ as $x \to -\infty$, $F(x)$ has such a density if and only if F is absolutely continuous, and then $f(x) = F'(x)$ m-a.e. Since such $f(x)$ is always Lebesgue integrable by definition, the Riemann-Lebesgue lemma is always satisfied in this case of absolutely continuous $F(x)$:

$$|\phi_F(t)| \to 0 \text{ as } t \to \pm\infty .$$

Proposition 7.15 (On $\left|\widehat{f}(t)\right|$ as $t \to \pm\infty$) *Given a Lebesgue measurable function $f(x)$:*

1. ***(Riemann-Lebesgue lemma)*** *If $f(x)$ is integrable, then:*

$$\left|\widehat{f}(t)\right| \to 0 \text{ as } t \to \pm\infty . \tag{7.17}$$

2. *If $f^{(k)}(x)$ exists and is integrable for $k \leq n$, then:*

$$\left|\widehat{f}(t)\right| = o(|t|^{-n}) \text{ as } t \to \pm\infty . \tag{7.18}$$

Proof. *For the Riemann-Lebesgue lemma, assume that $f(x)$ is nonnegative and integrable, as the general case follows from applying this result to $f^{+}(x)$ and $f^{-}(x)$ of Definition 3.37.*

By Proposition 2.33, there is an increasing sequence of simple functions $\{\varphi_n(x)\}_{j=1}^{\infty}$, each defined on a finite collection of disjoint right semi-closed intervals, each equal to 0 outside a set of finite Lebesgue measure, and $\varphi_n(x) \to f(x)$ for all x.

Hence, by Lebesgue's dominated convergence theorem:

$$\int_{-\infty}^{\infty} \varphi_n(x)dm \to \int_{-\infty}^{\infty} f(x)dm,$$

and for any $\epsilon > 0$, there is an M so that for $n \geq M$:

$$\int_{-\infty}^{\infty} |f(x) - \varphi_n(x)|\, dm < \epsilon/2. \tag{1}$$

Since $||a| - |b|| \leq |a - b|$ for $a, b \in \mathbb{R}$, it follows from (1), the triangle inequality, and (7.6), that for any $n \geq M$ and all t :

$$\begin{aligned} &\left|\left|\int_{-\infty}^{\infty} f(x)e^{itx}dm\right| - \left|\int_{-\infty}^{\infty} \varphi_n(x)e^{itx}dm\right|\right| \\ \leq\ &\left|\int_{-\infty}^{\infty} f(x)e^{itx}dm - \int_{-\infty}^{\infty} \varphi_n(x)e^{itx}dm\right| \\ <\ &\epsilon/2. \end{aligned} \tag{2}$$

For fixed $n \geq M$, if $\varphi_n(x) = \sum_{j=1}^{N_n} c_j \chi_{(a_j,b_j]}(x)$, then:

$$\int_{-\infty}^{\infty} \varphi_n(x)e^{itx}dm = \sum\nolimits_{j=1}^{N_n} c_j \left[e^{ita_j} - e^{itb_j}\right]/it,$$

and by (7.6):

$$\left|\int_{-\infty}^{\infty} \varphi_n(x)e^{itx}dm\right| \leq 2\sum\nolimits_{j=1}^{N_n} |c_j|/t.$$

From (2), $\left|\int_{-\infty}^{\infty} f(x)e^{itx}dm\right|$ can be made arbitrarily close to $\left|\int_{-\infty}^{\infty} \varphi_n(x)e^{itx}dm\right|$, which in turn converges to 0 as $t \to \infty$, and this obtains the Riemann-Lebesgue lemma.

For (7.18), we apply Lebesgue integration by parts from Proposition III.3.64. To justify this application, it follows from item 5 of Remark III.3.56 that the existence of $f^{(k)}(x)$ for $k \leq n$ implies that $f^{(k)}(x)$ is absolutely continuous for $0 \leq k \leq n-1$. Similarly, as

e^{itx} is infinitely differentiable as a function of t, it is also absolutely continuous and so by Proposition III.3.64:

$$\begin{aligned}\widehat{f}(t) &= \int_{-\infty}^{\infty} f(x)e^{itx}dm \\ &= \frac{-1}{it}\int_{-\infty}^{\infty} f'(x)e^{itx}dm \\ &\;\vdots \\ &= \frac{1}{(-it)^n}\int_{-\infty}^{\infty} f^{(n)}(x)e^{itx}dm. \end{aligned} \tag{3}$$

For this derivation, the Riemann-Lebesgue lemma and the integrability of $f^{(k)}(x)$ for $0 \leq k \leq n-1$ imply that the limit terms at $\pm\infty$ equal 0.

The integrability of $f^{(n)}(x)$ then obtains:

$$\left|\widehat{f}(t)\right| = |t|^{-n}\left|\int_{-\infty}^{\infty} f^{(n)}(x)e^{itx}dm\right|,$$

which by integrability of $f^{(n)}(x)$ is a big-O estimate:

$$\left|\widehat{f}(t)\right| = O(|t|^{-n}).$$

But then one last application of the Riemann-Lebesgue lemma to integrable $f^{(n)}(x)$ improves this result to $\left|\widehat{f}(t)\right| = o(|t|^{-n})$. ■

The above proof reveals an interesting identity that is worth noting.

Corollary 7.16 (On $\widehat{f^{(k)}}(x)$) *If $f^{(k)}(x)$ is an integrable function for $k \leq n$, then for any such k:*

$$\widehat{f^{(k)}}(x) = (-it)^k\,\widehat{f}(t). \tag{7.19}$$

Proof. *This is a restatement of (3) in the above proof.* ■

The final result of this section states that Fourier transforms work naturally with convolutions. It makes sense to contemplate $\widehat{f * g}(t)$, the Fourier transform of the convolution of integrable f and g, since $f * g(x)$ is then integrable by Proposition 6.17, and hence $\widehat{f * g}(t)$ is well-defined.

Proposition 7.17 (On $\widehat{f * g}(t)$) *For integrable functions $f(x)$ and $g(x)$:*

$$\widehat{f * g}(t) = \widehat{f}(t)\widehat{g}(t). \tag{7.20}$$

Proof. *Not surprisingly, this result requires an application of Fubini's theorem. First by (7.4):*

$$\begin{aligned}\widehat{f * g}(t) &= \int_{-\infty}^{\infty} f * g(x)e^{itx}dm \\ &\equiv \int_{-\infty}^{\infty}\left[\int_{-\infty}^{\infty} f(y)g(x-y)dm(y)\right]e^{itx}dm(x).\end{aligned}$$

Proposition 6.17 proved the Lebesgue integrability of $f(y)g(x-y)$ in $\mathbb{R}^2$, and since e^{itx} is absolutely bounded by 1 by (7.6), this integrand is Lebesgue integrable on $\mathbb{R}^2$.

An application of Fubini's theorem justifies that the order of these iterated integrals can be reversed and thus:

$$\begin{aligned}\widehat{f*g}(t) &= \int_{-\infty}^{\infty}\left[\int_{-\infty}^{\infty} g(x-y)e^{itx}dm(x)\right] f(y)dm(y) \\ &= \int_{-\infty}^{\infty}\left[\int_{-\infty}^{\infty} g(x-y)e^{it(x-y)}dm(x)\right] f(y)e^{ity}dm(y).\end{aligned}$$

The final step is to use Proposition 4.24 on the x*-integral:*

$$\int_{-\infty}^{\infty} g(x-y)e^{it(x-y)}dm(x) = \int_{-\infty}^{\infty} g(w)e^{itw}dm(w) = \widehat{g}(t),$$

and the result follows. ■

7.4 Fourier-Stieltjes Inversion of $\phi_F(t)$

The previous section demonstrated that when considered in pairs: $(\mu_F, \phi_F(t))$ or $(f(x), \widehat{f}(t))$, certain properties of μ_F or $f(x)$ predict properties of the associated $\phi_F(t)$ or $\widehat{f}(t)$, and conversely. The question addressed in this section relates to the invertibility of the Fourier transform, or the Fourier-Stieltjes transform. In other words, given $\phi_F(t)$ or $\widehat{f}(t)$, do these uniquely determine the associated μ_F or $f(x)$, and if "yes," is there a constructive method for finding μ_F or $f(x)$? When inversion is possible, the uniqueness of the pairings $(\mu_F, \phi_F(t))$ or $(f(x), \widehat{f}(t))$ will have important probability theory applications in Book VI.

To prepare for the statement and proof of the main result, we require a small modification of the result in (5.17) from Proposition 5.27, which proved that with:

$$S(t) \equiv \int_0^t \frac{\sin x}{x} dx,$$

that:

$$\lim_{t\to\infty} S(t) = \frac{\pi}{2}.$$

To this end, define:

$$S_\theta(t) \equiv \int_0^t \frac{\sin \theta x}{x} dx, \text{ for } \theta \in \mathbb{R}.$$

It follows by substitution that $S_\theta(t) = S(t\theta)$. Recalling Remark 3.8, we eliminate the ambiguity as to whether θ is positive or negative by writing:

$$S_\theta(t) = sgn(\theta) S(t\,|\theta|).$$

Here the **sign function** or **signum function** $sgn(\theta)$ is defined by:

$$sgn(\theta) = \begin{cases} 1, & \theta > 0, \\ 0, & \theta = 0, \\ -1, & \theta < 0. \end{cases}$$

It follows from (5.17) that:

$$\lim_{t\to\infty} S_\theta(t) = \frac{\pi}{2} sgn(\theta). \tag{7.21}$$

We are now ready for the main result. The limit in (7.22) below is called the **inverse Fourier-Stieltjes transform of** $\phi_F(t)$, since this limit essentially reproduces the measure μ_F on which the Fourier-Stieltjes transform is defined. By "essentially" is meant that this limit obtains the measure μ_F on bounded open intervals, with adjustments for the measures of the endpoints.

By Exercise I.5.22, this limit can also be expressed:

$$\mu_F[[a,b]] - \frac{1}{2}\mu_F[\{a,b\}],$$

or:

$$\mu_F[(a,b]] + \frac{1}{2}\mu_F[a] - \frac{1}{2}\mu_F[b].$$

When $F(x)$ is continuous at a and/or b, then $\mu_F[a] = 0$ and/or $\mu_F[b] = 0$ by item 4 of Proposition I.5.7.

Proposition 7.18 (Inverse Fourier-Stieltjes transform: $\phi_F(t) \to \mu_F$) *If μ_F is a finite Borel measure on $\mathbb{R}$ and $\phi_F(t)$ its Fourier-Stieltjes transform, then for $b > a$,*

$$\lim_{T\to\infty} \frac{1}{2\pi} \int_{-T}^{T} \frac{e^{-iat} - e^{-ibt}}{it} \phi_F(t) dm = \mu_F[(a,b)] + \frac{1}{2}\mu_F[\{a,b\}], \tag{7.22}$$

where $\mu_F[\{a,b\}]$ denotes the Borel measure of these points:

$$\mu_F[\{a,b\}] \equiv \mu_F[a] + \mu_F[b].$$

Proof. *First note that the integrand in (7.22) does not have a singularity at $t = 0$ since:*

$$\frac{e^{-iat} - e^{-ibt}}{it} = \int_a^b e^{-itx} dx,$$

and this is bounded in absolute value by $b - a$ by (7.6).

By definition of $\phi_F(t)$:

$$\int_{-T}^{T} \frac{e^{-iat} - e^{-ibt}}{it} \phi_F(t) dm = \int_{-T}^{T} \int_{-\infty}^{\infty} \frac{e^{i(x-a)t} - e^{i(x-b)t}}{it} d\mu_F(x) dm(t).$$

This integrand is continuous on $[-T,T] \times \mathbb{R}$ and bounded by $b-a$ as noted above. Hence since μ_F is a finite Borel measure, this function is integrable on the product measure space $([-T,T] \times \mathbb{R}, \sigma([-T,T] \times \mathbb{R}), m \times \mu_F)$.

Thus the order of the iterated integrals can be reversed by Fubini's theorem, and then by Euler's formula in (7.4):

$$\begin{aligned}\int_{-T}^{T} \frac{e^{i(x-a)t} - e^{i(x-b)t}}{it} dm(t) &= \int_{-T}^{T} \frac{\cos[(x-a)t] - \cos[(x-b)t]}{it} dm(t) \\ &\quad + \int_{-T}^{T} \frac{\sin[(x-a)t] - \sin[(x-b)t]}{t} dm(t).\end{aligned}$$

The first integral equals 0 because the integrand is an odd function of t, meaning $g(-t) = -g(t)$, and so $\int_{-T}^{0} g(t)dt = -\int_0^T g(t)dt$. The integrand in the second integral is an even function, meaning $h(-t) = h(t)$, so $\int_{-T}^{0} h(t)dt = \int_0^T h(t)dt$.

Combining:

$$\int_{-T}^{T} \frac{e^{-iat} - e^{-ibt}}{it} \phi_F(t) dm = 2 \int_{-\infty}^{\infty} \int_0^T \frac{\sin[(x-a)t] - \sin[(x-b)t]}{t} dm(t) d\mu_F(x),$$

and with $S(t)$ defined above:

$$\frac{1}{2\pi}\int_{-T}^{T}\frac{e^{-iat}-e^{-ibt}}{it}\phi_F(t)dm = \frac{1}{\pi}\int_{-\infty}^{\infty}\left[S_{(x-a)}(T) - S_{(x-b)}(T)\right]d\mu_F(x).$$

By (7.21), this integrand is bounded and converges for all x as $T \to \infty$ to the μ_F-integrable function $\lambda_{a,b}(x)$ defined by:

$$\lambda_{a,b}(x) = \begin{cases} 0, & x < a, \\ \frac{1}{2}, & x = a, \\ 1, & a < x < b, \\ \frac{1}{2}, & x = b, \\ 0, & x > b. \end{cases}$$

Because μ_F is a finite measure and the integrand is bounded, Lebesgue's dominated convergence theorem of Proposition 7.4 can be applied to derive:

$$\begin{aligned} \lim_{T\to\infty}\frac{1}{2\pi}\int_{-T}^{T}\frac{e^{-iat}-e^{-ibt}}{it}\phi_F(t)dm &= \int_{-\infty}^{\infty}\lambda_{a,b}(x)d\mu_F(x) \\ &= \mu_F[(a,b)] + \frac{1}{2}\mu_F[\{a,b\}]. \end{aligned}$$

■

Exercise 7.19 (On $F(x)$) *If $F(x)$ is the distribution function induced by the finite measure μ_F noted in Notation 7.7, show that (7.22) can be expressed:*

$$\lim_{T\to\infty}\frac{1}{2\pi}\int_{-T}^{T}\frac{e^{-iat}-e^{-ibt}}{it}\phi_F(t)dm = \frac{F(b)+F(b^-)}{2} - \frac{F(a)+F(a^-)}{2}. \tag{7.23}$$

Hint: Exercise I.5.22.

Remark 7.20 (Cauchy principal value) *For any T, the integral in (7.22) is also defined as a Riemann integral by Proposition III.1.15, since the integrand is continuous by the proof of Proposition 7.18, and for such bounded Riemann integrable functions, Proposition III.2.18 applies.*

The use of a limit in (7.22) and of the symmetrical integral over $[-T,T]$ is essential, in general. This is because the integrand need not be integrable over $\mathbb{R}$, and hence the limit needs not exist if more generally defined.

This approach to the evaluation of an improper integral $\int_{-\infty}^{\infty} f dm$ is called the ***Cauchy principal value of the improper integral****, named for* ***Augustin-Louis Cauchy*** *(1789–1857). Cauchy introduced this and related evaluations of improper integrals that were not formally definable by traditional means. Of course, this limit produces the correct value if the integral over $\mathbb{R}$ exists as an improper integral.*

It is not difficult to develop examples of functions for which the improper integrals exist in the sense of this principal value, but not formally as improper integrals. For example, $f(x) = \sin x$ has limit 0 since as an odd function, $\int_{-T}^{T} f dm = 0$ for all T, but this integral does not exist over $\mathbb{R}$ more generally. The same is true of many even functions, for example, $f(x) = \cos x$.

The next example illustrates that the Cauchy principal value in (7.22) is also necessary, since the integrand there needs not be integrable.

Example 7.21 ($F(x) = \chi_{[0,\infty)}(x)$) *Let $F(x) = 0$ for $x < 0$ and $F(x) = 1$ otherwise, so the induced μ_F is the finite measure with $\mu_F(A) = 1$ if $0 \in A$ and $\mu_F(A) = 0$ otherwise.*

Then:

$$\phi_F(t) = \int_{-\infty}^{\infty} e^{itx} d\mu_F(x) = 1,$$

which is certainly not Lebesgue integrable.

However:

$$\int_{-T}^{T} \frac{e^{-iat} - e^{-ibt}}{it} dm(t) = \int_{-T}^{T} \int_{a}^{b} e^{-itx} dx dm(t),$$

and this integrand is continuous and bounded and hence integrable on the bounded domain $[-T,T] \times [a,b]$. By Fubini's theorem and (7.8):

$$\begin{aligned}\int_{-T}^{T} \frac{e^{-iat} - e^{-ibt}}{it} dm(t) &= \int_{a}^{b} \int_{-T}^{T} e^{-itx} dm(t) dx \\ &= \int_{a}^{b} \frac{e^{iTx} - e^{-iTx}}{ix} dx \\ &= 2\int_{Ta}^{Tb} \frac{\sin y}{y} dy.\end{aligned}$$

Thus since $\phi_F(t) = 1$:

$$\frac{1}{2\pi} \int_{-T}^{T} \frac{e^{-iat} - e^{-ibt}}{it} \phi_F(t) dm = \frac{1}{\pi} \int_{Ta}^{Tb} \frac{\sin y}{y} dy.$$

If $a, b > 0$ or $a, b < 0$, this integral converges to 0 by (5.17), which agrees with the value of $\mu_F[(a,b)]$. However, if $a < 0 < b$, then:

$$\frac{1}{\pi} \int_{Ta}^{Tb} \frac{\sin y}{y} dy = \frac{1}{\pi} \left[\int_{0}^{Tb} \frac{\sin y}{y} dy + \int_{0}^{-Ta} \frac{\sin y}{y} dy \right],$$

which converges to 1, again the value of $\mu_F[(a,b)]$. In all such cases, $\mu_F[a] + \mu_F[b] = 0$, and thus these results agree with (7.22).

The reader is invited to investigate the cases where $a = 0$ or $b = 0$ to again confirm (7.22), which in this case obtains a limit of $1/2$.

Exercise 7.22 *Let $F(x)$ be defined as in Example 7.21. Show that:*

$$\lim_{S,T\to\infty} \frac{1}{2\pi} \int_{-S}^{T} \frac{e^{-iat} - e^{-ibt}}{it} \phi_F(t) dm,$$

does not in general exist. Specifically, show this limit is not independent of how S and T approach ∞.

As noted in the introduction to this section, inversion obtains the important consequence of the uniqueness of the Fourier-Stieltjes transform.

Corollary 7.23 (Uniqueness of the Fourier-Stieltjes transform) *If μ_F and μ_G are finite Borel measures on $\mathbb{R}$ and the associated Fourier-Stieltjes transforms satisfy $\phi_F(t) = \phi_G(t)$ for all t, then on $\mathcal{B}(\mathbb{R})$:*

$$\mu_F = \mu_G.$$

Proof. *There is no loss of generality to assume that F and G, the right continuous increasing functions underlying these finite measures, are equal to $\bar{F}$ and $\bar{G}$ in the notation of Proposition I.5.7, and thus satisfy $\bar{F}(-\infty) = \bar{G}(-\infty) = 0$. We now resume the original notation.*

As summarized in Section 1.1.3, if $(a,b] \in \mathcal{A}'$, the semi-algebra of right semi-closed intervals:

$$\mu_F\left[(a,b]\right] = F(b) - F(a), \qquad \mu_G\left[(a,b]\right] = G(b) - G(a). \tag{1}$$

If it can be shown that $\mu_F(A) = \mu_G(A)$ for all $A \in \mathcal{A}$, the algebra generated by $\mathcal{A}'$, then the uniqueness theorem of Proposition 1.22 assures that $\mu_F(A) = \mu_G(A)$ for all $A \in \sigma(\mathcal{A})$, where $\sigma(\mathcal{A})$ denotes the smallest sigma algebra that contains $\mathcal{A}$. Since $\sigma(\mathcal{A}) = \mathcal{B}(\mathbb{R})$ by Proposition I.8.1, the proof will be complete.

To show that $\mu_F(A) = \mu_G(A)$ for all $A \in \mathcal{A}$, it is enough to show this for $A \in \mathcal{A}'$. Recall that $\mathcal{A}$ is the collection of finite disjoint unions of $\mathcal{A}'$-sets by Exercise I.6.10, to which the set functions in (1) are extended additively in Proposition I.5.11.

If $A = (a,b]$ is a right semi-closed interval for which both a and b are continuity points of both $F(x)$ and $G(x)$, then $\mu_F[\{a,b\}] = \mu_G[\{a,b\}] = 0$ by Proposition I.5.7. In addition, $\mu_F[(a,b]] = \mu_F[(a,b)]$ and $\mu_G[(a,b]] = \mu_G[(a,b)]$ by Exercise I.5.22. Thus the assumption that $\phi_F(t) = \phi_G(t)$ implies by (7.22) that $\mu_F[(a,b]] = \mu_G[(a,b]]$ for all such intervals $(a,b]$.

Since F and G can have at most countably many discontinuities by Proposition I.5.8, for any set $A = (a,b] \in \mathcal{A}'$, define decreasing sequences of continuity points of both F and G, $a_j \to a$ and $b_j \to b$, where $a_j > a$ and $b_j > b$ for all j. Then:

$$(a,b] = \bigcup_{j=1}^{\infty} (a_{j+1}, a_j] \bigcup (a_1, b],$$

and by countable additivity:

$$\mu_F\left[(a,b]\right] = \sum\nolimits_{j=1}^{\infty} \mu_F\left[(a_{j+1}, a_j]\right] + \mu_F\left[(a_1, b]\right],$$

with an identical statement for $\mu_G\left[(a,b]\right]$. These summations contain only $\mathcal{A}'$-sets with endpoints on which both $F(x)$ and $G(x)$ are continuous, and so by the above argument:

$$\sum\nolimits_{j=1}^{\infty} \mu_F\left[(a_{j+1}, a_j]\right] = \sum\nolimits_{j=1}^{\infty} \mu_G\left[(a_{j+1}, a_j]\right]. \tag{2}$$

By construction, $(a_1, b] = \bigcap_{j=1}^{\infty} (a_1, b_j]$, so by finiteness of measures and Proposition 1.29 on continuity from above:

$$\mu_F\left[(a_1, b]\right] = \lim_{j\to\infty} \mu_F\left[(a_1, b_j]\right],$$

with an identical statement for $\mu_G\left[(a_1, b]\right]$. Since all endpoints are continuity points of F and G, $\mu_F\left[(a_1, b_j]\right] = \mu_G\left[(a_1, b_j]\right]$ for all j, and thus:

$$\lim_{j\to\infty} \mu_F\left[(a_1, b_j]\right] = \lim_{j\to\infty} \mu_G\left[(a_1, b_j]\right].$$

In other words, $\mu_F\left[(a_1, b]\right] = \mu_G\left[(a_1, b]\right]$, and the proof is complete with (2). ■

Corollary 7.24 (On $\phi_F(t) \equiv 0$) *If μ_F is a finite Borel measure on $\mathbb{R}$ and $\phi_F(t) = 0$ for all t, then $\mu_F(A) = 0$ for all $A \in \mathcal{B}(\mathbb{R})$.*

Proof. *Define $G(x) \equiv 0$ and use the prior proposition.* ■

7.5 Fourier Inversion of Integrable $\phi_F(t)$

As seen in Example 7.21, $\phi_F(t)$ need not be integrable for a finite measure μ_F. Yet the Cauchy principal value of the Lebesgue integral in (7.22) converges to a limit as $T \to \infty$, and this integral is also definable as a Riemann integral due to the continuity of the integrand. This limit then provides the value of $\mu_F[(a,b]]$ for all such intervals where a and b are continuity points of $F(x)$ and more generally provides the value of $\mu_F[(a,b)] + \frac{1}{2}\mu_F[\{a,b\}]$.

In this section, we investigate the implications for μ_F in the special case when $\phi_F(t)$ is in fact Lebesgue integrable, meaning:

$$\int_{-\infty}^{\infty} |\phi_F(t)|\, dm < \infty.$$

Extending the connections between properties of a measure and those of its Fourier-Stieltjes transform, the main result of this section states that integrability of $\phi_F(t)$ assures that the associated distribution function $F(x)$ has a continuous density function $f(x)$ for which $F(x) = \int_{-\infty}^{x} f(y)dy$ as a Riemann integral.

This result is related to the Riemann-Lebesgue lemma and generalization given in Proposition 7.15. Recalling Remark 7.14, the Riemann-Lebesgue lemma states that if the finite measure μ_F is given by a density function $f(x)$, which is therefore Lebesgue integrable, then $|\phi_F(t)| \to 0$ as $t \to \pm\infty$. The generalization states that if this density function has a Lebesgue integrable first derivative, then $|\phi_F(t)| = o(|t|^{-1})$ as $t \to \pm\infty$, and if $f(x)$ has a Lebesgue integrable *nth* derivative, then $|\phi_F(t)| = o(|t|^{-n})$ as $t \to \pm\infty$. Hence for $n \geq 2$, continuity of $\phi_F(t)$ by Proposition 7.10 and this estimate at $\pm\infty$ assure the Riemann integrability of $\phi_F(t)$, and Lebesgue integrability by Proposition III.2.56.

The main result of this section reverses this implication to state that if $\phi_F(t) \to 0$ as $t \to \pm\infty$ fast enough and/or appropriately enough to ensure Lebesgue integrability, then $F(x)$ has an associated density function $f(x)$ that is a continuous function.

Before developing this result, the next section investigates relationships between integrability and growth at infinity, extending Exercise 7.12.

7.5.1 Integrability vs. Decay at $\pm\infty$

First, integrability of a function $h(x)$ does not ensure that $h(x) \to 0$ as $x \to \pm\infty$, nor even that $h(x)$ is bounded.

Example 7.25 *Define $h(x) = n$ on $[n - 1/2^{n+1}, n + 1/2^{n+1}]$, for $n = 1, 2, 3...$, and $h(x) = 0$ otherwise. Then $h(x)$ is Lebesgue and Riemann integrable with $\int h(x)dm = 1$. But $h(x) \nrightarrow 0$ as $x \to +\infty$, and indeed, $h(x)$ is unbounded. As an exercise, this function can also be modified to be and even function, and/or to be continuous, with the same conclusion.*

Since $\phi_F(t)$ is in fact uniformly continuous by Proposition 7.10, and the above function $h(x)$ is at best modifiable to be continuous, perhaps integrability of a uniformly continuous function $h(x)$ ensures $h(x) \to 0$ as $x \to \pm\infty$.

The next result is **Barbălat's lemma**, from a 1959 paper by **I. Barbălat**. It proves that integrability plus uniform continuity imply decay at $\pm\infty$.

Proposition 7.26 (Barbălat's lemma) *If $h(x)$ is uniformly continuous and integrable on $\mathbb{R}$, then $|h(x)| \to 0$ as $x \to \pm\infty$.*

Proof. *Let $h(x)$ be uniformly continuous and integrable on $\mathbb{R}$ and assume for example that $|h(x)| \nrightarrow 0$ as $x \to +\infty$. Then for some $\epsilon > 0$, there is an unbounded increasing sequence $\{x_n\}_{n=1}^{\infty}$ so that $|h(x_n)| \geq \epsilon$ for all n.*

By uniform continuity, there is a δ so that if $|x - x'| < \delta$:

$$||h(x)| - |h(x')|| \leq |h(x) - h(x')| < \epsilon/2,$$

and hence $|h(x)| \geq \epsilon/2$ on $[x_n - \delta, x_n + \delta]$ for all n. This contradicts integrability of $h(x)$. ■

Conclusion 7.27 ***Barbălat:*** *Integrability of uniformly continuous $h(x)$ implies that $|h(x)| \to 0$ as $t \to \pm\infty$.*

Example 7.28 (Improving Barbălat's lemma?) *For an integrable, uniformly continuous function $h(x)$, Barbălat's lemma asserts that $|h(x)| \to 0$ as $x \to \pm\infty$. In this example, we investigate the possibility that there is a more quantifiable implication for the degree of decay at $\pm\infty$ for such a function.*

Claim: *Given any strictly decreasing function $g(x)$ on $[x_0, \infty)$ for $x_0 > 0$ with $g(x) \to 0$ as $x \to \infty$, there is an integrable and uniformly continuous function $h(x)$ with $h(x) = O(g(|x|))$ as $x \to \pm\infty$.*

Proof of Claim: *First let $g(x) = 1/\ln x$ on $[e, \infty)$, and define $h_0(x) = 1/\ln|x|$ on:*

$$\left[\bigcup\nolimits_{n=1}^{\infty}[e^{n^2}, e^{n^2}+1]\right] \bigcup \left[\bigcup\nolimits_{n=1}^{\infty}[-e^{n^2}-1, -e^{n^2}]\right],$$

and $h_0(x) = 0$ otherwise. Then $h_0(x) = O(g(|x|))$ as $x \to \pm\infty$, and $h_0(x)$ is integrable since:

$$\int_{-e^{n^2}-1}^{-e^{n^2}} h_0(x)dm = \int_{e^{n^2}}^{e^{n^2}+1} h_0(x)dm < e/n^2.$$

Now modify $h_0(x)$ to be continuous as follows. If $[a_n, b_n] = [e^{n^2}, e^{n^2}+1]$, define $h(x)$ on $[a_n - \epsilon_n, b_n + \epsilon_n]$ so that $h(x) = h_0(x)$ on $[a_n, b_n]$, with $h(x)$ extending $h_0(x)$ linearly to 0 on $[a_n - \epsilon_n, a_n]$ and $[b_n, b_n + \epsilon_n]$. Choose $\{\epsilon_n\}_{n=1}^{\infty}$ so that $\{[a_n - \epsilon_n, b_n + \epsilon_n]\}_{n=1}^{\infty}$ are disjoint, and for all n:

$$\int_{a_n-\epsilon_n}^{a_n} h(x)dm = \int_{b_n}^{b_n+\epsilon_n} h(x)dm < 1/n^2.$$

Repeating the same construction and definition of $h(x)$ for $[a_n, b_n] = [-e^{n^2}-1, -e^{n^2}]$ obtains that $h(x)$ is uniformly continuous, integrable, and $h(x) = O(g(|x|))$ as $x \to \pm\infty$.

In the general case, since $1/g(x)$ is strictly increasing and unbounded, replace e^{n^2} in the above example by a_n where $1/g(a_n) = n^2$. That is, $h_0(x) = g(|x|)$ on:

$$\left[\bigcup\nolimits_{n=1}^{\infty}[a_n, a_n+1]\right] \bigcup \left[\bigcup\nolimits_{n=1}^{\infty}[-a_n-1, -a_n]\right],$$

and then extend $h_0(x)$ continuously to $h(x)$ as above.

Conclusion 7.29 *Integrability of uniformly continuous $h(x)$ does not imply any decay rate at infinity stronger than $|h(x)| \to 0$ as $x \to \pm\infty$.*

Proof: *Given any proposed decay rate, choose $g(x)$ to have a slower decay rate. Then the associated integrable and uniformly continuous function $h(x) = O(g(|x|))$ as $x \to \pm\infty$ contradicts the proposed decay rate.*

Remark 7.30 (On Conclusion 7.29) *This second conclusion is perhaps initially surprising, because one often thinks of the examples $h(x) = |x|^{-1-\epsilon}$ as integrable on $|x| \geq 1$ for $\epsilon > 0$, and $h(x) = |x|^{-1+\epsilon}$ as not integrable for $\epsilon \geq 0$. The key insight is that a function's order of magnitude at $\pm\infty$ is not changed by redefining the function to be zero on even infinitely many intervals, as long as the original definition is preserved for infinitely large x. On the other hand, the value and even existence of the integral of a function can be materially modified by such redefinitions. The above construction shows that one can redefine the function enough to make it integrable without changing the original order of magnitude.*

Looked at a different way, Lebesgue measure is translation invariant, while the notion of order of magnitude is clearly not.

For example, with $g(x) = 1/\ln x$, define $k(x)$ to have the same values on $[n, n+1]$ as does $h_0(x)$ on $[e^{n^2}, e^{n^2}+1]$. In other words, $k(n) = 1/n^2$ and in general:

$$k(n+\lambda) = \left[n^2 + \ln(1+\lambda e^{-n^2})\right]^{-1}, \ 0 \leq \lambda \leq 1.$$

While the integral of $k(x)$ equals that of $h_0(x)$, we now see that $k(x) = O\left(|x|^{-2}\right)$ while $h_0(x) = O\left(1/\ln x\right)$ as $x \to \infty$.

7.5.2 Fourier Inversion: From Integrable $\phi_F(t)$ to $f(x)$

The main inversion result for Lebesgue integrable $\phi_F(t)$ is summarized in the next proposition. Namely, the associated measure μ_F is then induced by an increasing function F with continuous and necessarily nonnegative density function $f(x)$ with:

$$F(x) = \int_{-\infty}^{x} f(y)dm.$$

Since μ_F is finite, F is bounded and hence f is Lebesgue integrable. It then follows by continuity and Propositions III.1.15 and III.2.18 that f is also Riemann integrable. Extending (7.22), such f is recoverable from ϕ_F.

When $\phi_F(t)$ is integrable, the formula in (7.25) below is called the **inverse Fourier transform of** $\phi_F(t)$. Because $\phi_F(t) = \widehat{f}(t)$ in this case, $f(x)$ is the **inverse Fourier transform of** $\widehat{f}(t)$, since this integral reproduces the function f on which the Fourier transform was defined:

$$f(x) = \frac{1}{2\pi}\int_{-\infty}^{\infty} \widehat{f}(t)e^{-ixt}dm.$$

Proposition 7.31 (Inverse Fourier transform: Integrable $\phi_F(t) \to f(x)$) *Let μ_F be a finite measure and $\phi_F(t)$ its Fourier-Stieltjes transform. If $\phi_F(t)$ is Lebesgue integrable, then there is a continuous function $f(x)$ so that:*

$$\mu_F[A] = \int_A f(x)dm, \ A \in \mathcal{B}(\mathbb{R}), \tag{7.24}$$

and thus $\phi_F(t) = \widehat{f}(t)$.

In addition, $f(x)$ is given by:

$$f(x) = \frac{1}{2\pi}\int_{-\infty}^{\infty} \phi_F(t)e^{-ixt}dm. \tag{7.25}$$

Proof. *By (7.9) and (7.6):*

$$\left|\frac{e^{-iat} - e^{-ibt}}{it}\right| = \frac{\left|e^{i(b-a)t} - 1\right|}{|t|} \leq (b-a).$$

Since $\phi_F(t)$ is Lebesgue integrable, (7.22) can be rewritten as:

$$\frac{1}{2\pi}\int_{-\infty}^{\infty}\frac{e^{-iat}-e^{-ibt}}{it}\phi_F(t)dm = \mu_F[(a,b)] + \frac{1}{2}\mu_F[\{a,b\}].$$

The triangle inequality and above bound obtain:

$$\left|\int_{-\infty}^{\infty}\frac{e^{-iat}-e^{-ibt}}{it}\phi_F(t)dm\right| \leq (b-a)\int_{-\infty}^{\infty}|\phi_F(t)|\,dm \leq c(b-a). \tag{1}$$

Letting $b=a+1/n$ in (7.22), it follows from (1):

$$\mu_F[(a,a+1/n)] + \frac{1}{2}\left(\mu_F[\{a\}] + \mu_F[\{a+1/n\}]\right) \leq c/(2\pi n) \to 0.$$

Hence, μ_F has no point masses, meaning $\mu_F[\{a\}]=0$ for all a.

By Proposition I.5.7, define the distribution function of finite μ_F by $F(x) \equiv \mu_F\left[(-\infty,x]\right]$. We claim that $F'(x)$ exists and is continuous.

Since μ_F has no point masses, $F(x+h)-F(x)=\mu_F[(x,x+h)]$ for $h\neq 0$, and thus:

$$\frac{F(x+h)-F(x)}{h} = \frac{1}{2\pi}\int_{-\infty}^{\infty}\frac{e^{-ixt}-e^{-i(x+h)t}}{ith}\phi_F(t)dm.$$

Because $\left|\frac{e^{-ixt}-e^{-i(x+h)t}}{ith}\right| \leq 1$ by (7.9), we can apply Lebesgue's dominated convergence theorem of Proposition 7.4 with any sequence $h_n\to 0$ and conclude that $F(x)$ is differentiable:

$$\begin{aligned} F'(x) &= \frac{1}{2\pi}\int_{-\infty}^{\infty}\lim_{h\to 0}\left[\frac{e^{-ixt}-e^{-i(x+h)t}}{ith}\right]\phi_F(t)dm \\ &= \frac{1}{2\pi}\int_{-\infty}^{\infty}\phi_F(t)e^{-ixt}dm. \end{aligned}$$

Defining $f(x)=F'(x)$ obtains (7.25), and $f(x)\geq 0$ by Corollary III.3.13.

For continuity of $f(x)$:

$$\begin{aligned} |f(x+h)-f(x)| &\leq \frac{1}{2\pi}\int_{-\infty}^{\infty}\left|e^{-i(x+h)t}-e^{-ixt}\right||\phi_F(t)|\,dm \\ &\leq \frac{1}{\pi}\int_{-\infty}^{\infty}|\phi_F(t)|\,dm. \end{aligned}$$

Since $e^{-i(x+h)t}-e^{-ixt}\to 0$ as $h\to 0$ for all x,t, Lebesgue's dominated convergence theorem of Proposition 7.4 assures that $|f(x+h)-f(x)|\to 0$ as $h\to 0$, and $f(x)$ is continuous.

With $f(x) = F'(x)$ continuous, the fundamental theorem of calculus of Proposition III.1.30 yields for any interval $[a,b]$:

$$(\mathcal{R})\int_a^b f(x)dx = F(b)-F(a).$$

Since μ_F is a finite measure and $F(x)\equiv\mu_F\left[(-\infty,x]\right]$, this formula holds for all finite and infinite intervals. By Proposition III.2.18, $f(x)$ is Lebesgue integrable on these intervals with the same values. Hence for all $(a,b]\in\mathcal{A}'$, where $\mathcal{A}'$ is the semi-algebra of right semi-closed intervals:

$$\begin{aligned} \int_a^b f(x)dm &= F(b)-F(a) \\ &\equiv \mu_F[(a,b]]. \end{aligned}$$

Define $\widetilde{\mu}_F(A)$ by (7.24) for $A \in \mathcal{B}(\mathbb{R})$:

$$\widetilde{\mu}_F(A) = \int_A f(x)dm.$$

Then for all $A \in \mathcal{A}$, the algebra of finite disjoint unions of $\mathcal{A}'$-sets:

$$\widetilde{\mu}_F(A) = \mu_F(A).$$

This is true on $\mathcal{A}'$ by definition of $f(x)$, and extends to $\mathcal{A}$ by finite additivity. Since $\mathcal{A}$ generates $\mathcal{B}(\mathbb{R})$ by Proposition I.8.1, and $\mathbb{R}$ is σ-finite under m, it follows by the uniqueness theorem of Proposition 1.22 that $\widetilde{\mu}_F(A) = \mu_F(A)$ for all $A \in \mathcal{B}(\mathbb{R})$, proving (7.24).

Finally, as noted in item 3 of Remark 7.8, since μ_F is given by the density function $f(x)$, it follows that $\phi_F(t) = \widehat{f}(t)$. ■

Remark 7.32 (Fourier transform notational conventions) *When both $f(x)$ and $\widehat{f}(t)$ are integrable, there is a great deal of symmetry between the formula for the Fourier transform, and that for the inverse Fourier transform given in (7.10) and (7.25), respectively:*

$$\widehat{f}(t) = \int_{-\infty}^{\infty} f(x)e^{itx}dm, \qquad f(x) = \frac{1}{2\pi}\int_{-\infty}^{\infty} \phi_F(t)e^{-ixt}dm.$$

The symmetry observed motivates a variety of notational conventions in general Fourier analysis.

While it is always the case that the exponential switches between positive and negative exponents in these formulas, in some texts, the Fourier transform is defined with the negative exponent, and then the inverse transform obtains the positive exponent.

Also, because there is an $(2\pi)^{-1}$ factor in one formula and not the other, it is not uncommon in Fourier analysis to see the factor of $(2\pi)^{-1/2}$ in both formulas, adding to the apparent symmetry. It is also possible to eliminate this coefficient entirely by defining the Fourier transform with a $\pm 2\pi itx$ in the exponential, and then the inverse transform will have $\mp 2\pi itx$ in the exponential and with no coefficient in either case.

The approach taken in this text is consistent with the ultimate application we have in mind, which is to use Fourier methods in the context of probability theory. In this case, the Fourier transform is called the ***characteristic function.*** *The notational convention for this latter function is nearly universally consistent with that used here for the Fourier transform, whereby this transform reflects the positive exponent itx and with no numerical coefficient.*

However, it is always a good idea to verify the notational conventions for Fourier analysis used by any given reference.

7.6 Continuity Theorem for Fourier Transforms

The last topic investigated in this chapter is the "continuity property" of the Fourier-Stieltjes transform. Namely, if $\{\mu_{F_n}\}_{n=1}^{\infty}$ is a sequence of finite measures for which $\mu_{F_n} \to \mu_F$ in some sense, must $\phi_{F_n}(t) \to \phi_F(t)$, and if so, in what sense? Conversely, if $\phi_{F_n}(t) \to \phi(t)$, again in some sense, what if anything does this tell us about the convergence of the associated measure sequence, $\{\mu_{F_n}\}_{n=1}^{\infty}$? In particular, must $\mu_{F_n} \to \mu_F$ in some sense for a measure μ_F, and if so, must the associated $\phi_F(t) = \phi(t)$?

Introduced in Book II and studied further in Book IV, the notion of **weak convergence of probability measures** is a natural one to investigate for this purpose. Specifically, we investigate the implications of $\mu_{F_n} \Rightarrow \mu_F$, meaning $\{\mu_{F_n}\}_{n=1}^{\infty}$ converges weakly to μ_F, for the pointwise convergence of $\phi_{F_n}(t) \to \phi_F(t)$. We also investigate the reverse implication.

Recall Definition II.8.2:

Definition 7.33 (Weak convergence - Distributions and measures on $\mathbb{R}$) *Weak convergence has two characterizations:*

1. *A sequence of distribution functions on $\mathbb{R}$, $\{F_n(x)\}_{n=1}^{\infty}$, is said to **converge weakly** to a distribution function $F(x)$, denoted $F_n \Rightarrow F$, if $F_n(x) \to F(x)$ for every continuity point of $F(x)$.*

2. *A sequence of probability measures on $\mathbb{R}$, $\{\mu_n\}_{n=1}^{\infty}$, is said to **converge weakly** to a probability measure μ, denoted $\mu_n \Rightarrow \mu$, if $\mu_n((-\infty, x])$ converges to $\mu((-\infty, x])$ for all x for which $\mu[x] = 0$.*

By Exercise II.8.4, these notions are equivalent. If F_n and the probability measure μ_n are related in the usual way:

$$F_n(x) \equiv \mu_n((-\infty, x]),$$

meaning either can be used to define the other, and similarly:

$$F(x) \equiv \mu((-\infty, x]),$$

then $\mu_n \Rightarrow \mu$ if and only if $F_n \Rightarrow F$.

That weak convergence of measures has implications for convergence of certain associated integrals was already seen in Propositions IV.4.72 and IV.4.74 in the context of convergence of moments and moment generating functions. More general results on convergence of such integrals will be seen in Book VI in the so-called **portmanteau theorem.** These results were proved in 1940 by **Aleksandr Aleksandrov** (1912–1999) and in some references identified with his name.

The following continuity theorem provides a very strong result on the above questions but is not self-contained, requiring one result from the Book VI portmanteau theorem. We also recall the notion of a **tight sequence of probability measures** from Definition II.8.16.

Definition 7.34 (Tight sequence) *A sequence of probability measures $\{\mu_{F_n}\}_{n=1}^{\infty}$ is **tight** if for any $\epsilon > 0$ there is a finite interval $(a, b]$ so that $\mu_{F_n}((a, b]) > 1 - \epsilon$ for all n.*

Equivalently, for any $\epsilon > 0$ there is a $c > 0$ so that $\mu_{F_n}(\{|x| \geq c\}) \leq \epsilon$ for all n.

Proposition 7.35 (Fourier-Stieltjes transform continuity theorem) *Let $\{\mu_{F_n}\}_{n=1}^{\infty}$, μ_F be probability measures and $\{\phi_{F_n}\}_{n=1}^{\infty}$, ϕ_F the associated Fourier-Stieltjes transforms.*

Then $\mu_{F_n} \Rightarrow \mu_F$ if and only if $\phi_{F_n}(t) \to \phi_F(t)$ for all t.

Proof. *Assume that $\mu_{F_n} \Rightarrow \mu_F$. By the portmanteau theorem of Book VI, for every bounded, continuous, real-valued function $g(x)$:*

$$\int g(x)d\mu_n \to \int g(x)d\mu. \tag{1}$$

Since $\cos tx$ *and* $\sin tx$ *are bounded and continuous for any* t*, we have by (7.4) and (1):*

$$\begin{aligned}
\phi_{F_n}(t) &= \int_{-\infty}^{\infty} e^{itx} d\mu_{F_n} \\
&= \int_{-\infty}^{\infty} \cos tx d\mu_{F_n} + i\int_{-\infty}^{\infty} \sin tx d\mu_{F_n} \\
&\to \int_{-\infty}^{\infty} \cos tx d\mu_F + i\int_{-\infty}^{\infty} \sin tx d\mu_F \\
&= \phi_F(t).
\end{aligned}$$

Thus $\phi_{F_n}(t) \to \phi_F(t)$ *for all* t.

Conversely, assume that $\phi_{F_n}(t) \to \phi_F(t)$ *for all* t*. We first show that* $\{\mu_{F_n}\}_{n=1}^{\infty}$ *is tight and then apply a Book II result. To this end, since* $\mu_{F_n}(\{|x| \geq 2/|u|\}) = \int_{|x|\geq 2/|u|} d\mu_{F_n}$ *and* $|x| \geq 2/|u|$ *implies that* $1 \leq 2(1 - 1/|ux|)$*:*

$$\begin{aligned}
\mu_{F_n}(\{|x| \geq 2/|u|\}) &\leq 2\int_{|x|\geq 2/|u|} (1 - 1/|ux|)\, d\mu_{F_n} \\
&\leq 2\int_{|x|\geq 2/|u|} \left(1 - \frac{\sin ux}{ux}\right) d\mu_{F_n} \\
&\leq 2\int_{-\infty}^{\infty} \left(1 - \frac{\sin ux}{ux}\right) d\mu_{F_n}.
\end{aligned}$$

The second step follows since if $|xu| \geq c > 0$*, the inequality* $1 - 1/|ux| \leq 1 - \frac{\sin ux}{ux}$ *is equivalent to* $\sin ux \leq 1$.

Now:

$$2\left(1 - \frac{\sin ux}{ux}\right) = \frac{1}{u}\int_{-u}^{u} (1 - e^{itx})dt,$$

and an application of Fubini's theorem is justified since μ_{F_n} *is a probability measure and thus* $1 - e^{itx}$ *is* $m \times \mu_{F_n}$*-integrable on* $[-u, u] \times \mathbb{R}$ *. Hence:*

$$\begin{aligned}
\mu_{F_n}(\{|x| \geq 2/|u|\}) &\leq \frac{1}{u}\int_{-u}^{u}\int_{-\infty}^{\infty} (1 - e^{itx})d\mu_{F_n} dt \\
&\leq \frac{1}{u}\int_{-u}^{u} (1 - \phi_{F_n}(t))\, dt. \qquad (2)
\end{aligned}$$

Since $\phi_F(t)$ *is continuous by Proposition 7.10 and* $\phi_F(0) = 1$*:*

$$\left|\frac{1}{u}\int_{-u}^{u} (1 - \phi_F(t))\, dt\right| \leq 2\sup_{[-u,u]} |1 - \phi_F(t)|,$$

and this upper bound converges to 0 *as* $u \to 0$.

So given $\epsilon > 0$ *there is a* u *so that:*

$$\frac{1}{u}\int_{-u}^{u} (1 - \phi_F(t))\, dt < \epsilon. \qquad (3)$$

But $\phi_{F_n}(t) \to \phi_F(t)$ *for all* t*, and since* $|(1 - \phi_{F_n}(t))| \leq 2$ *by the triangle inequality, the bounded convergence theorem obtains that for all* u*:*

$$\frac{1}{u}\int_{-u}^{u} (1 - \phi_{F_n}(t))\, dt \to \frac{1}{u}\int_{-u}^{u} (1 - \phi_F(t))\, dt.$$

Thus given $\epsilon > 0$ and u in (3), there is an N so that for $n \geq N$:

$$\frac{1}{u}\int_{-u}^{u}\left(1-\phi_{F_n}(t)\right)dt < 2\epsilon.$$

Since all $\phi_{F_n}(t)$ are continuous and $\phi_{F_n}(0) = 1$, this bound will remain valid for all $n < N$ by reducing u. Hence, for this u and all n :

$$\mu_{F_n}\left(\{|x| \geq 2/|u|\}\right) < 2\epsilon.$$

Letting $c \equiv \frac{2}{|u|}$ proves that $\{\mu_{F_n}\}_{n=1}^{\infty}$ is tight.

By Corollary II.8.22, if it can be shown that any subsequence of $\{\mu_{F_n}\}_{n=1}^{\infty}$ that converges weakly in fact converges weakly to μ_F, then $\mu_{F_n} \Rightarrow \mu_F$.

Let $\{\mu_{F_{n_k}}\}_{k=1}^{\infty}$ be such a subsequence and assume that $\mu_{F_{n_k}} \Rightarrow \upsilon$. Then by the first part of this theorem, $\phi_{F_{n_k}}(t) \to \phi(t)$ where $\phi(t)$ is the Fourier transform of υ. But by assumption, $\phi_{F_n}(t) \to \phi_F(t)$, and hence $\phi(t) = \phi_F(t)$ for all t. Thus by Corollary 7.23, $\upsilon = \mu_F$. ■

The next corollary almost certainly provides the most important application of the continuity theorem in probability theory. It requires Helly's selection theorems of Book II.

For this application, we recall Remark II.8.11 which noted that any of the results of that Section II.8.2 could be stated in terms of weakly convergent distribution functions or the associated weakly convergent measures. While the results there were presented in the terminology of distribution functions, true to this remark, we apply those results here in the context of measures.

Corollary 7.36 (On convergence of $\{\phi_{F_n}\}_{n=1}^{\infty}$) *Let $\{\mu_{F_n}\}_{n=1}^{\infty}$ be a collection of probability measures and $\{\phi_{F_n}\}_{n=1}^{\infty}$ the associated Fourier-Stieltjes transforms.*

If $\phi_{F_n}(t) \to \phi_0(t)$ for all t with $\phi_0(t)$ continuous at $t = 0$, then there exists a probability measure μ so that $\mu_{F_n} \Rightarrow \mu$, and $\phi_0(t)$ is the Fourier transform of μ.

Proof. *By the proof of the above proposition, the assumed continuity of $\phi_0(t)$ at $t = 0$ is enough to derive that $\{\mu_{F_n}\}_{n=1}^{\infty}$ is tight, since $\phi_0(0) = 1$ follows from $\phi_{F_n}(0) = 1$ for all n. By Helly's selection theorems of Propositions II.8.14 and II.8.20, there exists a subsequence $\{\mu_{F_{n_k}}\}_{k=1}^{\infty}$ and a probability measure μ with $\mu_{F_{n_k}} \Rightarrow \mu$.*

By Proposition 7.35, $\phi_{F_{n_k}}(t) \to \phi(t)$ for all t where $\phi(t)$ is the Fourier transform of μ. Since $\phi_{F_n}(t) \to \phi_0(t)$ by assumption, it follows that $\phi(t) = \phi_0(t)$ and $\phi_0(t)$ is the Fourier transform of μ.

Moreover, $\phi_{F_n}(t) \to \phi_0(t)$ for all t assures that any subsequence of $\{\mu_{F_n}\}_{n=1}^{\infty}$, which converges weakly must converge to this same μ. Corollary II.8.22 then obtains that $\mu_{F_n} \Rightarrow \mu$.

■

Remark 7.37 (Characteristic functions) *Jumping ahead to Book VI, the* ***characteristic function of a distribution function*** *F will be defined as the Fourier-Stieltjes transform of the associated finite measure μ_F. As will be seen, the continuity result above will provide a powerful tool for investigations into weak convergence of distribution functions, $F_n \Rightarrow F$, or equivalently, the convergence in distribution of the underlying random variable sequence, $X_n \Rightarrow X$.*

And indeed this will be a much more powerful tool for such results than that offered by ***moment generating functions*** *developed in Book IV.*

Characteristic functions exist for every probability measure and hence the above results are applicable to any distribution function sequence. By Propositions IV.4.25, IV.4.27, and IV.4.59, a moment generating function $M(t)$ exists if and only if the distribution function *has infinitely many moments* $\{\mu'_n\}_{n=1}^{\infty}$, *and* $\sum_{n=0}^{\infty}\mu'_n t^n/n!$ *converges absolutely on* $(-t_0,t_0)$ *for some* $t_0>0$. *Then* $M(t)=\sum_{n=0}^{\infty}\mu'_n t^n/n!$.

The **method of moments** *results for moment generating functions provided similar continuity conclusions to those above, though with somewhat more requirements because of the existence problem:*

1. ***Proposition IV.4.74:*** *If* $F_n \Rightarrow F$, *and the associated* $M_n(t)$ *and* $M(t)$ *exist on a common interval* $(-t_0,t_0)$ *with* $t_0>0$, *and* $\{M_n(t)\}_{n=1}^{\infty}$ *is bounded on this interval for each* t, *then* $M_n(t)\to M(t)$ *all* $t\in(-t_0,t_0)$.

2. ***Proposition IV.4.77:*** *Given* $\{F_n\}_{n=1}^{\infty}$ *and* F *with associated* $M_n(t)$ *and* $M(t)$, *which exist on a common interval* $(-t_0,t_0)$ *with* $t_0>0$, *if* $M_n(t)\to M(t)$ *for all* $t\in(-t_0,t_0)$, *then* $F_n\Rightarrow F$.

Example 7.38 (Poisson limit theorem: MGF approach) *Proposition IV.6.5 presenbreak ted a moment generating function proof of the* **Poisson limit theorem** *using Proposition IV.4.77. This result states that the distribution function of a properly calibrated binomial random variable sequence converged weakly to the Poisson distribution function. Proposition II.1.11 provided an alternative proof using explicit calculations with density functions and limiting arguments.*

The general binomial random variable X_{B_n} *has parameters* $n\in\mathbb{N}$, *and* p *with* $0<p<1$, *and has range* $[0,1,...,n]$ *with:*

$$\Pr\{X_{B_n}=j\}=\binom{n}{j}p^j(1-p)^{n-j}.$$

For this notation, if X_{B_n} *is defined on a probability space* $(\mathcal{S},\mathcal{E},\lambda)$, *then for any Borel set* $A\in\mathcal{B}(\mathbb{R})$:

$$\Pr\{X_{B_n}\in A\}\equiv\lambda\left[X_{B_n}^{-1}(A)\right].$$

The induced probability measure on $\mathbb{R}$ *is then defined by:*

$$\mu_{B_n}(j)=\binom{n}{j}p^j(1-p)^{n-j},\qquad j=0,1,..,n,$$

where the **binomial coefficient** $\binom{n}{j}$ *is defined for* $0\le j\le n$ *by:*

$$\binom{n}{j}=\frac{n!}{j!(n-j)!}.\tag{7.26}$$

The Poisson random variable X_P, *named for* **Siméon-Denis Poisson** *(1781–1840), is characterized by a single parameter* $\lambda>0$ *and has range on the nonnegative integers with:*

$$\Pr\{X_P=j\}=e^{-\lambda}\frac{\lambda^j}{j!},$$

and induced probability measure:

$$\mu_P(j)=e^{-\lambda}\frac{\lambda^j}{j!},\qquad j=0,1,2,...$$

The Poisson limit theorem states that for $\lambda=np$ *fixed, that as* $n\to\infty$:

$$\mu_{B_n}\Rightarrow\mu_P.$$

In this limit, $p\to0$ *as* $n\to\infty$ *so that* np *has constant value* λ.

By Definition 7.33, this means that $F_{B_n}(x) \to F_P(x)$ *for every* ***continuity point*** *of* F_P, *so for all* $x \in CP(F_P)$, *where:*

$$CP(F_P) = (-\infty, 0) \bigcup \cup_{n=0}^{\infty} (n, n+1).$$

We now prove this result again using the above Fourier transform tools. In Book VI, these tools will be applied in more general situations where moment generating functions do not exist.

Proposition 7.39 (Poisson limit theorem) *For* $\lambda = np$ *fixed, let* μ_{B_n} *denote the binomial measure defined on* $j = 0, 1, .., n$, *and* μ_P *the Poisson measure. Then as* $n \to \infty$:

$$\mu_{B_n} \Rightarrow \mu_P. \tag{7.27}$$

Proof. *By the Fourier-Stieltjes transform continuity theorem, (7.27) follows by proving that for all* t, $\phi_{F_n}(t) \to \phi_{F_P}(t)$, *where to simplify notation let* $F_n \equiv F_{B_n}$.

By definition,

$$\phi_{F_n}(t) = \int_{-\infty}^{\infty} e^{itx} d\mu_{B_n}.$$

With μ_{B_n} *defined above and supported on the integers* $0 \leq j \leq n$, *it follows that with* $p = \lambda/n$,

$$\phi_{F_n}(t) = \sum_{j=0}^{n} \binom{n}{j} \left(\frac{\lambda}{n}\right)^j \left(1 - \frac{\lambda}{n}\right)^{n-j} e^{itj}.$$

With $\phi_{F_1}(t) = (1 - \lambda/n) + (\lambda/n) e^{it}$, *a calculation shows that:*

$$\phi_{F_{n+1}}(t) = \phi_{F_n}(t)\phi_{F_1}(t),$$

and so by induction:

$$\begin{aligned} \phi_{F_n}(t) &= \left[\left(1 - \frac{\lambda}{n}\right) + \left(\frac{\lambda}{n}\right) e^{it}\right]^n \\ &= \left[1 + \frac{\lambda\left(e^{it} - 1\right)}{n}\right]^n. \end{aligned} \tag{1}$$

Now for all y:

$$\left(1 + \frac{y}{n}\right)^n \to e^y, \tag{7.28}$$

which is derived by taking n *large enough so that* $-1 < \frac{y}{n}$, *taking logarithms, and then considering the derivative of* $f(y) = \ln(1 + y)$ *at* $y = 0$. *This limit and* (1) *obtain for all* t:

$$\phi_{F_n}(t) \to \exp\left[\lambda\left(e^{it} - 1\right)\right], \text{ as } n \to \infty. \tag{2}$$

The final step is to evaluate $\phi_{F_P}(t)$ *for which we recall (7.3):*

$$\begin{aligned} \phi_{F_P}(t) &= \int_{-\infty}^{\infty} e^{itx} d\mu^P \\ &= \sum_{j=0}^{\infty} e^{-\lambda} \frac{\lambda^j}{j!} e^{itj} \\ &= e^{-\lambda} \sum_{j=0}^{\infty} \frac{\left(\lambda e^{it}\right)^j}{j!} \\ &= \exp\left[\lambda\left(e^{it} - 1\right)\right]. \end{aligned} \tag{3}$$

Hence, $\phi_{F_n}(t) \to \phi_{F_P}(t)$ *for all* t *by* (2) *and* (3), *and (7.27) follows from the continuity theorem of Proposition 7.35.* ■

8

General Measure Relationships

In this chapter, we investigate ways in which one measure can be decomposed relative to another measure. In the basic set-up, we are given a measure space $(X, \sigma(X), \mu)$, as well as a second measure υ on this space, meaning that υ is defined on the same sigma algebra $\sigma(X)$. The question is: Can υ be decomposed:

$$\upsilon = \upsilon_1 + \upsilon_2,$$

where υ_1 is in some sense closely related to μ, while υ_2 is in some sense quite unrelated to μ?

Once the notions of "closely related" and "quite unrelated" have been formalized, the **Lebesgue decomposition theorem** will address this. Named for **Henri Lebesgue** (1875–1941), it will state that if μ and υ are σ-finite measures, then this decomposition exists and is unique.

It will also turn out that υ_1, the measure closely related to μ, can in fact be defined in terms of the μ-integral of a μ-measurable function f. That is, for all $A \in \sigma(X)$:

$$\upsilon_1(A) = \int_A f d\mu.$$

This result is called the **Radon-Nikodým theorem,** named for **Johann Radon** (1887–1956), who proved it when $X = \mathbb{R}^n$, and **Otto Nikodým** (1887–1974), who generalized Radon's result to all σ-finite measure spaces.

As we have already developed much of this theory in the special case of Borel measures μ defined on $(\mathbb{R}, \mathcal{B}(\mathbb{R}), m)$, the next section summarizes these results and sets the stage for the more general development that follows.

8.1 Decomposition of Borel Measures on $(\mathbb{R}, \mathcal{B}(\mathbb{R}), m)$

Let μ be a Borel measure defined on $\mathbb{R}$ with associated distribution function $F_\mu(x)$ as in I.(5.1), or when μ is finite, equivalently given by $F_\mu(x) \equiv \mu\left[(-\infty, x]\right]$ as in I.(5.3). By Proposition I.5.7, as summarized in Section 1.1.3, $F_\mu(x)$ is increasing and thus Borel measurable, is right continuous, and for any right semi-closed interval $(a, b]$:

$$\mu\left[(a, b]\right] = F_\mu(b) - F_\mu(a). \tag{1}$$

As an increasing function, $F_\mu(x)$ is differentiable m-a.e. by Proposition III.3.12, $F'_\mu(x) \geq 0$ m-a.e. by Corollary III.3.13, and $F'_\mu(x)$ is Lebesgue integrable on bounded intervals by Proposition III.3.19.

Define $F_1(x)$ by:

$$F_1(x) = \int_0^x F'_\mu(y) dm,$$

DOI: 10.1201/9781003264576-8

recalling Remark 3.8 that $F_1(x) \equiv -\int_x^0 F'_\mu(y)dm$ for $x < 0$. Since $F'_\mu(x) \geq 0$ m-a.e., $F_1(x)$ is an increasing function.

Proposition III.3.39 states that if $x \geq 0$:

$$F'_1(x) = F'_\mu(x), \ m\text{-a.e.}$$

This identity is also true for $x < 0$ by the same result since if constant $a < x$:

$$-\int_x^0 F'_\mu(y)dm = -\int_a^0 F'_\mu(y)dm + \int_a^x F'_\mu(y)dm.$$

Then by Proposition III.3.19, if $a < b$:

$$F_1(b) - F_1(a) = \int_a^b F'_\mu(y)dm \leq F_\mu(b) - F_\mu(a). \tag{2}$$

Next define:

$$F_2(x) = F_\mu(x) - F_1(x).$$

Then $F_2(x)$ is Borel measurable, and increasing by (2) since if $x < x'$:

$$\begin{aligned} F_2(x') - F_2(x) &\equiv [F_\mu(x') - F_1(x')] - [F_\mu(x) - F_1(x)] \\ &= [F_\mu(x') - F_\mu(x)] - [F_1(x') - F_1(x)] \\ &\geq 0. \end{aligned}$$

Hence, $F_2(x)$ is again differentiable m-a.e., but since $F'_1(x) = F'_\mu(x)$ m-a.e. as noted above:

$$\begin{aligned} F'_2(x) &= F'_\mu(x) - F'_1(x) \\ &= 0, \qquad m\text{-a.e.} \end{aligned}$$

Summarizing, the function F_μ induced from the Borel measure μ can be decomposed as a sum of increasing functions:

$$\begin{aligned} F_\mu(x) &= F_1(x) + F_2(x), \\ F'_1(x) &= F'_\mu(x), \ m\text{-a.e.}, \\ F'_2(x) &= 0, \ m\text{-a.e.} \end{aligned} \tag{3}$$

Now define set functions on right semi-closed intervals $(a, b] \in \mathcal{A}'$:

$$\begin{aligned} \upsilon_{F_1}\left[(a, b]\right] \equiv F_1(b) - F_1(a), \\ \upsilon_{F_2}\left[(a, b]\right] \equiv F_2(b) - F_2(a), \end{aligned} \tag{4}$$

and let υ_1 and υ_2 be defined as the extensions of υ_{F_1} and υ_{F_2} to $\mathcal{B}(\mathbb{R})$ as summarized in Section 1.1.3. In brief, υ_{F_j} is extended additively from $\mathcal{A}'$ to the algebra $\mathcal{A}$ of finite disjoint unions of $\mathcal{A}'$-sets, and then to the complete sigma algebra of sets which are **Carathéodory measurable with respect to** $\upsilon^*_{F_j}$, the outer measure induced by υ_{F_j}. On this complete sigma algebra $\mathcal{M}_{\upsilon_{F_j}}(\mathbb{R})$, as well as on the Borel sigma algebra $\mathcal{B}(\mathbb{R}) \subset \mathcal{M}_{\upsilon_{F_j}}(\mathbb{R})$, υ_j is defined to equal $\upsilon^*_{F_j}$.

Then υ_j extends υ_{F_j} in the sense that $\upsilon_j = \upsilon_{F_j}$ on $\mathcal{A}'$. Further, as $\mathcal{B}(\mathbb{R})$ is the smallest sigma algebra that contains $\mathcal{A}'$, the extension to $\mathcal{B}(\mathbb{R})$ is unique, meaning that if there are measures υ'_j with $\upsilon'_j = \upsilon_j$ on $\mathcal{A}'$, then also $\upsilon'_j = \upsilon_j$ on $\mathcal{B}(\mathbb{R})$.

By (1), (3), and (4) it follows that on the semi-algebra $\mathcal{A}'$:

$$\mu = \upsilon_1 + \upsilon_2,$$

and this extends additively to the algebra $\mathcal{A}$. We now prove that this identity extends to $\mathcal{B}(\mathbb{R})$ by investigating the associated outer measures.

If $A \subset \mathbb{R}$, then by the definition of outer measure in (1.8) and the remarks following this equation:

$$\begin{aligned}
\mu^*(A) &= \inf\left\{\sum\nolimits_n \mu(A_n) \mid A \subset \bigcup A_n,\ \{A_n\} \subset \mathcal{A}' \text{ disjoint}\right\} \\
&= \inf\left\{\sum\nolimits_n [\upsilon_{F_1}(A_n) + \upsilon_{F_2}(A_n)] \mid A \subset \bigcup A_n,\ \{A_n\} \subset \mathcal{A}' \text{ disjoint}\right\} \\
&\geq \inf\left\{\sum\nolimits_n \upsilon_{F_1}(A_n) \mid A \subset \bigcup A_n\right\} + \inf\left\{\sum\nolimits_n \upsilon_{F_2}(A'_n) \mid A \subset \bigcup A'_n,\right\} \\
&= \upsilon^*_{F_1}(A) + \upsilon^*_{F_2}(A).
\end{aligned}$$

Conversely, if $A \subset \bigcup A_n$ and $A \subset \bigcup A'_n$, then $A \subset (\bigcup A_n) \bigcap (\bigcup A'_n)$ and this induces another disjoint countable union of right semi-closed intervals with $A \subset \bigcup A''_n$ where each $A''_n = A_{n_1} \cap A'_{n_2}$. Thus since υ_{F_1} and υ_{F_2} are monotonic:

$$\begin{aligned}
\upsilon^*_{F_1}(A) + \upsilon^*_{F_2}(A) &= \inf\left\{\sum\nolimits_n \upsilon_{F_1}(A_n) \mid A \subset \bigcup A_n\right\} + \inf\left\{\sum\nolimits_n \upsilon_{F_2}(A'_n) \mid A \subset \bigcup A'_n,\right\} \\
&\geq \inf\left\{\sum\nolimits_n \upsilon_{F_1}(A''_n) \mid A \subset \bigcup A''_n\right\} + \inf\left\{\sum\nolimits_n \upsilon_{F_2}(A''_n) \mid A \subset \bigcup A''_n\right\} \\
&= \inf\left\{\sum\nolimits_n [\upsilon_{F_1}(A''_n) + \upsilon_{F_2}(A''_n)] \mid A \subset \bigcup A''_n,\ \{A''_n\} \subset \mathcal{A}' \text{ disjoint}\right\} \\
&= \mu^*(A).
\end{aligned}$$

In summary:

$$\mu^* = \upsilon^*_{F_1} + \upsilon^*_{F_2},$$

and since $\mu = \mu^*$ on $\mathcal{B}(\mathbb{R})$ and similarly for υ_1 and υ_2, this analysis produces the **Lebesgue decomposition** in (8.1) of a Borel measure μ on $(\mathbb{R}, \mathcal{B}(\mathbb{R}), m)$.

Proposition 8.1 (Lebesgue decomposition of Borel μ on $(\mathbb{R}, \mathcal{B}(\mathbb{R}), m)$) *Let μ be a Borel measure on $(\mathbb{R}, \mathcal{B}(\mathbb{R}), m)$. Then there exists Borel measures υ_1 and υ_2 so that:*

$$\mu = \upsilon_1 + \upsilon_2. \tag{8.1}$$

Further:

1. *$\upsilon_1[A] = 0$ for all $A \in \mathcal{B}(\mathbb{R})$ with $m(A) = 0$.*

2. *There exists $E \in \mathcal{B}(\mathbb{R})$ so that:*

$$\upsilon_2[E] = m[\widetilde{E}] = 0,$$

and thus for all $A \in \mathcal{B}(\mathbb{R})$:

$$\mu[A] = \upsilon_1\left[A \bigcap E\right] + \upsilon_2\left[A \bigcap \widetilde{E}\right]. \tag{8.2}$$

Proof. *The decomposition in (8.1) was derived by the above construction.*

Since $\upsilon_1 = \upsilon_{F_1}$ on $\mathcal{A}'$, it follows that for all $(a, b]$:

$$\upsilon_1[(a, b]] = \int_a^b F'_\mu(y)dm.$$

By the nonnegativity and Lebesgue measurability of $F'_\mu(y)$ *and Proposition 4.3, define a Borel measure* υ *by:*

$$\upsilon[A] \equiv \int_A F'_\mu(y)dm, \;\; A \in \mathcal{B}(\mathbb{R}).$$

By definition, $\upsilon = \upsilon_1$ *on* $\mathcal{A}'$ *and then also on the algebra* $\mathcal{A}$ *of finite disjoint unions of* $\mathcal{A}'$*-sets. Thus by the uniqueness theorem of Proposition 1.22,* $\upsilon = \upsilon_1$ *on* $\sigma(\mathcal{A})$, *the smallest sigma algebra that contains* $\mathcal{A}$. *Since* $\sigma(\mathcal{A}) = \mathcal{B}(\mathbb{R})$ *by Proposition I.8.1, it follows that* $\upsilon = \upsilon_1$ *on* $\mathcal{B}(\mathbb{R})$, *and so:*

$$\upsilon_1[A] \equiv \int_A F'_\mu(y)dm, \;\; A \in \mathcal{B}(\mathbb{R}). \tag{1}$$

Thus by (1) *and item 9 of Proposition 3.42,* $\upsilon_1[A] = 0$ *for all* $A \in \mathcal{B}(\mathbb{R})$ *with* $m(A) = 0$, *proving item 1.*

By the above construction, $F'_2(x)$ *exists* m*-a.e. and* $F'_2(x) = 0$ m*-a.e. Let* D *denote the set on which* $F'_2(x) = 0$, *and thus* $m(\widetilde{D}) = 0$. *While* D *and* $\widetilde{D}$ *are Lebesgue measurable, they need not be Borel sets and this is required for the next step. Applying Proposition I.2.42, let* $\widetilde{E}$ *be a Borel set with* $\widetilde{D} \subset \widetilde{E}$ *and* $m(\widetilde{E} - \widetilde{D}) = 0$, *recalling that* $\widetilde{E} \in \mathcal{G}_\delta$, *the class of countable intersections of open sets. By finite additivity* $m(\widetilde{E}) = 0$, *and defining* E *as the complement of* $\widetilde{E}$, *it follows again by finite additivity that for any* $A \in \mathcal{B}(\mathbb{R})$*:*

$$\mu[A] = \mu\left[A \bigcap E\right] + \mu\left[A \bigcap \widetilde{E}\right]. \tag{2}$$

Note that this decomposition required E, $\widetilde{E} \in \mathcal{B}(\mathbb{R})$, *and would not have been valid with Lebesgue measurable* D *and* $\widetilde{D}$ *since by assumption,* μ *is only defined on* $\mathcal{B}(\mathbb{R})$.

Now $m(\widetilde{E}) = 0$ *implies* $\upsilon_1\left[\widetilde{E}\right] = 0$ *by the first result. More generally,* $m\left[A \bigcap \widetilde{E}\right] = 0$ *for all* $A \in \mathcal{B}(\mathbb{R})$ *obtains:*

$$\upsilon_1\left[A \bigcap \widetilde{E}\right] = 0.$$

Hence by (8.1):

$$\mu\left[A \bigcap \widetilde{E}\right] = \upsilon_2\left[A \bigcap \widetilde{E}\right]. \tag{3}$$

Now $E \subset D$ *and so* $A \bigcap E \subset D$ *for any Borel set* A. *Since* $D = \{F'_2(x) = 0\}$, *Proposition I.5.30 obtains:*

$$\upsilon_2\left[A \bigcap E\right] = 0.$$

Hence $\upsilon_2[E] = 0$ *and:*

$$\mu\left[A \bigcap E\right] = \upsilon_1\left[A \bigcap E\right]. \tag{4}$$

Combining (3) *and* (4) *into* (2) *obtains (8.2).* ■

We now summarize the general results that are forthcoming in this chapter from the more limited point of view of the above result.

Summary 8.2 *The above discussion has demonstrated that for any Borel measure* μ *defined on the Lebesgue measure space* $(\mathbb{R}, \mathcal{B}(\mathbb{R}), m)$*:*

1. ***Lebesgue decomposition theorem:*** *There exist Borel measures υ_1 and υ_2, with:*

   (a) $\mu = \upsilon_1 + \upsilon_2$, *where:*

   i. υ_1 *is* ***absolutely continuous with respect to*** m, *to be denoted* $\upsilon_1 \ll m$, *and which means that* $\upsilon_1(A) = 0$ *for all* $A \in \mathcal{B}(\mathbb{R})$ *with* $m(A) = 0$.

   ii. υ_2 *and* m *are* ***mutually singular***, *to be denoted* $\upsilon_2 \perp m$, *and which means that there is a set* $E \in \mathcal{B}(\mathbb{R})$, *for which* $m(\widetilde{E}) = \upsilon_2(E) = 0$.

   (b) *For all* $A \in \mathcal{B}(\mathbb{R})$, *and the set* E *identified in item 1.a.ii:*

$$\upsilon_1[A] = \upsilon_1\left[A\bigcap E\right],$$
$$\upsilon_2[A] = \upsilon_2\left[A\bigcap \widetilde{E}\right],$$

and hence

$$\mu[A] = \upsilon_1\left[A\bigcap E\right] + \upsilon_2\left[A\bigcap \widetilde{E}\right].$$

The Lebesgue decomposition theorem will be generalized in Proposition 8.26 to all σ-finite measure spaces.

2. *Another investigation of this chapter, but presented in the current context, is the Radon-Nikodým theorem.*

   Radon-Nikodým theorem: *Given a Borel measure* υ *which is* ***absolutely continuous with respect to*** m, *there is a nonnegative* m*-measurable function* f *so that:*

$$\upsilon[A] = \int_A f dm.$$

While it may appear that this result has only been proved for $\upsilon = \upsilon_1$ *constructed above for which* $f(x) \equiv F'_\mu(x)$, *the more general statement is obtained as follows.*

The above decomposition can be applied again to υ *to derive:*

$$\upsilon = \omega_1 + \omega_2,$$

where ω_1 *is absolutely continuous with respect to* υ, ω_2 *and* υ *are mutually singular, and* $\omega_1[A]$ *is given by the above integral representation with* $f(x) \equiv F'_\upsilon(x)$. *By item 1.a.ii, there is a set* $E \in \mathcal{B}(\mathbb{R})$ *for which* $m(\widetilde{E}) = \omega_2(E) = 0$. *But the absolute continuity of* υ *with respect to* m *assures that* $\omega_2(\widetilde{E}) = 0$, *and this along with* $\omega_2(E) = 0$ *proves that* $\omega_2 \equiv 0$ *and thus* $\upsilon \equiv \omega_1$.

The Radon-Nikodým theorem will be seen to apply in all σ-finite measure spaces in Proposition 8.23. Properties of signed measures will need to be derived for this result.

3. *In the special case of Borel measures, we can in fact refine the Lebesgue decomposition by splitting the* ***singular component*** υ_2 *into:*

$$\upsilon_2 = \upsilon_2^D + \upsilon_2^C.$$

Here υ_2^D *is a discrete Borel measure that allocates non-zero measure to at most countably many points, and* υ_2^C *is a Borel measure induced by a continuous function. In essence, this is accomplished by splitting* F_2, *the monotonic function that induces* υ_2, *into the sum of a "singular" function and a "saltus" function. See Section IV.1.2. for details on this construction.*

8.2 Decomposition of σ-Finite Measures

In this section, we generalize the **Lebesgue decomposition theorem** of the above section to the decomposition of a σ-finite measure υ defined on a σ-finite measure space $(X, \sigma(X), \mu)$. We also generalize the **Radon-Nikodým theorem** in the same setting when υ is absolutely continuous with respect to μ.

We begin by formalizing the definitions introduced above, and then pursue a necessary digression into the notion of a **signed measure**, which also played a role in the development of integration by parts for Lebesgue-Stieltjes integrals.

Definition 8.3 (Absolutely continuous; equivalent; mutually singular) *Given a measure space $(X, \sigma(X), \mu)$ and measures υ_1 and υ_2 defined on $\sigma(X)$:*

1. *υ_1 is* ***absolutely continuous with respect to*** *μ, denoted:*
$$\upsilon_1 \ll \mu,$$
if $\upsilon_1(A) = 0$ for all $A \in \sigma(X)$ with $\mu(A) = 0$.

2. *υ_1 is* ***equivalent to*** *μ, denoted:*
$$\upsilon_1 \sim \mu,$$
if:
$$\upsilon_1 \ll \mu, \text{ and, } \mu \ll \upsilon_1.$$

3. *υ_2 and μ are* ***mutually singular,*** *denoted:*
$$\upsilon_2 \perp \mu,$$
if there is a measurable set $E_\mu \in \sigma(X)$ for which:
$$\mu(\widetilde{E}_\mu) = \upsilon_2(E_\mu) = 0.$$

Remark 8.4 (On absolute continuity; singular; support of a measure) *The above terminology appears in a new context but has been encountered before.*

1. *The notion of* ***absolute continuity*** *was introduced in Definition III.3.54 as a property of a function. In the section on the Radon-Nikodým theorem, this earlier definition and that related to measures above will be reconciled in Proposition 8.18.*

2. *The notion of a* ***singular function*** *was introduced in Definition III.3.49. In the section on the Lebesgue decomposition theorem, this earlier definition and that related to measures above will be reconciled in Proposition 8.29.*

3. *Given a measure space $(X, \sigma(X), \mu)$, a measurable set E_μ is called* ***a support of*** *μ if for all $A \in \sigma(X)$:*
$$\mu(A) = \mu\left(A \bigcap E_\mu\right), \tag{8.3}$$
and thus $\mu(\widetilde{E}_\mu) = 0$. In this case, it is also said that μ is ***concentrated on*** *E_μ.*

   In item 3 of Definition 8.3, E_μ is a support of μ, and $E_{\upsilon_2} \equiv \widetilde{E}_\mu$ is a support of υ_2. Consequently, in this definition, the defining relationship can be stated as:
$$\upsilon_2 \perp \mu \iff \upsilon_2 \text{ and } \mu \text{ have disjoint supports.}$$

   The support of a measure is not unique. If $E_\mu \subset E'_\mu$ with E'_μ measurable, then E'_μ is also a support of μ since (8.3) is again satisfied, and $\mu(\widetilde{E}'_\mu) = 0$ since $\widetilde{E}'_\mu \subset \widetilde{E}_\mu$.

8.2.1 Signed Measures

In the development of this section, we will have the need to contemplate properties of the difference of two measures μ_1 and μ_2 defined on the sigma algebra $\sigma(X)$:

$$\nu \equiv \mu_1 - \mu_2.$$

To avoid ever having the expressions $\infty - \infty$ or $-\infty + \infty$, it must always be assumed that at least one of these measures is finite.

In general, the set function ν will not be nonnegative, and hence will not be a measure on $\sigma(X)$ unless $\mu_1(A) > \mu_2(A)$ for all $A \in \sigma(X)$. But this set function has the important property of countable additivity, as well as $\nu(\emptyset) = 0$, and so it is called a **signed measure.** Such measures were encountered in the study of Lebesgue-Stieltjes integration by parts.

More formally:

Definition 8.5 (Signed measure) *A set function ν defined on a sigma algebra $\sigma(X)$ is a **signed measure** if the range of ν includes at most one of $\pm\infty$, and if $\{A_i\} \subset \sigma(X)$ is a finite or countable disjoint collection:*

$$\nu\left[\bigcup_i A_i\right] = \sum_i \nu\left[A_i\right].$$

Here it is assumed that the summation is absolutely convergent if $\nu\left[\bigcup_i A_i\right]$ is finite, and divergent to one of $\pm\infty$ otherwise.

Remark 8.6 (On $\nu \equiv \mu_1 - \mu_2$.) *While initially defined as a more general notion than that of a difference of two measures, it will be seen below that every signed measure can in fact be expressed as a difference of measures. See the **Hahn decomposition** and **Jordan decomposition** theorems in Section 8.2.2.*

The essential question we need to answer for the development of the next section is as follows. Given a signed measure ν, can X be decomposed into measurable sets A^+ and A^-:

$$X = A^+ \bigcup A^-,$$

so that:

1. $A^+ \bigcap A^- = \emptyset$,
2. $\nu(A) \geq 0$ for all measurable $A \subset A^+$, and,
3. $\nu(A) \leq 0$ for all measurable $A \subset A^-$.

Such a decomposition is not unique in general, since if A^0 is a measurable set for which $\nu(A) = 0$ for all $A \subset A^0$, then A^0 can be part of either of the A^+ and A^- sets, or measurably split and divided between these sets.

Momentarily ignoring the requirement that $X = A^+ \bigcup A^-$, we introduce some terminology.

Definition 8.7 (Positive, negative, null sets) *Given $\sigma(X)$ and a signed measure ν, a measurable set A^+ is called **a positive set for** ν if:*

$$\nu(A) \geq 0 \textit{ for all measurable } A \subset A^+.$$

*Similarly, a measurable set A^- is called **a negative set for** ν if:*

$$\nu(A) \leq 0 \textit{ for all measurable } A \subset A^-,$$

*and a measurable set A^0 is called **a null set for** ν if:*

$$\nu(A) = 0 \textit{ for all measurable } A \subset A^0.$$

Example 8.8 *In the case where $\nu \equiv \mu_1 - \mu_2$, a difference of measures as in Exercise 6.7, this splitting can be understood as:*

1. A^+ : the set on which $\mu_1(A) \geq \mu_2(A)$ for all measurable $A \subset A^+$.
2. A^- : the set on which $\mu_1(A) \leq \mu_2(A)$ for all measurable $A \subset A^-$.
3. A^0 : the set on which $\mu_1(A) = \mu_2(A)$ for all measurable $A \subset A^0$.

It is apparent by definition that a subset of a positive set is positive, and the intersection of a collection of positive sets is a positive set. Assigned as an exercise below is the proof that the union of a collection of positive sets is a positive set. Analogously, the same statements are true of negative sets and null sets.

Unfortunately, the terminology for these sets can be misleading in that it is natural to assume that if A^+ is a positive set, then $\widetilde{A^+}$ is a negative set, and analogously for negative sets. But this needs not be true and is also assigned as Exercise 8.10.

Example 8.9 ($\nu(A) = \int_A f dm$, **measurable** f) *Let f be a Lebesgue measurable function and $f = f^+ - f^-$ its decomposition into positive and negative parts as in Definition 3.37. Assume that at least one of these components is Lebesgue integrable, and define:*

$$\nu(A) = \int_A f dm.$$

Then ν is countably additive by Corollary 3.49 if both component functions are Lebesgue integrable over the unioned set $A \equiv \bigcup A_i$.

In the general case, if one of the component functions is not integrable over A, say $\int_A f dm = \infty$, then it follows that $\int_A f^+ dm = \infty$ and thus $\sum_i \int_{A_i} f^+ dm = \infty$ by Lebesgue's monotone convergence theorem.

In detail, define:

$$f_n^+(x) = f^+(x) \sum_{i=1}^n \chi_{A_i}(x),$$

and so by Proposition 3.29 and Definition 3.7:

$$\int f_n^+(x) dm = \sum_{i=1}^n \int_{A_i} f^+ dm.$$

Then $f_n^+(x) \to f^+(x)\chi_A(x)$ monotonically, and so by monotone convergence:

$$\sum_{i=1}^n \int_{A_i} f^+ dm \to \int f^+(x)\chi_A(x) dm \equiv \int_A f^+ dm = \infty.$$

This then implies that $\sum_i \int_{A_i} f dm = \infty$ since $\int_A f^- dm < \infty$ by the assumption that at least one component function is integrable.

For such examples, it is not difficult to identify a positive set and negative set for ν. For example, $A^+ \equiv \{x | f \geq 0\}$ is a positive set and $A^- \equiv \{x | f < 0\}$ a negative set. But any measurable subsets of these are also valid candidates, as are definitions that allocate $A^0 = \{x | f = 0\}$ differently, in whole or measurably in part.

If we seek to also have $A^+ \bigcup A^- = \mathbb{R}$, then the definitions posed are best possible subject to the ambiguity on the placement of measurable splittings of A^0.

Exercise 8.10 (On positive, negative sets) *If ν is a signed measure defined on a sigma algebra $\sigma(X)$, prove that if $\{A_i^+\}_{i=1}^\infty$ is a countable collection of positive sets, then $\bigcup_{i=1}^\infty A_i^+$ is a positive set. This result is also true for negative and null sets by the same proof. Hint: If*

$A \subset \bigcup_{i=1}^{\infty} A_i^+$, *the goal is to express* $A = \bigcup_{i=1}^{\infty} A_i$ *where* $\{A_i\}_{i=1}^{\infty}$ *are disjoint and* $A_i \subset A_i^+$ *for all* i. *Consider:*

$$A_i \equiv A \bigcap A_i^+ \bigcap \widetilde{A_{i-1}^+} \bigcap ... \bigcap \widetilde{A_1^+}.$$

Then show by example that if A^+ *is a positive set, then* $\widetilde{A^+}$ *may be a negative set, but need not be a negative set.*

8.2.2 The Hahn and Jordan Decompositions

The following result is known as the **Hahn decomposition theorem** and is named for **Hans Hahn** (1879–1934). It states that one can always determine a positive set and negative set for a signed measure. Further, except for the ambiguity related to the location of null sets denoted A^0 above, this result asserts that this positive set and negative set are "maximal" in that:

$$A^+ \bigcup A^- = X, \text{ and, } A^+ \bigcap A^- = \emptyset.$$

For its proof, we require a result that at first seems apparent, that every measurable set of positive measure contains a positive subset of positive measure. However, its demonstration is subtle because to be deemed a positive set requires a verification of $\nu \geq 0$ on all measurable subsets.

Proposition 8.11 (On existence of A^+) *Given a signed measure* ν *on a sigma algebra* $\sigma(X)$, *if* $A \in \sigma(X)$ *satisfies* $0 < \nu(A) < \infty$, *then there exists a positive set* $A^+ \subset A$ *with* $\nu(A^+) > 0$.

Proof. *Given such* A, *the strategy is to define* A^+ *as the set that is left over after removing all measurable subsets of* A *with negative measure.*

If there is no subset of negative measure, then A *is a positive set and we are done. Otherwise, let* k_1 *be the smallest positive integer so that there is a set* $A_1 \subset A$ *with* $\nu(A_1) < -1/k_1$. *Inductively, once* $\{A_i\}_{i=1}^{n-1}$ *are chosen let* k_n *be the smallest positive integer so that there is a measurable set* $A_n \subset A - \bigcup_{i=1}^{n-1} A_i$ *with* $\nu(A_n) < -1/k_n$. *By construction,* $k_{n+1} \geq k_n$ *for all* n. *Now define:*

$$A^+ = A - \bigcup_{i=1}^{\infty} A_i,$$

noting that this union could be finite if there are no sets of negative measure after the N*th step.*

Since $A = A^+ \bigcup (\bigcup_{i=1}^{\infty} A_i)$ *and this union is disjoint, it follows by countable additivity that:*

$$\nu(A) = \nu(A^+) + \sum_{i=1}^{\infty} \nu(A_i). \tag{1}$$

If this summation contains a finite number of terms, then $0 < \nu(A)$ *by assumption and* $\nu(A_i) < 0$ *for all* i *by construction assures that* $\nu(A^+) > 0$. *If this is an infinite sum, this series converges absolutely by definition since* $\nu(A) < \infty$. *Then again,* $0 < \nu(A)$ *and absolute convergence assure that* $\nu(A^+) > 0$.

The final step is to show that A^+ *so defined is a positive set. This is true by definition if the summation in* (1) *is finite, since by construction there were no additional sets of negative measure. For the countable sum case, we show that given* $\epsilon > 0$, A^+ *can contain no measurable set with* ν*-measure less than* $-\epsilon$.

To this end, absolute convergence of the sum in (1) *obtains that* $k_n \to \infty$, *so given* $\epsilon > 0$ *there is an* n *with* $1/(k_n - 1) < \epsilon$. *Now by construction,* $A - \bigcup_{i=1}^{n} A_i$ *can contain no set with measure less that* $-1/(k_n - 1)$, *because at step* n *we defined* k_n *as the smallest integer with* $\nu(A_n) < -1/k_n$. *Since* $A^+ \subset A - \bigcup_{i=1}^{n} A_i$, *it follows that* A^+ *can contain no set with* *measure less that $-1/(k_n - 1)$. But $-\epsilon < -1/(k_n - 1)$, and hence A^+ can contain no set with measure less that $-\epsilon$.*

Since ϵ was arbitrary, it follows that A^+ contains no measurable set with negative measure, and thus A^+ is a positive set. ■

As is seen next, this technical result is the key ingredient to prove a main result of this section. For its statement, recall the definition of a **symmetric set difference**:

$$A \triangle B \equiv (A - B)\bigcup(B - A), \tag{8.4}$$

and note that by definition:

$$A \triangle B = B \triangle A.$$

Proposition 8.12 (Hahn decomposition theorem) *Let ν be a signed measure on a sigma algebra $\sigma(X)$. Then there is a positive set A^+ and a negative set A^- for ν, for which:*

$$A^+\bigcup A^- = X, \text{ and, } A^+\bigcap A^- = \emptyset.$$

Further, this decomposition is unique up to null sets. Specifically, if $B^+\bigcup B^- = X$ is another such decomposition, then:

$$A^+ \triangle B^+ = A^- \triangle B^-,$$

and this set is a null set.

Proof. *For specificity, assume that $-\infty \leq \nu(A) < \infty$ for all $A \in \sigma(X)$. Then given $-\infty < \nu(A) \leq \infty$, this proved result applies to the signed measure $-\nu$, and we can then simply switch positive and negative sets to obtain the desired result for the given ν.*

Let

$$P \equiv \{A \in \sigma(X) | A \text{ is a positive set}\},$$

noting that P is not empty since $\emptyset \in P$. Define:

$$\rho = \sup\{\nu(A) | A \in P\},$$

then $\rho \geq 0$ since $\nu(\emptyset) = 0$.

If $\rho = 0$, then P is a null set, and then $A^- = X - P$ and $A^+ = P$ is one decomposition of X. Here A^- so defined must be a negative set since if $A \subset A^-$ with $\nu(A) > 0$, this would contradict the construction of P by Proposition 8.11.

So assume $\rho > 0$ and let $\{A_i\}_{i=1}^{\infty} \subset P$ be chosen so that $\rho = \lim_{i\to\infty} \nu(A_i)$. This collection can be assumed to be nested and increasing since given arbitrary $\{A_i'\}_{i=1}^{\infty} \subset P$, let $A_i = \bigcup_{j=1}^{i} A_j'$. Now define $A^+ = \bigcup_{i=1}^{\infty} A_i$, and we show that this set satisfies the requirements of the proposition.

First, A^+ is indeed a positive set by Exercise 8.10 since it is the countable union of positive sets. Since $A^+ \in P$ it follows by definition of ρ that $\nu(A^+) \leq \rho$. But for any j, $A^+ - A_j = \bigcup_{i\neq j} A_i$ is a positive set by the same argument, and:

$$\nu(A^+) = \nu(A_j) + \nu(A^+ - A_j) \geq \nu(A_j).$$

Thus $\nu(A^+) \geq \nu(A_j)$ for all j implies that $\nu(A^+) \geq \rho = \lim_{i\to\infty} \nu(A_i)$ and thus $\nu(A^+) = \rho$.

Now define $A^- = \widetilde{A^+}$, and we show A^- is a negative set. Assume that there is a set $B \subset A^-$ with $\nu(B) > 0$. Then by the Proposition 8.11, there is a positive set $B' \subset B$ with $\nu(B') > 0$, and B' is disjoint from A^+ by construction. But then $A^+\bigcup B'$ is a positive set, and by disjointedness:

$$\nu\left(A^+\bigcup B'\right) = \nu(A^+) + \nu(B') > \rho,$$

contradicting the definition of ρ. Hence no such B can exist and A^- is thus a negative set.

For uniqueness, $\widetilde{B^+} = B^-$ and so $A^+ - B^+ \equiv A^+ \bigcap B^-$ is a positive set and a negative set and thus a null set. Similarly for $B^+ - A^+$ and then $A^+ \triangle B^+$. ■

We next turn to an important result known as the **Jordan decomposition theorem,** named for **Camille Jordan** (1838–1922). This result can be derived directly using the tools underlying the proof of the Hahn decomposition theorem, but now that the Hahn result is proved, we can obtain Jordan's result as a corollary.

Given a signed measure ν and the measurable sets A^+ and A^- defined in the Hahn decomposition theorem, define for $A \in \sigma(X)$:

$$\mu^+(A) = \nu\left(A\bigcap A^+\right), \qquad \mu^-(A) = -\nu\left(A\bigcap A^-\right).$$

Then μ^+ and $\mu^-$are measures, supported on A^+ and A^-, respectively, and:

$$\nu = \mu^+ - \mu^-.$$

Further, since $A^+ \bigcap A^- = \emptyset$, these measures have disjoint supports, and so $\mu^+$and μ^- are mutually singular.

This decomposition of a signed measure ν into a difference of measures is far from unique, since for any finite measure λ:

$$\nu = (\mu^+ + \lambda) - (\mu^- + \lambda).$$

The Jordan decomposition theorem states that a decomposition of ν into a difference of mutually singular measures is unique.

Proposition 8.13 (Jordan decomposition theorem) *Let ν be a signed measure on a sigma algebra $\sigma(X)$. Then there are mutually singular measures μ^+ and μ^- defined on $\sigma(X)$ with $\nu = \mu^+ - \mu^-$.*

Further, this decomposition is unique. If μ_0^+ and μ_0^- are mutually singular measures with $\nu = \mu_0^+ - \mu_0^-$, then $\mu^+ = \mu_0^+$ and $\mu^- = \mu_0^-$.

Proof. *Existence was demonstrated in the introductory remarks using the Hahn decomposition theorem.*

For uniqueness, let A^+ and A^- be the respective supports for μ^+ and μ^- as above, and A_0^+ and A_0^- the supports of μ_0^+ and μ_0^-. In other words,

$$A^+\bigcup A^- = X, \ A^+\bigcap A^- = \emptyset, \ \mu^+(A^-) = \mu^-(A^+) = 0,$$

and similarly for A_0^+ and A_0^-. Since $\mu^+(A) = \mu^+(A\bigcap A^+)$ for any $A \in \sigma(X)$, and similarly for the other measures:

$$\begin{aligned} \nu(A) &= \mu^+\left(A\bigcap A^+\right) - \mu^-\left(A\bigcap A^-\right), \\ &= \mu_0^+\left(A\bigcap A_0^+\right) - \mu_0^-\left(A\bigcap A_0^-\right). \end{aligned} \tag{1}$$

Letting $A = A^+ \bigcap A_0^-$ in (1) yields $\mu^+(A^+ \bigcap A_0^-) = -\mu_0^-(A^+ \bigcap A_0^-)$, and since measures are nonnegative:

$$\mu^+\left(A^+\bigcap A_0^-\right) = \mu_0^-\left(A^+\bigcap A_0^-\right) = 0,$$

and so μ^+ is supported on $A^+ \bigcap A_0^+$, and μ_0^- are supported on $A^- \bigcap A_0^-$.

Letting $A = A^- \bigcap A_0^+$ *in* (1) *similarly obtains:*

$$\mu^-\left(A^- \bigcap A_0^+\right) = \mu_0^+\left(A^- \bigcap A_0^+\right) = 0,$$

and so μ_0^+ *is supported on* $A^+ \bigcap A_0^+$, *and* μ^- *is supported on* $A^- \bigcap A_0^-$.

Then as in (1), *for any* $A \in \sigma(X)$:

$$\begin{aligned}\nu(A) &= \mu^+\left(A \bigcap A^+ \bigcap A_0^+\right) - \mu^-\left(A \bigcap A^- \bigcap A_0^-\right) \\ &= \mu_0^+\left(A \bigcap A^+ \bigcap A_0^+\right) - \mu_0^-\left(A \bigcap A^- \bigcap A_0^-\right).\end{aligned}$$

Hence, if $A \subset A^+ \bigcap A_0^+$, *then* $\mu^+(A) = \mu_0^+(A)$. *This obtains that* $\mu^+ = \mu_0^+$ *since for general* A:

$$\mu^+(A) = \mu^+\left(A \bigcap A^+ \bigcap A_0^+\right), \qquad \mu_0^+(A) = \mu_0^+\left(A \bigcap A^+ \bigcap A_0^+\right).$$

Similarly, $\mu^- = \mu_0^-$. ■

8.2.3 The Radon-Nikodým Theorem

As noted in the introduction, the **Radon-Nikodým theorem** is named for **Johann Radon** (1887–1956), who proved this result for $X = \mathbb{R}^n$, and **Otto Nikodým** (1887–1974), who generalized Radon's result to all σ-finite measure spaces. This theorem was introduced in Remark IV.1 21 in the context of existence of density functions.

To set the stage, let $(X, \sigma(X), \mu)$ be a σ-finite measure space and υ a σ-finite measure defined on $\sigma(X)$, which is absolutely continuous with respect to μ:

$$\upsilon \ll \mu.$$

The classic example of such absolutely continuous υ is next.

Example 8.14 ($\upsilon \ll \mu$) *If* $f : X \to \mathbb{R}$ *is a nonnegative* μ*-measurable function, define:*

$$\upsilon(A) = \int_A f d\mu.$$

Then υ *is a measure by Proposition 4.5 and is absolutely continuous with respect to* μ *by item 3 of Proposition 3.29.*

If f *is* μ*-integrable, then* υ *is a finite measure and thus* σ*-finite by definition.*

If f *is bounded, then* υ *is again a* σ*-finite measure by* σ*-finiteness of* μ. *In detail, if* $X = \bigcup_{j=1}^{\infty} B_j$ *for* $\{B_j\}_{j=1}^{\infty} \subset \sigma(X)$ *with* $\mu(B_j) < \infty$ *for all* j, *then* $\upsilon(B_j) < \infty$ *for all* j *by item 6 of Proposition 3.29. More generally,* f *need not be bounded to obtain that* υ *is* σ*-finite, only "locally" integrable in the sense that* $\int_{C_j} f d\mu < \infty$ *for* $\{C_j\}_{j=1}^{\infty} \subset \sigma(X)$ *with* $X = \bigcup_{j=1}^{\infty} C_j$.

The goal of the Radon-Nikodým theorem is to reverse this conclusion to state that every σ-finite measure υ that is **absolutely continuous** with respect to μ arises in exactly this way. The associated function μ-measurable function f is called the **Radon-Nikodým derivative of** υ **with respect to** μ, and will prove to be unique μ-a.e. In other words, if also $\upsilon(A) = \int_A g d\mu$, then $f = g$ except on a set of μ-measure 0.

Notation 8.15 *The* ***Radon-Nikodým derivative of*** υ ***with respect to*** μ *is often denoted* $\left[\frac{d\upsilon}{d\mu}\right]$ *or simply* $\frac{d\upsilon}{d\mu}$, *reflecting both by terminology and notation the notion of a derivative from calculus.*

This also allows the notationally compelling re-interpretation of (4.6) of Proposition 4.10. A function $g(x)$ is υ-integrable if and only if $g(x)\frac{d\upsilon}{d\mu}$ is μ-integrable, and when integrable, for all $A \in \sigma(X)$:

$$\int_A g(x)d\upsilon = \int_A g(x)\frac{d\upsilon}{d\mu}d\mu. \tag{8.5}$$

Remark 8.16 ($f \equiv \frac{d\upsilon}{d\mu}$ as a "derivative") *It is natural to wonder in what way the function $f \equiv \frac{d\upsilon}{d\mu}$ can be understood to be a derivative.*

Assume that $(X, \sigma(X), \mu) = (\mathbb{R}, \mathcal{B}(\mathbb{R}), m)$, the Borel measure space with Lebesgue measure, and let υ_F be a Borel measure induced by an increasing, right continuous function F, as summarized in Section 1.1.3. Equivalently, let υ be any Borel measure. Then by the discussion in Section I.5.3, $\upsilon = \upsilon_F$ for such a function F as defined in I.(5.1).

We consider two perspectives.

1. *From Proposition III.3.12, F is differentiable m-a.e. In the special case where F is absolutely continuous by Definition 8.17, Proposition III.3.62 obtains that F can be recovered by the Lebesgue integral of F'. In particular, it follows that on any interval $[a, b]$ on which F is absolutely continuous,*

$$F(x) = F(a) + \int_a^x F'(y)dm.$$

Since υ_F is defined on the semi-algebra $\mathcal{A}'$ by $\upsilon_F[(a, b]] = F(b) - F(a)$, this obtains that:

$$\upsilon_F[(a, b]] = \int_a^b F'(y)dm.$$

This characterization of υ_F generalizes to $A \in \mathcal{A}$, the algebra of finite disjoint unions of $\mathcal{A}'$-sets, by item 6 of Proposition III.2.49, and then to all $A \in \mathcal{B}(\mathbb{R})$:

$$\upsilon_F[A] = \int_A F'(y)dm. \tag{1}$$

This extension to $\mathcal{B}(\mathbb{R})$ follows from the uniqueness result of Proposition 1.22. The set function in (1), temporarily called $\widetilde{\upsilon}_F$, is a measure on $\mathcal{B}(\mathbb{R})$ by Proposition 4.3, and $\widetilde{\upsilon}_F = \upsilon_F$ on the algebra $\mathcal{A}$. Since υ_F is σ-finite, the uniqueness theorem obtains that $\widetilde{\upsilon}_F = \upsilon_F$ on $\sigma(\mathcal{A})$, the smallest sigma algebra that contains $\mathcal{A}$. Finally, $\sigma(\mathcal{A}) = \mathcal{B}(\mathbb{R})$ by Proposition I.8.1.

Note, however, that the representation in (1) is not unique. If $f = F'$ m-a.e., then f works equally well in this representation of υ_F, again by Proposition III.3.62.

Summary: *In the special case where υ_F is a Borel measure induced by an increasing, absolutely continuous function F, the function f identified in the Radon-Nikodým theorem satisfies $f = F'$ m-a.e. In other words, in this special case:*

$$\frac{d\upsilon_F}{dm} = \frac{dF}{dx}, \text{ m-a.e.}$$

Hence m-a.e., the Radon-Nikodým derivative $\frac{d\upsilon_F}{dm}$ equals $\frac{dF}{dx}$, the derivative of the distribution function F that defines υ_F.

2. *Another insight to the notation $\frac{d\upsilon}{dm}$ as a calculus derivative is a corollary to item 1.*

If $F'(x)$ is continuous on (a, b), then for any $x \in (a, b)$:

$$\lim_{\epsilon_n, \epsilon_n' \to 0} \frac{\upsilon_F\left[(x - \epsilon_n, x + \epsilon_n']\right]}{m\left[(x - \epsilon_n, x + \epsilon_n']\right]} = F'(x). \tag{2}$$

To see this, recall that the Lebesgue and Riemann integrals agree for continuous $F'(x)$ by Propositions III.1.15 and III.2.18, and thus by the mean value theorem for Riemann integrals of Corollary III.1.24:

$$\frac{\upsilon_F\left[(x - \epsilon_n, x + \epsilon_n']\right]}{m\left[(x - \epsilon_n, x + \epsilon_n']\right]} \equiv \frac{1}{\epsilon_n + \epsilon_n'} \int_{x - \epsilon_n}^{x + \epsilon_n'} F'(y) dy = F'(y_0),$$

where $y_0 \in [x - \epsilon_n, x + \epsilon_n']$. The result in (2) now follows by continuity of $F'(x)$.

The limit on the left in (2) is reasonably interpreted as the derivative $\frac{d\upsilon_F}{dm}$. If υ_F is a finite measure, then by I.(5.3):

$$\upsilon_F\left[(x - \epsilon_n, x + \epsilon_n']\right] = \upsilon_F\left[(-\infty, x + \epsilon_n']\right] - \upsilon_F\left[(-\infty, x - \epsilon_n']\right],$$

while $m\left[(x - \epsilon_n, x + \epsilon_n']\right] = \epsilon_n + \epsilon_n'$. For general υ_F, a similar representation is possible using I.(5.1) but depends on the sign of x.

*For more general measurable $F'(x)$, the result in (2) is again satisfied, but now only m-a.e. See for example, Theorem 8.6 in **Rudin** (1974).*

Before continuing to the main result of the Radon-Nikodým theorem, we take a short detour to address the perhaps surprising double use of the terminology "absolutely continuous." This notion was first introduced in the context of a function in Definition III.3.54, and now again in Definition 8.3 in the context of a measure. The final conclusion derived below is that if F is an increasing, right continuous function, then F is absolutely continuous by Definition III.3.54 if and only if $\upsilon_F \ll m$ by Definition 8.3.

For completeness, we recall the earlier notion.

Definition 8.17 (Absolute continuity) *A real-valued function $F(x)$ defined on $[a, b]$ is **absolutely continuous** if for any $\epsilon > 0$ there is a δ so that:*

$$\sum_{i=1}^{n} |F(x_i) - F(x_i')| < \epsilon,$$

for any finite collection of disjoint subintervals $\{(x_i', x_i)\}_{i=1}^{n} \subset [a, b]$ with:

$$\sum_{i=1}^{n} |x_i - x_i'| < \delta.$$

A function is absolutely continuous if the above condition is satisfied for all $\{(x_i', x_i)\}_{i=1}^{n} \subset \mathbb{R}$.

Thus if an increasing, right continuous function F is absolutely continuous, then given $\epsilon > 0$ there is a δ so that with υ_F denoting the induced Borel measure:

$$\sum_{i=1}^{n} \upsilon_F\left[(x_i', x_i]\right] < \epsilon,$$

for any finite collection of disjoint intervals $\{(x_i', x_i]\}_{i=1}^{n}$, with:

$$\sum_{i=1}^{n} m\left[(x_i', x_i]\right] < \delta.$$

So absolute continuity of an increasing right continuous function induces a property between the measures υ_F and m. This representation implies, that, at least for collections of intervals, we can make their υ_F-measure arbitrarily small by making the m-measure small. This result is approaching $\upsilon_F \ll m$, where $m(A) = 0$ implies that $\upsilon_F(A) = 0$.

The next result formalizes this connection.

Proposition 8.18 (Absolute continuity definitions) *Let υ_F be a Borel measure on $(\mathbb{R}, \mathcal{B}(\mathbb{R}), m)$ induced by an increasing, right continuous function F.*

Then F is absolutely continuous by Definition 8.17 if and only if $\upsilon_F \ll m$.

Proof. *If F is absolutely continuous, then by item 1 of Remark 8.16, for all $A \in \mathcal{B}(\mathbb{R})$:*

$$\upsilon_F [A] = \int_A F'(y)dm,$$

and so $\upsilon_F \ll m$ by item 9 of Proposition 3.42.

If $\upsilon_F \ll m$, the Radon-Nikodým theorem of Proposition 8.23 can be applied since the Borel measure υ_F is finite on compact sets by definition and thus is a σ-finite measure. This theorem proves existence of a Lebesgue measurable function f so that for all $A \in \mathcal{B}(\mathbb{R})$:

$$\upsilon_F [A] = \int_A f(y)dm.$$

Since υ_F is finite on compact sets, $f(y)$ is Lebesgue integrable on all bounded sets.

Letting $A = (a, x]$ obtains by Section 1.1.3:

$$\upsilon_F [A] = F(x) - F(a) = \int_a^x f(y)dm.$$

As an indefinite integral of Lebesgue integrable $f(x)$, the function $F(x)$ is absolutely continuous by Proposition III.3.58. ■

The Radon-Nikodým theorem requires that $(X, \sigma(X), \mu)$ be a σ-finite measure space, and υ a σ-finite measure on $\sigma(X)$ with $\upsilon \ll \mu$. In the next example, we illustrate the need for σ-finiteness of the underlying space.

Example 8.19 (Need for measure space σ-finiteness) *Let $(X, \sigma(X), \mu) = ((0,1), \mathcal{B}(0,1), \mu)$, and define μ as the counting measure where $\mu(A)$ is defined as the number of points in A if finite, and $\mu(A) = \infty$ otherwise. Then μ is not a σ-finite measure since $(0,1)$ has uncountably many points, and thus cannot equal a countable union of sets of finite measure.*

Let $\upsilon = m$, Lebesgue measure, so m is finite and hence σ-finite. In addition, $m \ll \mu$ since $\mu(A) = 0$ implies $A = \emptyset$ and hence $m(A) = 0$.

Assume that there exists a μ-measurable and thus Borel measurable function f so that for all $A \in \mathcal{B}(0,1)$:

$$m(A) = \int_A f d\mu. \tag{1}$$

Express $f = f^+ - f^-$ as in Definition 3.37 and apply the following argument to each nonnegative function.

Denoting either by f for simplicity, it follows from (3.9) that:

$$\int_A f(x)d\mu = \sup_{\varphi \leq f} \int_A \varphi(x)d\mu,$$

where each $\varphi(x) = \sum_{i=1}^{n} a_i \chi_{A_i}(x)$ is a simple function, $\{A_i\}_{i=1}^n$ a disjoint collection of μ-measurable sets, and by (3.1):

$$\int_A \varphi(x)d\mu = \sum_{i=1}^{n} a_i \mu(A_i).$$

Letting $A = \{a\}$ in (1), the integral equals $f(a)$, while $m(A) = 0$. Thus for all a :

$$f(a) = 0,$$

and (1) obtains that $m(A) = 0$ for all $A \in \mathcal{B}(0,1)$, a contradiction.

In summary, even though $m \ll \mu$, there is no μ-measurable function f that satisfies (1).

The proof of the Radon-Nikodým theorem is constructive, and for this proof a technical result will be needed on the existence of measurable functions with specified level sets defined as:

$$A_\alpha \equiv \{x | f(x) \leq \alpha\}.$$

If f is measurable on $(X, \sigma(X), \mu)$, then $A_\alpha \in \sigma(X)$ for all α by definition. In addition,

$$A_\alpha \subset A_\beta \text{ for } \alpha < \beta,$$

and:

$$f > \alpha \text{ on } X - A_\alpha.$$

The next results seek to reverse the above logic. Specifically, given a countable collection of sets $\{A_\alpha\}_\alpha \subset \sigma(X)$, when does an associated measurable f exist for which $\{A_\alpha\}_\alpha$ are its level sets, and what are the properties of such f?

- The first lemma provides the existence result for such a measurable function if $A_\alpha \subset A_\beta$ when $\alpha < \beta$, and proves that $f \leq \alpha$ on A_α, and $f \geq \alpha$ on $X - A_\alpha$.
- The second lemma generalizes this to the case where the assumption on sets for $\alpha < \beta$ is weakened from $A_\alpha - A_\beta = \emptyset$ to $\mu(A_\alpha - A_\beta) = 0$. In other words, if $\alpha < \beta$ then $A_\alpha \subset A_\beta$ except for a subset of A_α of μ-measure 0. The conclusion then becomes $f \leq \alpha$ on A_α, and $f \geq \alpha$, μ-a.e. on $X - A_\alpha$.

Recall that extended real value functions were discussed in Section I.3.1.

Lemma 8.20 (Strongly nested $\{A_\alpha\}_\alpha$) *If $\{A_\alpha\}_\alpha \subset \sigma(X)$ is a countable collection of sets with $A_\alpha \subset A_\beta$ for $\alpha < \beta$, then there is a $\sigma(X)$-measurable, extended real-valued function f on X with $f \leq \alpha$ on A_α and $f \geq \alpha$ on $X - A_\alpha$.*

Proof. *Given $x \in X$, define:*

$$f(x) = \inf\{\alpha' | x \in A_{\alpha'}\},$$

where $\inf \emptyset \equiv \infty$.

Hence, if $x \in A_\alpha$:

$$f(x) \equiv \inf\{\alpha' | x \in A_{\alpha'}\} \leq \alpha,$$

while if $x \in X - A_\alpha$:

$$f(x) = \inf\{\alpha' | x \in A_{\alpha'}\} \geq \alpha,$$

since $\{\alpha' | x \in A_{\alpha'}\} \subset (\alpha, \infty)$

To show that f is measurable, note that if $f(x) < \alpha$, then $x \in A_\beta$ for some $\beta < \alpha$, while if $x \in A_\beta$ for some $\beta < \alpha$, then $f(x) \leq \beta < \alpha$. Hence, as a countable union:

$$\{x | f(x) < \alpha\} = \bigcup_{\beta < \alpha} A_\beta \in \sigma(X),$$

and f is measurable. ∎

Lemma 8.21 (Weakly nested $\{A_\alpha\}_\alpha$) *If $\{A_\alpha\}_\alpha \subset \sigma(X)$ is a countable collection of sets with $\mu(A_\alpha - A_\beta) = 0$ for $\alpha < \beta$, then there is a $\sigma(X)$-measurable, extended real-valued function f on X with $f \leq \alpha$ on A_α, and $f \geq \alpha$, μ-a.e. on $X - A_\alpha$.*

Proof. *Let $C \equiv \bigcup_{\alpha<\beta}(A_\alpha - A_\beta)$, a countable union of sets of measure zero, so $\mu(C) = 0$, and define:*

$$A'_\alpha = A_\alpha \bigcup C. \tag{1}$$

We claim that $A'_\alpha \subset A'_\beta$ if $\alpha < \beta$.

By De Morgan's laws (Exercise I.2.2), given $\alpha < \beta$:

$$\begin{aligned} A'_\alpha - A'_\beta &= \left(A_\alpha \bigcup C\right) \bigcap \left(\widetilde{A}_\beta \bigcap \widetilde{C}\right) \\ &= \left(A_\alpha \bigcap \widetilde{A}_\beta \bigcap \widetilde{C}\right) \bigcup \left(C \bigcap \widetilde{A}_\beta \bigcap \widetilde{C}\right) \\ &= (A_\alpha - A_\beta) - C = \emptyset. \end{aligned}$$

This last set is empty since α and β are among the countable collection of pairs that define C.

By Lemma 8.20, there is a measurable function f on X with $f \leq \alpha$ on A'_α and $f \geq \alpha$ on $X - A'_\alpha$. By (1) this obtains that $f \leq \alpha$ on A_α. Further, since:

$$X - A'_\alpha = X \bigcap \widetilde{A}_\beta \bigcap \widetilde{C} = (X - A_\alpha) - C,$$

$f \geq \alpha$ on $X - A'_\alpha$ obtains:

$$f \geq \alpha \text{ on } (X - A_\alpha) - C.$$

Thus except possibly on C, a set of measure 0, $f \geq \alpha$ on $X - A_\alpha$, and the result follows. ■

Before developing the main result, we outline how the proof of the Radon-Nikodým theorem will utilize the results of the above lemmas.

Remark 8.22 (On the Radon-Nikodým proof) *The following proof identifies a countable collection of "weakly nested" sets, $\{A^-_{r_k}\}$ for rational r_k, so that if $r_k < r_j$ then $\mu[A^-_{r_k} - A^-_{r_j}] = 0$. Using Lemma 8.21, a measurable function f is then defined on X so that $f \leq r_k$ on $A^-_{r_k}$, and $f \geq r_k$ μ-a.e. on $X - A^-_{r_k} = A^+_{r_k}$.*

With this construction, it will follow that for any N and $r_k = k/N$, that if $A_k \subset A^-_{r_{k+1}} - A^-_{r_k}$, then:

$$r_k\mu[A_k] \leq \upsilon[A_k] \leq r_{k+1}\mu[A_k].$$

In addition, on any such A_k we have by construction that $r_k \leq f \leq r_{k+1}$, and so:

$$r_k\mu[A_k] \leq \int_{A_k} f d\mu \leq r_{k+1}\mu[A_k].$$

These results bound the υ measures for any set $A_k \subset A^-_{r_{k+1}} - A^-_{r_k}$ in terms of $\mu[A_k]$ and also bound the integrals $\int_{A_k} f d\mu$ similarly.

It will then be shown that any $A \in \sigma(X)$ can be expressed as a disjoint union of such A_k-sets where $A_k \subset A^-_{r_{k+1}} - A^-_{r_k}$ if $k < \infty$, and a set $A_\infty \subset X_\infty \equiv X - \bigcup_{k=0}^\infty A^-_{r_k}$, so the above analysis will lead to estimates for A. But as it cannot be assumed that $X = \bigcup_k \left(A^-_{r_{k+1}} - A^-_{r_k}\right)$, this construction must address the υ-measure of the set $X_\infty = X - \bigcup_{k=0}^\infty A^-_{r_k}$.

While it will be relatively straightforward to show that $\mu[A_\infty] = 0$ for any A, the value of $\upsilon[A_\infty]$ is in general not constrained by this construction. But in the special case of this theorem where $\upsilon \ll \mu$, this will ensure that $\upsilon[A_\infty] = 0$. So then for every component of the disjoint union for A, the υ measures can be estimated based on the μ integrals of f.

With these technical results and the above introduction in hand, we are now ready for the statement and proof of the Radon-Nikodým theorem. It should be noted that the ideas underlying this theorem and the Lebesgue decomposition theorem of the following section are intimately related, as are the tools used in their proofs. It is largely a matter of taste whether one proves Radon-Nikodým first and derives the Lebesgue decomposition or uses the opposite logic. In fact, some texts combine the results as the Lebesgue-Radon-Nikodým theorem.

Proposition 8.23 (Radon-Nikodým theorem) *Let $(X, \sigma(X), \mu)$ be a σ-finite measure space and υ a σ-finite measure on $\sigma(X)$ that is absolutely continuous with respect to μ, so $\upsilon \ll \mu$.*

Then there exists a nonnegative $\sigma(X)$-measurable function $f : X \to \mathbb{R}$, also denoted $f \equiv \frac{\partial \upsilon}{\partial \mu}$, so that for all $A \in \sigma(X)$:

$$\upsilon(A) = \int_A f d\mu. \tag{8.6}$$

Further, f is unique μ-a.e. If g is a $\sigma(X)$-measurable function so that (8.6) is satisfied with g, then $g = f$, μ-a.e.

Proof.

1. ***Reduction of Existence Proof to Finite Measures:***

   Both μ and υ are σ-finite measures, and thus there is a disjoint collection $\{A_n\}_{n=1}^{\infty} \subset \sigma(X)$ so that $\bigcup_{n=1}^{\infty} A_n = X$, and both $\mu(A_n) < \infty$ and $\upsilon(A_n) < \infty$ for all n. This follows because by Proposition I.2.20, such a collection exists for each measure separately, so we can by monotonicity of measures define the A_n-sets in terms of intersections of these collections. For $A \in \sigma(X)$, define finite measures $\mu_n(A) \equiv \mu(A \bigcap A_n)$ and $\upsilon_n(A) \equiv \nu(A \bigcap A_n)$, and note that then $\upsilon_n \ll \mu_n$.

   Now assume that the existence statement of the theorem has been proven for finite measures. Then for each n, there is a nonnegative $\sigma(X)$-measurable function, $f_n : X \to \mathbb{R}$, so that for $A \in \sigma(X)$:

   $$\upsilon_n(A) = \int_A f_n d\mu_n = \int_A f_n \chi_{A_n} d\mu.$$

   This identity between the μ_n-integral and the μ-integral of f_n is true for all simple functions by (3.1) and generalizes by Definition 3.11.

   By countable additivity and disjointedness of $\{A_n\}_{n=1}^{\infty}$:

   $$\begin{aligned} \upsilon(A) &= \sum \upsilon_n(A) \\ &= \sum \int_A f_n \chi_{A_n} d\mu \\ &= \int_A \sum f_n \chi_{A_n} d\mu. \end{aligned}$$

   Hence, the existence statement of the theorem is true in the σ-finite case with $f = \sum f_n \chi_{A_n}$.

2. ***Existence Proof for Finite Measures:***

   a. **Construction of** f : *Define* $A_0^+ = X$ *and* $A_0^- = \emptyset$, *and for any positive rational* r_k, *let* $\{A_{r_k}^+, A_{r_k}^-\}$ *be the positive and negative sets of the Hahn decomposition of the signed measure* $\upsilon - r_k\mu$, *so:*

$$A_{r_k}^+ \bigcup A_{r_k}^- = X, \quad A_{r_k}^+ \bigcap A_{r_k}^- = \emptyset.$$

   For any $r_j \neq r_k$, $A_{r_k}^- - A_{r_j}^- = A_{r_k}^- \bigcap A_{r_j}^+$ *and hence this is a positive set for* $\upsilon - r_j\mu$ *and a negative set for* $\upsilon - r_k\mu$. *But if* $r_k < r_j$, *then for any* $A \in \sigma(X)$:

$$(\upsilon - r_j\mu)[A] = (\upsilon - r_k\mu)[A] - (r_j - r_k)\mu[A] \leq (\upsilon - r_k\mu)[A].$$

   Hence, when $r_k < r_j$, *if* A *is a positive set for* $\upsilon - r_j\mu$ *and a negative set for* $\upsilon - r_k\mu$ *then* $\mu[A] = 0$. *Letting* $A = A_{r_k}^- - A_{r_j}^-$, *it follows that if* $r_k < r_j$ *then* $\mu[A_{r_k}^- - A_{r_j}^-] = 0$.
   By Lemma 8.21, let f *be the measurable function on* X *so that* $f \leq r_k$ *on* $A_{r_k}^-$, *and* $f \geq r_k$ μ*-a.e. on* $X - A_{r_k}^- = A_{r_k}^+$. *Since* $A_0^- = \emptyset$, *it follows that* $f \geq 0$ μ*-a.e. on* X.

   b. μ**-Integral of** f : *Given integer* N, *consider the subset of rationals with* $r_k = k/N$, $k \geq 0$. *Given* $A \in \sigma(X)$, *define:*

$$A_k = A \bigcap \left(A_{r_{k+1}}^- - A_{r_k}^-\right), \qquad A_\infty = A - \bigcup\nolimits_{k=0}^{\infty} A_{r_k}^-.$$

   Then:

$$A = A_\infty \bigcup \bigcup\nolimits_{k=0}^{\infty} A_k, \tag{1}$$

   is a disjoint union, and so:

$$\mu[A] = \mu[A_\infty] + \sum\nolimits_{k=0}^{\infty} \mu[A_k].$$

   Since:

$$A_k \subset A_{r_{k+1}}^- - A_{r_k}^- = A_{r_{k+1}}^- \bigcap A_{r_k}^+,$$

   part 2.a obtains that $k/N \leq f \leq (k+1)/N$ μ*-a.e. on* A_k, *and so by item 5 of Proposition 3.42:*

$$\frac{k}{N}\mu[A_k] \leq \int_{A_k} f d\mu \leq \frac{k+1}{N}\mu[A_k]. \tag{2}$$

   As A_k *is by definition a positive set for* $\upsilon - r_k\mu$ *and a negative set for* $\upsilon - r_{k+1}\mu$, *it follows that* $(\upsilon - r_k\mu)[A_k] \geq 0$ *and* $(\upsilon - r_{k+1}\mu)[A_k] \leq 0$. *This obtains that* $r_k\mu[A_k] \leq \upsilon[A_k] \leq r_{k+1}\mu[A_k]$, *and from* (2) *and the definition of* r_k *is derived that for all* $k < \infty$:

$$\upsilon[A_k] - \frac{1}{N}\mu[A_k] \leq \int_{A_k} f d\mu \leq \upsilon[A_k] + \frac{1}{N}\mu[A_k]. \tag{3}$$

   Now by construction $f = \infty$ μ*-a.e. on* A_∞. *If* $\mu[A_\infty] > 0$, *then* $\upsilon[A_\infty] = \infty$ *since* $(\upsilon - r_k\mu)[A_\infty] \geq 0$ *for all* k, *contradicting the assumption that* υ *is a finite measure. So it must be the case that* $\mu[A_\infty] = 0$ *and hence* $\upsilon[A_\infty] = 0$ *since* $\upsilon \ll \mu$, *and thus by Definition 3.11:*

$$\upsilon[A_\infty] = \int_{A_\infty} f d\mu = 0. \tag{4}$$

Given the disjointedness of the union in (1) for A, the above estimates in (3) and (4) can be added together by Corollary 3.32 to obtain for all $A \in \sigma(X)$:

$$\upsilon[A] - \frac{1}{N}\mu[A] \leq \int_A f d\mu \leq \upsilon[A] + \frac{1}{N}\mu[A]. \tag{5}$$

As this is now true for every subset of rationals of the form $r_k = k/N$, $k \geq 0$, it follows that for all $A \in \sigma(X)$:

$$\int_A f d\mu = \upsilon[A].$$

3. ***Uniqueness of*** f *: Let g be another $\sigma(X)$-measurable function that satisfies (8.6). Then with $h = f - g$, it follows that for all $A \in \sigma(X)$:*

$$\int_A h d\mu = 0.$$

For any rational $a, b > 0$, let $A_{a,b} = \{x | a \leq h \leq b\}$. Then by item 5 of Proposition 3.42:

$$a\mu\left[A_{a,b}\right] \leq \int_A h d\mu \leq b\mu\left[A_{a,b}\right],$$

and hence $\mu\left[A_{a,b}\right] = 0$. A similar conclusion is found for any rational $a, b < 0$. So $\{x | h \neq 0\}$ is a countable union of sets of μ-measure zero.

■

The following corollary was promised in Remark 4.11.

Corollary 8.24 (Radon-Nikodým theorem) *Let $(X, \sigma(X), \mu)$ be a σ-finite measure space and υ a σ-finite measure on $\sigma(X)$ that is absolutely continuous with respect to μ, so $\upsilon \ll \mu$.*

Then there exists a nonnegative $\sigma(X)$-measurable function $f : X \to \mathbb{R}$ so that for all $\sigma(X)$-measurable functions $g : X \to \mathbb{R}$:

$$\int_X g d\upsilon = \int_X g f d\mu, \tag{8.7}$$

although both integrals may be infinite.

However, g is υ-integrable if and only if gf is μ-integrable, and when integrable:

$$\int_A g d\upsilon = \int_A g f d\mu,$$

for all $A \in \sigma(X)$.

Proof. *Since the measure υ is given by (8.6), this result is an application of Proposition 4.10.* ■

For the next result, note that while the "a.e." in (8.8) seems ambiguous, by assumption $\mu(A) = 0$ if and only if $\upsilon(A) = 0$. Thus a.e. means both μ-a.e. and υ-a.e.

Corollary 8.25 (Radon-Nikodým theorem) *Let $(X, \sigma(X), \mu)$ be a σ-finite measure space and υ a σ-finite measure on $\sigma(X)$ for $\upsilon \sim \mu$, meaning both $\upsilon \ll \mu$ and $\mu \ll \upsilon$. Then:*

$$\frac{\partial \upsilon}{\partial \mu} = \left[\frac{\partial \mu}{\partial \upsilon}\right]^{-1}, \ a.e. \tag{8.8}$$

Proof. *By Proposition 8.23, there exists nonnegative $\sigma(X)$-measurable functions $f_\upsilon \equiv \frac{\partial \upsilon}{\partial \mu}$ and $f_\mu \equiv \frac{\partial \mu}{\partial \upsilon}$ defined on $X \to \mathbb{R}$ so that for all $A \in \sigma(X)$:*

$$\upsilon(A) = \int_A f_\nu d\mu, \qquad \mu(A) = \int_A f_\mu d\upsilon.$$

Letting $g = f_\mu$ in (8.7) obtains that for all $A \in \sigma(X)$:

$$\mu(A) = \int_A f_\mu d\upsilon = \int_A f_\mu f_\nu d\mu.$$

Since both functions are nonnegative, $f_\mu f_\nu = 1$ a.e. ■

8.2.4 The Lebesgue Decomposition Theorem

The Lebesgue decomposition theorem is named for **Henri Lebesgue** (1875–1941) and can be proved based on a small adaptation of the constructive proof of the Radon-Nikodým theorem. Briefly, we are again starting with a σ-finite measure space $(X, \sigma(X), \mu)$ and a σ-finite measure υ defined on $\sigma(X)$. The Lebesgue decomposition of υ obtains two measures, one absolutely continuous with respect to μ, the other mutually singular with μ.

This result was introduced in Remark IV.1.23 in the discussion of the Proposition IV.1.18 decomposition of an arbitrary distribution function on $\mathbb{R}$ into distribution functions with certain characteristic properties. Of course, there is an intimate connection between Borel measures and distribution functions, so it should be no surprise that Lebesgue's result was introduced in that context. While the decomposition of a distribution function obtained three component distribution functions, two of these combine into what is Lebesgue's singular component, as noted in item 3 of Summary 8.2.

The reader may want to review the earlier development as a less abstract introduction to this result.

Proposition 8.26 (Lebesgue decomposition theorem) *Let $(X, \sigma(X), \mu)$ be a σ-finite measure space and υ a σ-finite measure defined on $\sigma(X)$.*

Then there are measures υ_{ac} and υ_s defined on $\sigma(X)$, so that for all $A \in \sigma(X)$:

$$\upsilon(A) = \upsilon_{ac}(A) + \upsilon_s(A), \tag{8.9}$$

where:

$$\upsilon_{ac} \ll \mu, \quad \upsilon_s \perp \mu.$$

Further, this decomposition is unique in that if $\upsilon(A) = \upsilon'_{ac}(A) + \upsilon'_s(A)$ with $\upsilon'_{ac} \ll \mu$ and $\upsilon'_s \perp \mu$, then $\upsilon'_{ac} = \upsilon_{ac}$ and $\upsilon'_s = \upsilon_s$.

Proof.

1. ***Existence:*** *Recalling the construction in the proof of the Radon-Nikodým theorem, for given N let $r_k = k/N$, $k \geq 0$. With $A^-_{r_k}$ denoting the negative set of the Hahn decomposition for the signed measure $\upsilon - r_k\mu$, define the $\sigma(X)$-measurable sets:*

$$E_{ac} = \bigcup\nolimits_{k=0}^{\infty} A^-_{r_k}, \qquad E_s = X - \bigcup\nolimits_{k=0}^{\infty} A^-_{r_k}.$$

Define υ_{ac} and υ_s on $A \in \sigma(X)$ by:

$$\upsilon_{ac}(A) = \upsilon\left(A \bigcap E_{ac}\right), \qquad \upsilon_s(A) = \upsilon\left(A \bigcap E_s\right).$$

By Exercise II.1.29, υ_{ac} and υ_s are measures, and (8.9) is satisfied since $X = E_{ac} \bigcup E_s$ as a disjoint union.

To prove that $\upsilon_{ac} \ll \mu$, (5) of the Radon-Nikodým proof obtains the existence of $\sigma(X)$-measurable $f(x)$ so that for any N:

$$\upsilon\left(A \bigcap E_{ac}\right) - \frac{1}{N}\mu\left(A \bigcap E_{ac}\right) \leq \int_{A \bigcap E_{ac}} f d\mu \leq \upsilon\left(A \bigcap E_{ac}\right) + \frac{1}{N}\mu\left(A \bigcap E_{ac}\right).$$

As N is arbitrary, this implies that for all $A \in \sigma(X)$:

$$\upsilon_{ac}(A) = \int_{A \bigcap E_{ac}} f d\mu,$$

and consequently $\upsilon_{ac}(A) = 0$ if $\mu(A) = 0$ by item 9 of Proposition 3.42.

To prove that $\upsilon_s \perp \mu$, the Radon-Nikodým proof yielded that $\mu(E_s) = 0$. But then:

$$\upsilon_s(\widetilde{E}_s) \equiv \upsilon\left(\widetilde{E}_s \bigcap E_s\right) = 0,$$

and the result follows by definition.

2. ***Uniqueness:*** *Assume that $\upsilon(A) = \upsilon'_{ac}(A) + \upsilon'_s(A)$, and so for all $A \in \sigma(X)$:*

$$\upsilon_{ac}(A) + \upsilon_s(A) = \upsilon'_s(A) + \upsilon'_{ac}(A),$$

which we would like to express as:

$$\upsilon_{ac}(A) - \upsilon'_{ac}(A) = \upsilon'_s(A) - \upsilon_s(A). \tag{1}$$

To avoid A-sets that would produce $\infty - \infty$, we need to reduce this proof to the case of finite measures.

As in the proof of the Radon-Nikodým theorem, σ-finiteness of μ and υ allows the identification of disjoint $\{A_n\}_{n=1}^{\infty} \subset \sigma(X)$ with $\bigcup_{n=1}^{\infty} A_n = X$, and where each A_n has finite μ- and υ-measure. Define the finite measure μ_n on $\sigma(X)$ by:

$$\mu_n(A) = \mu(A \bigcap A_n),$$

and similarly for υ_n. It now follows from part 1 that for all n, $\upsilon_n(A) = \upsilon_{ac_n}(A) + \upsilon_{s_n}(A)$ with $\upsilon_{ac_n} \ll \mu_n$ and $\upsilon_{s_n} \perp \mu_n$.

Now if also $\upsilon(A) = \upsilon'_{ac}(A) + \upsilon'_s(A)$, then this is also true with subscripts n defined as above, and once again $\upsilon'_{ac_n} \ll \mu_n$ and $\upsilon'_{s_n} \perp \mu_n$. Thus the proof will be complete if we can prove that $\upsilon'_{ac_n} = \upsilon_{ac_n}$ and $\upsilon'_{s_n} = \upsilon_{s_n}$. Equivalently, we can now drop the notationally burdensome n-subscripts and complete the proof assuming all measures are finite, and thus the subtractions in (1) are well-defined.

First, $(\upsilon_{ac} - \upsilon'_{ac}) \ll \mu$ by definition, meaning that if $\mu(A) = 0$ then both $\upsilon_{ac}(A) = 0$ and $\upsilon'_{ac}(A) = 0$, and thus the signed measure $(\upsilon_{ac} - \upsilon'_{ac})(A) = 0$. Since $\upsilon_s \perp \mu$ and $\upsilon'_s \perp \mu$, let E_s be defined as above with $\mu(E_s) = \upsilon_s(\tilde{E}_s) = 0$, and let E'_s be the analogously defined set for υ'_s, so that $\mu(E'_s) = \upsilon'_s(\tilde{E}'_s) = 0$. Defining $E''_s = E_s \bigcup E'_s$ obtains $\mu(E''_s) = 0$.

Thus $\mu(A) = 0$ for all measurable $A \subset E''_s$, and since $(\upsilon_{ac} - \upsilon'_{ac}) \ll \mu$, and this implies that $(\upsilon_{ac} - \upsilon'_{ac})(A) = 0$ for all such $A \subset E''_s$. By (1), this in turn obtains $(\upsilon'_s - \upsilon_s)(A) = 0$.

This implies that $\upsilon_{ac} = \upsilon'_{ac}$ and $\upsilon'_s = \upsilon_s$ on all measurable subsets of E''_s. In addition, $\upsilon_s(\tilde{E}''_s) = \upsilon'_s(\tilde{E}''_s) = 0$ since $\tilde{E}''_s = \tilde{E}_s \bigcap \tilde{E}'_s$, and so $\upsilon_s(A) = \upsilon'_s(A) = 0$ for all measurable $A \subset \tilde{E}''_s$, and by (1), $\upsilon_{ac}(A) = \upsilon'_{ac}(A) = 0$ for all such A.

Combining obtains $\upsilon_{ac} = \upsilon'_{ac}$ and $\upsilon'_s = \upsilon_s$ on $\sigma(X)$ and the proof of uniqueness is complete.

■

Proposition 8.18 investigated the notion of **absolutely continuous**, both as this terms applies to a given function and as a property between two measures. The next result addresses the notion of **singular,** again as this term applies to a function and as a property between two measures. But as noted in item 3 of Summary 8.2, singular as a notion between measures reflected two kinds of singularity of functions. For the result here, we introduce the notion of a **weakly singular function** to facilitate the discussion.

Definition 8.27 (Weakly singular function) *A function $f(x)$ will be said to be* ***weakly singular*** *if $f'(x) = 0$ m-a.e.*

There are many examples of such functions, with $f(x) = c$ the simplest.

Example 8.28 (Weakly singular functions) *Other categories of weakly singular functions introduced in earlier books:*

1. ***Saltus functions:*** *By Definition IV.1.15, given $\{x_n\}_{n=1}^{\infty} \subset \mathbb{R}$ and real sequences $\{u_n\}_{n=1}^{\infty}$, $\{v_n\}_{n=1}^{\infty}$, which are absolutely convergent:*

$$\sum\nolimits_{n=1}^{\infty} |u_n| < \infty, \ \sum\nolimits_{n=1}^{\infty} |v_n| < \infty,$$

a ***saltus function*** *$f(x)$ is defined as:*

$$f(x) = \sum\nolimits_{x_n \leq x} u_n + \sum\nolimits_{x_n < x} v_n.$$

If $\{x_n\}_{n=1}^{\infty}$ has no accumulation points, then $f'(x) = 0$ outside $\{x_n\}_{n=1}^{\infty}$, and thus $f(x)$is weakly singular. Further, $f(x)$ is right continuous if and only if $v_n = 0$ for all n, and is then also increasing if and only if $u_n > 0$ for all n.

2. ***Singular functions:*** *By Definition III.3.49, a function $f(x)$ is* ***singular*** *on the interval $[a, b]$ if $f(x)$ is continuous, monotonically increasing with $f(b) > f(a)$, and $f'(x) = 0$ almost everywhere. A function is* ***singular*** *if singular on every interval $[a, b]$.*

   The existence of a singular function on $[0, 1]$ was proved in Proposition III.3.53 with the ***Cantor function,*** *named for* ***Georg Cantor*** *(1845–1918), the discoverer of the* ***Cantor ternary set*** *of Example III.3.50 on which his function is defined. Since $f(0) = 0$ and $f(1) = 1$, this function can be concatenated to a singular function $\widetilde{f}(x)$ on $\mathbb{R}$ by defining $\widetilde{f}(x) = n + f(x)$ on $[n, n+1]$.*

   By definition, $\widetilde{f}(x)$ is weakly singular, as is $-\widetilde{f}(x)$, as is $\widetilde{f}(x) = a_n + (-1)^{b_n} f(x)$ on $[n, n+1]$ for any collections $\{a_n\}_{n=1}^{\infty} \subset \mathbb{R}$ and $\{b_n\}_{n=1}^{\infty} \subset \{0, 1\}$.

The conclusion below is that if F is an increasing, right continuous function, then F is weakly singular by Definition 8.27 if and only if $\upsilon_F \perp m$.

This result may seem surprising initially. A function F is weakly singular if $F'(x) = 0$ almost everywhere, while $\upsilon_F \perp m$ means that there is a set A so that $m(A) = \upsilon_F(\widetilde{A}) = 0$. In other words, m and υ_F have disjoint supports with union equal to X, or in the special case here, with union equal to $\mathbb{R}$.

But note that a weakly singular function $F(x)$ could have been called **Lebesgue weakly singular** and the defining property stated as $m[A] = 0$ where $A = \{x | F'(x) \neq 0\}$. So indeed, weak singularity of a function will obtain mutual singularity if it can be shown that $\upsilon_F(\widetilde{A}) = 0$, where υ_F is the Borel measure induced by F.

Proposition 8.29 (Singular definitions) *Let υ_F be a Borel measure on $(\mathbb{R}, \mathcal{B}(\mathbb{R}), m)$ induced by an increasing, right continuous function F. Then F is a weakly singular function if and only if $\upsilon_F \perp m$.*

Proof. *If F is a weakly singular function, then $F'(x) = 0$ m-a.e., so let $\widetilde{A} = \{x|F'(x) = 0\}$ and $A = \{x|F'(x) \neq 0\}$. As A and $\widetilde{A}$ need not be Borel sets, recall Proposition I.2.42 that there exists a Borel set G with $A \subset G$ and $m(G-A) = 0$. Then $m[G] = 0$ by finite additivity, and so $\upsilon_F \perp m$ will follow from a proof that $\upsilon_F\left[\widetilde{G}\right] = 0$.*

To this end, recall Proposition I.5.30 that proved that if $F'(x) \leq \epsilon$ on a Borel set B, then $\upsilon_F(B) \leq \epsilon m(B)$. Now $F'(x) = 0$ on $\widetilde{G} \subset \widetilde{A}$ implies that for any interval $[-a, a]$ and all $\epsilon > 0$:

$$\upsilon_F\left[\widetilde{G}\bigcap[-a,a]\right] \leq 2a\epsilon,$$

and so $\upsilon_F\left[\widetilde{G}\bigcap[-a,a]\right] = 0$ for all a. Since these sets can be nested:

$$\widetilde{G} = \bigcup\nolimits_{n=1}^{\infty}\left(\widetilde{G}\bigcap[-n,n]\right),$$

continuity from below of Proposition 1.29 obtains that $\upsilon_F\left[\widetilde{G}\right] = 0$, and hence, $\upsilon_F \perp m$.

If $\upsilon_F \perp m$, then for some Borel set A, $m(A) = \upsilon_F(\widetilde{A}) = 0$. Since $F(x)$ is increasing, $F'(x)$ exists m-a.e. by Proposition III.3.12, $F'(x) \geq 0$ m-a.e. by Corollary III.3.13, and $F'(x)$ is Lebesgue measurable by Corollary III.3.14. Given $\epsilon > 0$, consider the Lebesgue measurable set $\{x|F'(x) \geq \epsilon\}$. Proposition I.5.30 proved that if $F'(x) \geq \epsilon$ on a Borel set B, then $\upsilon_F(B) \geq \epsilon m(B)$. Since $m(A) = 0$, this result and monotonicity obtain:

$$\begin{aligned} m\left[\{x|F'(x) \geq \epsilon\}\right] &= m\left[\widetilde{A}\bigcap\{F'(x) \geq \epsilon\}\right] \\ &\leq \frac{1}{\epsilon}\upsilon_F(\widetilde{A}) \\ &= 0. \end{aligned}$$

Hence, $m\left[\{x|F'(x) \geq \epsilon\}\right] = 0$ for all $\epsilon > 0$.

Since these sets are nested, $\{x|F'(x) \geq \epsilon'\} \subset \{x|F'(x) \geq \epsilon\}$ for $\epsilon < \epsilon'$:

$$\{x|F'(x) > 0\} = \bigcup\nolimits_{n=1}^{\infty}\{x|F'(x) \geq 1/n\},$$

and so $m\left[\{x|F'(x) > 0\}\right] = 0$ by continuity from below.

Since $m\left[\{x|F'(x) < 0\}\right] = 0$ by Corollary III.3.13, F is a weakly singular function. ■

9

The L_p Spaces

In this chapter, we introduce an important collection of function spaces called the L_p spaces, and sometimes denoted L^p spaces, and often with additional clarifying notation. While of interest generally in the development of measure spaces with special properties, these spaces will be crucial in the later books on stochastic integration.

The L_p spaces are examples of **Banach spaces,** named for **Stefan Banach** (1892–1945), who axiomatized this theory in the early 1920s. Other notable early contributors to this theory are **Frigyes Riesz** (1880–1956), **Ernst Fischer** (1875–1954), **Hans Hahn** (1879–1934), and **Norbert Wiener** (1894–1964).

See Section III.4.3 for a different example of a Banach space, and other results on bounded linear functionals.

9.1 Introduction to Banach Spaces

We begin with a collection of definitions. While the notion of a vector space initially appears very abstract, it is important to keep in mind that this definition effectively summarizes all the operations one takes for granted in the familiar vector space of $\mathbb{R}^n$, which is formally a vector space over the real field $\mathbb{R}$. The reader may find it useful to formally translate this abstract definition to the familiar context of $\mathbb{R}^n$ to provide concreteness.

Beyond the notion of vector space are the notions of norm, convergence of a sequence of points, and completeness, collectively providing a definition of Banach space. Again, $\mathbb{R}^n$ with Euclidean norm $\|x\| = |x|$ of (4.32) is the quintessential example of a Banach space.

The underlying field $\mathcal{F}$ for a vector space is notationally general in the definition, but in these books, $\mathcal{F}$ is always $\mathbb{R}$ or $\mathbb{C}$. The general definition will identify familiar properties of these special fields.

Definition 9.1 (Field) *A field $\mathcal{F}$ is a collection of elements on which is defined an* ***addition****, denoted $+$, and a* ***multiplication****, denoted $\cdot$ or simply by juxtaposition, with the following properties:*

1. ***Closure:*** *If $\alpha, \beta \in \mathcal{F}$ then $\alpha + \beta \in \mathcal{F}$ and $\alpha\beta \in \mathcal{F}$.*

2. ***Commutativity:*** *If $\alpha, \beta \in \mathcal{F}$, then $\alpha + \beta = \beta + \alpha$ and $\alpha\beta = \beta\alpha$.*

3. ***Units:*** *There exists $0, 1 \in \mathcal{F}$ so that for all $\alpha \in \mathcal{F}$, $\alpha + 0 = \alpha$ and $1\alpha = \alpha$.*

4. ***Inverses:*** *For all $\alpha \in \mathcal{F}$, there exists $-\alpha$ so that $-\alpha + \alpha = 0$. For $\alpha \neq 0$, there exists α^{-1} so that $\alpha^{-1}\alpha = 1$.*

5. ***Associativity:*** *If $\alpha, \beta, \gamma \in \mathcal{F}$, then $(\alpha + \beta) + \gamma = \alpha + (\beta + \gamma)$ and $(\alpha\beta)\gamma = \alpha(\beta\gamma)$.*

6. ***Distributivity:*** *If $\alpha, \beta, \gamma \in \mathcal{F}$, then $\alpha(\beta + \gamma) = \alpha\beta + \alpha\gamma$.*

DOI: 10.1201/9781003264576-9

As noted above, in the same way that the notion of a field abstracts the familiar properties of $\mathbb{R}$, a vector space abstracts the familiar properties of $\mathbb{R}^n$ when one identifies n-tuples $(x_1, ..., x_n)$ with the vector x. This is the view of $\mathbb{R}^n$ that is prominent in the subject of linear algebra.

Definition 9.2 (Vector space) *A space V is called a* ***vector space over a field*** $\mathcal{F}$ *or a* ***linear space over a field*** $\mathcal{F}$*, if it is a collection of* ***points,*** *which are often called* ***vectors,*** *on which is defined* ***vector addition*** *and* ***scalar multiplication*** *which satisfy:*

1. ***Vector Addition:*** *For all $x, y, z \in V$:*
   (a) ***Closure:*** $x + y \in V$.
   (b) ***Commutativity:*** $x + y = y + x$.
   (c) ***Associativity:*** $(x + y) + z = y + (x + z)$.
   (d) ***Zero Vector:*** *There is an element $\theta \in V$ so that $x + \theta = x$ for all $x \in V$. This element is often denoted* 0.
2. ***Scalar Multiplication:*** *For all $x, y \in V$, $\alpha, \beta \in \mathcal{F}$:*
   (a) $\alpha x \in V$.
   (b) $\alpha (x + y) = \alpha x + \alpha y$.
   (c) $(\alpha + \beta) x = \alpha x + \beta x$.
   (d) $\alpha (\beta x) = (\alpha\beta) x$.
3. *For all $x \in V$:*
   (a) $0x = \theta$, *where "0" denotes the additive unit in $\mathcal{F}$, meaning $0+\alpha = \alpha$ for all $\alpha \in \mathcal{F}$.*
   (b) $1x = x$, *where "1" denotes the multiplicative unit in $\mathcal{F}$, meaning $1\alpha = \alpha$ for all $\alpha \in \mathcal{F}$.*

Remark 9.3 *Note that the zero vector θ is unique by items 1.b and 1.d. Given θ and θ' :*

$$\theta' = \theta' + \theta = \theta + \theta' = \theta.$$

Also, vector spaces contain ***additive inverses,*** $-x \equiv -1x$, *where -1 is the additive inverse of 1 in $\mathcal{F}$. That $x + (-x) = \theta$ follows from items 2.c and 3.a.*

For a normed vector space, the field will be restricted to $\mathbb{R}$ or $\mathbb{C}$ since we require the familiar **absolute value.**

Definition 9.4 ($|\cdot|$ in $\mathbb{R}$ and $\mathbb{C}$) *For real or complex x, the* ***absolute value of*** *x, denoted $|x|$, is defined by:*

$$|x| \equiv \sqrt{x\bar{x}}, \tag{9.1}$$

where by convention, this is always taken as the positive square root.

Here $\bar{x}$ denotes the ***complex conjugate*** *of $x = a + bi$, defined as $\bar{x} = a - bi$, and thus $|x| \equiv \sqrt{a^2 + b^2}$, while for real x, $|x| = \sqrt{x^2}$.*

Definition 9.5 (Normed vector space) *A vector space V is called a* ***normed vector space*** *or* ***normed linear space over a field*** $\mathcal{F} \in \{\mathbb{R}, \mathbb{C}\}$ *if there exists a functional $\|\cdot\| : V \to \mathbb{R}^+$, which is called the* ***norm*** *on the vector space V, so that:*

4. $\|x\| = 0$ *if and only if* $x = 0$.

5. $\|\alpha x\| = |\alpha| \, \|x\|$ *for all* $x \in V$ *and* $\alpha \in \mathcal{F}$.

6. ***Triangle Inequality:*** $\|x + y\| \leq \|x\| + \|y\|$ *for all* $x, y \in V$.

A normed vector space is sometimes denoted $(V, \|\cdot\|)$.

Remark 9.6 (Reverse triangle inequality) *The triangle inequality implies what it known as the reverse triangle inequality:*

6.' ***Reverse triangle inequality:*** $|\|x\| - \|y\|| \leq \|x - y\|$ *for all* $x, y \in V$.

The triangle inequality obtains that both $\|x\| \leq \|y\| + \|x - y\|$ *and* $\|y\| \leq \|x\| + \|y - x\|$, *since* $\|x - y\| = \|y - x\|$ *by item 5, and the result follows.*

A norm on a vector space provides the means to define convergence of a sequence of points in the space. The notion of a Cauchy sequence is named for **Augustin-Louis Cauchy** (1789–1857) and is useful because it provides a criterion for convergence that does not depend on knowing the limit of the sequence.

Definition 9.7 (Cauchy sequence; completeness) *In a normed linear space, a sequence* $\{s_n\}_{n=1}^{\infty} \subset V$ *is a* ***Cauchy sequence*** *if for any* $\epsilon > 0$ *there is an* $N \in \mathbb{N}$ *so that* $\|s_n - s_m\| < \epsilon$ *for all* $n, m \geq N$.

A normed linear space is ***complete*** *if given any Cauchy sequence* $\{s_n\}_{n=1}^{\infty} \subset V$, *there is an* $s \in V$ *so that* $\|s_n - s\| \to 0$. *In other words, for any* $\epsilon > 0$ *there is an* $N \in \mathbb{N}$ *so that* $\|s_n - s\| < \epsilon$ *for all* $n \geq N$. *This is then denoted as* $s_n \to s$ *as well as* $s = \lim_{n \to \infty} s_n$.

Exercise 9.8 (Convergence ⇒ Cauchy) *In the definition of completeness,* $\{s_n\}_{n=1}^{\infty}$ *was specified to be a Cauchy sequence. Prove that if* $\{s_n\}_{n=1}^{\infty} \subset V$ *and there is an* $s \in V$ *so that* $\|s_n - s\| \to 0$, *then* $\{s_n\}_{n=1}^{\infty}$ *must be a Cauchy sequence.*

Finally, we arrive at the definition of a Banach space.

Definition 9.9 (Banach space) *A* ***Banach space*** *is a complete normed linear space. A* ***real Banach space*** *is a complete normed linear space over* $\mathcal{F} = \mathbb{R}$, *and analogously for a* ***complex Banach space.***

The next result shows that Cauchy sequences have Cauchy norm sequences in $\mathbb{R}$ and are thus bounded in norm. We demonstrate in this proof that such sequences converge and thus confirming that $\mathbb{R}$ is complete. See also Example 9.13.

Proposition 9.10 (Cauchy sequences have Cauchy norm sequences) *If* $\{s_n\}_{n=1}^{\infty}$ *is a Cauchy sequence in a normed linear space* V, *then* $\{\|s_n\|\}_{n=1}^{\infty}$ *is a Cauchy sequence in* $\mathbb{R}$.

Further, there exists λ, $K \in \mathbb{R}$ *so that as* $n \to \infty$:

$$\|s_n\| \to \lambda,$$

and for all n:

$$\|s_n\| \leq K.$$

If V *is complete and* $s \in V$ *is such that* $s_n \to s$, *then:*

$$\|s_n\| \to \|s\|,$$

and so $\lambda = \|s\|$.

Proof. *That* $\{\|s_n\|\}_{n=1}^{\infty}$ *is a Cauchy sequence in* $\mathbb{R}$ *follows from the reverse triangle inequality:*

$$\left|\|s_n\| - \|s_m\|\right| \leq \|s_n - s_m\|.$$

Thus given $\epsilon > 0$ *there exists* N *so that* $\|s_n - s_m\| < \epsilon$, *and so too* $\left|\|s_n\| - \|s_m\|\right| < \epsilon$, *for* $n, m \geq N$. *The existence of* λ *then follows from the completeness of* $\mathbb{R}$, *but* λ *can be constructed as follows.*

Taking $\epsilon < 10^{-M}$ *obtains that* $\{\|s_n\|\}_{n=N}^{\infty}$ *all have the same* M*-digit decimal expansions, so define the rational* λ_M *by these* M *decimals. Then* $\left|\|s_n\| - \lambda_M\right| < 10^{-M}$ *for* $n \geq N$, *and all the digits of* λ *are then defined by this process by letting* $M \to \infty$.

For boundedness of $\{\|s_n\|\}_{n=1}^{\infty}$, *let* $\epsilon = 1$. *Then there exists* N *so that* $\|s_n - s_m\| < 1$ *for* $n, m \geq N$. *By the triangle inequality it follows that for all* $n \geq N$:

$$\|s_n\| \leq \|s_n - s_N\| + \|s_N\| < 1 + \|s_N\|.$$

Thus $\|s_n\| \leq K$ *for all* n *with:*

$$K = \max\{\|s_1\|, ..., \|s_{N-1}\|, \|s_N\| + 1\}.$$

If V *is complete then there exists* $s \in V$ *with* $\|s_n - s\| \to 0$, *and then by the reverse triangle inequality,* $\|s_n\| \to \|s\|$, *so* $\lambda = \|s\|$. ■

Because addition is well-defined in a linear space, there are also notions of convergence of infinite series.

Definition 9.11 (Summable; absolutely summable) *A sequence* $\{x_n\}_{n=1}^{\infty} \subset V$ *in a normed linear space is said to be* ***summable*** *if there exists* $x \in V$ *such that* $\left\|\sum_{j=1}^{n} x_j - x\right\| \to 0$ *as* $n \to \infty$. *The sequence* $\{x_n\}_{n=1}^{\infty} \subset V$ *is said to be* ***absolutely summable*** *if* $\sum_{j=1}^{\infty} \|x_j\| < \infty$.

Given a sequence $\{x_n\}_{n=1}^{\infty}$, it will often be the case that one can evaluate absolute summability far more easily than summability since the former notion does not require the identification of x. Thus it is useful to know when absolute summability implies summability. The next result proves that completeness in a normed linear space is necessary and sufficient to assure this implication.

Proposition 9.12 (Completeness vs. summability) *A normed linear space* V *is complete if and only if every absolutely summable sequence is summable.*

Proof. *If* V *is complete and* $\{x_n\}_{n=1}^{\infty} \subset V$ *is an absolutely summable sequence, then for any* $\epsilon > 0$ *there is an* N *so that* $\sum_{j=N}^{\infty} \|x_j\| < \epsilon$. *Letting* $s_n = \sum_{j=1}^{n} x_j$, *then for* $n \geq m \geq N$, *the triangle inequality obtains:*

$$\|s_n - s_m\| \leq \sum_{j=m+1}^{n} \|x_j\| < \epsilon.$$

Thus $\{s_n\}_{n=1}^{\infty} \subset V$ *is a Cauchy sequence, and so converges to some point* $x \in V$ *by completeness. By definition of* s_n, $\{x_n\}_{n=1}^{\infty}$ *is a summable sequence with* $\left\|\sum_{j=1}^{n} x_j - x\right\| \to 0$ *as* $n \to \infty$.

Conversely, assume that every absolutely summable sequence is summable, and let $\{s_n\}_{n=1}^{\infty} \subset V$ *be a Cauchy sequence. By definition, for any* k *there is an* N_k *so that* $\|s_n - s_m\| \leq 1/2^k$ *for all* $n, m \geq n_k$. *We can by construction assure that* $\{n_k\}_{k=1}^{\infty}$ *is an increasing sequence, and so* $\{s_{n_k}\}_{k=1}^{\infty}$ *is a subsequence of* $\{s_n\}_{n=1}^{\infty}$. *Define* $x_1 = s_{n_1}$ *and* $x_k = s_{n_k} - s_{n_{k-1}}$ *for* $k > 1$. *Then* $\{x_k\}_{k=1}^{\infty}$ *is a sequence with partial sums given by*

$\sum_{j=1}^{k} x_j = s_{n_k}$. *Also, since* $\|x_j\| \leq 1/2^{j-1}$ *for* $j \geq 2$, *it follows that* $\sum_{j=1}^{k} \|x_j\| \leq \|s_{n_1}\| + 1$, *and so* $\{x_k\}_{k=1}^{\infty}$ *is an absolutely summable series which by assumption is summable to some* $x \in V$.

The subsequence $\{s_{n_k}\}_{k=1}^{\infty}$ *therefore satisfies* $\|s_{n_k} - x\| \to 0$. *But since* $\{s_n\}_{n=1}^{\infty}$ *is a Cauchy sequence, for any* $\epsilon > 0$ *there is an* N *so that* $\|s_n - s_m\| < \epsilon$ *for all* $n, m \geq N$. *For this same* ϵ *there is a* K *so that* $\|s_{n_k} - x\| < \epsilon$ *for* $k \geq K$. *Hence, if* $n \geq \max\{N, n_K\}$:

$$\|s_n - x\| \leq \|s_n - s_{n_k}\| + \|s_{n_k} - x\| < 2\epsilon,$$

and so $s_n \to x$ *and* V *is complete.* ■

Example 9.13 ($\mathbb{R}^n$, $\mathbb{C}^n$, and generalizations) *The classical example of a real Banach space is* ***Euclidean space*** $\mathbb{R}^n$ *under the* ***standard Euclidean norm*** *defined on* $x = (x_1, x_2, \ldots, x_n)$ *by (4.32):*

$$\|x\| \equiv \sqrt{\sum\nolimits_{j=1}^{n} x_j^2}.$$

The classical example of a complex Banach space is the ***complex Euclidean space*** $\mathbb{C}^n$ *with the standard norm defined on* $x = (x_1, x_2, \ldots, x_n)$ *by:*

$$\|x\| \equiv \sqrt{\sum\nolimits_{j=1}^{n} |x_j|^2},$$

where $|x_j| \equiv \sqrt{x_j \bar{x}_j}$ *by Definition 9.4. It is not uncommon to denote* $\|x\|$ *by* $|x|$ *as this rarely causes confusion.*

In both spaces, vector addition and scalar multiplication are defined component-wise:

$$x + y \equiv (x_1 + y_1, x_2 + y_2, \ldots, x_n + y_n),$$

$$\alpha x \equiv (\alpha x_1, \alpha x_2, \ldots, \alpha x_n).$$

Also, both spaces are complete by a modification of the derivation in the proof of Proposition 9.10.

For example, let $\{x_n\}_{n=1}^{\infty} \subset \mathbb{C}$ *be a Cauchy sequence, where* $x_n = a_n + ib_n$. *If* $\epsilon < 10^{-M}/2$, *then there exists* N *so that* $|x_n - x_m| < 10^{-M}/2$ *for* $n, m \geq N$. *Converting this* l_2*-norm bound to an* l_1*-norm bound by (4.42) obtains that all* $\{a_n\}_{n=N}^{\infty}$ *have the same* M*-digit decimal expansions, as do* $\{b_n\}_{n=N}^{\infty}$.

Define the complex rational $x_M = a_M + ib_M$ *with* a_M *and* b_M *defined by these decimal expansions. Then* $|x_n - x_M| < 10^{-M}$ *for* $n \geq N$, *and all the digits of* a *and* b *of* $x = a + ib$ *are then defined by this process by letting* $M \to \infty$. *It is an exercise to extend this to* $\mathbb{R}^n$ *and* $\mathbb{C}^n$, *noting that given* $\epsilon < 10^{-M}/k$ *for* $k = n$ *and* $k = 2n$, *respectively, determines the first* M*-digit decimal expansions of all components in the vectors.*

The Euclidean spaces $\mathbb{R}^n$ *and* $\mathbb{C}^n$ *also have* ***inner products*** *that are compatible with the above norms. See Definition 9.42 for properties of inner products. In these cases, if* $x = (x_1, x_2, \ldots, x_n)$ *and* $y = (y_1, y_2, \ldots, y_n)$, *then the inner product* (x, y) *is defined:*

$$(x, y) \equiv \sum\nolimits_{j=1}^{n} x_j y_j, \qquad x, y \in \mathbb{R}^n,$$

$$(x, y) \equiv \sum\nolimits_{j=1}^{n} x_j \bar{y}_j, \qquad x, y \in \mathbb{C}^n,$$

and in both cases:

$$\|x\| = \sqrt{(x, x)}.$$

The ***Cauchy-Schwarz inequality*** *then states:*

$$|(x, y)| \leq \|x\| \, \|y\|, \tag{9.2}$$

where the norms and inner products are defined in the respective spaces.

Introduced in Section IV.4.3.4, the Cauchy-Schwarz inequality was originally proved in 1821 by **Augustin-Louis Cauchy** *(1759–1857) in the context of the n-dimensional real Euclidean space* $\mathbb{R}^n$*, and generalized 25 years later to all inner product spaces by* **Hermann Schwarz** *(1843–1921).*

As it turns out, Banach spaces do not in general have such compatible inner products, though there is a special class of Banach spaces that do, called **Hilbert spaces,** *and named for* **David Hilbert** *(1862–1943). See Section 9.5 on the special case of* $p = 2$.

The classical examples of $\mathbb{R}^n$ *and* $\mathbb{C}^n$ *can be generalized in two directions, separately or together:*

1. ***Change to an*** l_p***-Norm***

   There are infinitely many norms that can be defined on $\mathbb{R}^n$ *and* $\mathbb{C}^n$*, and a popular collection is referred to as the* l_p*-norms of Remark 4.43. These norms are parametrized by* $1 \leq p \leq \infty$ *and defined on real or complex* $y = (y_1, y_2, \ldots, y_n)$ *as in (4.41) by:*

   $$\|y\|_p \equiv \begin{cases} \left(\sum_{j=1}^{n} |y_j|^p\right)^{1/p}, & 1 \leq p < \infty, \\ \max_j |y_j|, & p = \infty. \end{cases} \tag{9.3}$$

   Note that the standard norm is the l_2*-norm, and so:*

   $$\|y\| = \|y\|_2 .$$

   In this context, one sometimes sees the notation $\mathbb{R}^n_p$ *and* $\mathbb{C}^n_p$*, or* $l_p(\mathbb{R}^n)$ *and* $l_p(\mathbb{C}^n)$*, to denote that the* l_p*-norms are used in these linear spaces rather than the standard Euclidean norms.*

   All such norms are equivalent as noted in Remark 4.43.

2. ***Infinite Dimension with Standard Norm***

   There is an infinite dimensional variation of $\mathbb{R}^n$ *and* $\mathbb{C}^n$*, which we denote by* $\mathbb{R}^{\mathbb{N}}$ *and* $\mathbb{C}^{\mathbb{N}}$*, defined as the collection of infinite real or complex sequences:*

   $$y = (y_1, y_2, \ldots),$$

   which have finite norm:

   $$\|y\| \equiv \sqrt{\sum_{j=1}^{\infty} |y_j|^2} < \infty.$$

   Vector addition and scalar multiplication are again defined component-wise.

3. ***Infinite Dimension with*** l_p***-Norm***

   These spaces are defined identically to $\mathbb{R}^{\mathbb{N}}$ *and* $\mathbb{C}^{\mathbb{N}}$ *but with respect to the* l_p*-norms. Denoted* $\mathbb{R}^{\mathbb{N}}_p$ *and* $\mathbb{C}^{\mathbb{N}}_p$*, or* $l_p(\mathbb{R}^{\mathbb{N}})$ *and* $l_p(\mathbb{C}^{\mathbb{N}})$*, these spaces include all sequences* $y = (y_1, y_2, \ldots)$ *with finite* l_p*-norm,* $\|y\|_p < \infty$*, where generalizing (4.41):*

   $$\|y\|_p \equiv \begin{cases} \left(\sum_{j=1}^{\infty} |y_j|^p\right)^{1/p}, & 1 \leq p < \infty, \\ \sup_j |y_j|, & p = \infty. \end{cases}$$

Remark 9.14 (On $\mathbb{R}^{\mathbb{N}}$ **and** $\mathbb{C}^{\mathbb{N}}$**)** *As a collection of points, for any* p *with* $1 \leq p \leq \infty$*:*

$$\mathbb{R}^n_p = \mathbb{R}^n, \qquad \mathbb{C}^n_p = \mathbb{C}^n.$$

In other words, while the norms of points change, the collection of points does not.

However, for the infinite dimensional spaces, the value of p has an effect both on the norm of a sequence, as well as on the collection of points in the given space. For example, $y = (1, 1/2, 1/3, \ldots)$ *is an element of* l_p *for* $p > 1$ *but not a point in* l_1*. In general:*

$$l_{p'} \subsetneq l_p \qquad \text{for } p' < p. \tag{9.4}$$

Exercise 9.15 *Prove (9.4).*

9.2 The $L_p(X)$-Spaces

The l_p-spaces of sequences are generalized in this section to the L_p function spaces. We first introduce the necessary definitions, then develop some of the properties of these spaces. We restrict our definitions to real valued functions, since that is primarily what we need in later books. But as noted in Section 7.1, nothing significant changes for complex valued functions other than the definition of $|f(x)|$ as noted in Definition 9.4, and the use of $\mathbb{C}$ instead of $\mathbb{R}$ as the underlying field $\mathcal{F}$.

In the following definition, we call $\|f\|_p$ a norm, but this requires proof, which in turn requires a convention. See Remark 9.20.

Definition 9.16 ($L_p(X)$-space) *Given a measure space* $(X, \sigma(X), \mu)$*, define* $L_p(X)$***-space,*** *and sometimes denoted* $L_p(X, \mu)$***-space,*** *as the space of all extended real-valued,* μ*-measurable functions* f *on* X *with finite* $L_p(X)$***-norm:***

$$\|f\|_p < \infty.$$

When $1 \leq p < \infty$, $\|f\|_p$ *is defined:*

$$\|f\|_p = \left(\int_X |f|^p \, d\mu\right)^{1/p}. \tag{9.5}$$

For $p = \infty$, $\|f\|_\infty$ *denotes the* ***essential supremum of*** f*, denoted* $\operatorname{ess\,sup} f$*, and defined by:*

$$\|f\|_\infty = \operatorname{ess\,sup} f \equiv \inf\{\alpha | \mu\left[\{x| \, |f(x)| > \alpha\}\right] = 0\}. \tag{9.6}$$

When $(X, \sigma(X), \mu) = (\mathbb{R}^n, \mathcal{M}_L, m)$*, n-dimensional Lebesgue measure space, the* $L_p(X)$*-spaces are often called the* ***classical*** $L_p(\mathbb{R}^n)$***-spaces.***

Given index p with $1 < p < \infty$*, the index q is called the* ***conjugate index to*** *p if:*

$$\frac{1}{p} + \frac{1}{q} = 1.$$

In this case $1 < q < \infty$ *since* $q = \frac{p}{p-1}$*. The pair* (p, q) *are called* ***conjugate indexes,*** *and this definition is sometimes applied to the pair* $(1, \infty)$*, notationally defining* $\frac{1}{0} = \infty$*.*

The next exercise shows that in general, $L_p(X)$-spaces are not nested. See Exercise 9.26 for a very different conclusion then $\mu(X) < \infty$.

Exercise 9.17 (On $L_p(X)$-space relationships) *Let* $(X, \sigma(X), \mu) = ((0, \infty), \mathcal{B}(0, \infty), m)$*, the Borel measure space on* $(0, \infty)$ *with Lebesgue measure. Show by examples that for* $1 \leq p < p' \leq \infty$ *that neither* $L_p \subset L_{p'}$ *nor* $L_{p'} \subset L_p$ *need be true. Hint: Consider functions defined on* $(0, 1)$ *and* $[1, \infty)$*.*

Remark 9.18 (On $L_\infty(X)$) *Note that $\|f\|_\infty$ is well-defined. First, $\{x| \, |f(x)| > \alpha\} \in \sigma(X)$ for any α by the μ-measurability of f. Further, if $\mu[\{x| \, |f(x)| > \alpha\}] = 0$ then $\mu[\{x| \, |f(x)| > \alpha'\}] = 0$ for $\alpha' > \alpha$ since:*

$$\{x| \, |f(x)| > \alpha'\} \subset \{x| \, |f(x)| > \alpha\} \text{ if } \alpha < \alpha', \tag{1}$$

and so the infimum is well-defined.

Now if $\alpha > \|f\|_\infty$:

$$\mu[\{x| \, |f(x)| > \alpha\}] = 0.$$

This and the requirement that $\|f\|_\infty < \infty$ obtains that $L_\infty(X)$ is the space of ***μ-measurable functions*** *f on X that are bounded outside a set of* ***μ-measure*** *0. Such functions are sometimes referred to as* ***essentially bounded functions,*** *since by (1) and continuity from above:*

$$|f| \leq \|f\|_\infty, \ \mu\text{-a.e.} \tag{9.7}$$

Further, $\|f\|_\infty$ is the smallest real number with this property.

While $L_p(X)$-spaces are not in general nested, there is a predictable connection between $L_p(X)$-norms for functions that are in many spaces.

Exercise 9.19 (On $\|f\|_p \to \|f\|_\infty$ as $p \to \infty$.) *Given a measurable function on $(X, \sigma(X), \mu)$, define $\|f\|_p = \infty$ if $f \notin L_p(X)$.*

Prove that if $f \in L_r(X)$ for some $r < \infty$, then:

$$\|f\|_p \to \|f\|_\infty \text{ as } p \to \infty. \tag{9.8}$$

Hint: If $f \in L_r(X)$ for some $r < \infty$, then:

$$\limsup_{p\to\infty} \|f\|_p = \|f\|_\infty,$$

since $|f|^p \leq \|f\|_\infty^{p-r} |f|^r$, μ-a.e. for $r < p < \infty$. Further:

$$\liminf_{p\to\infty} \|f\|_p \geq \|f\|_\infty,$$

since given any $t < \|f\|_\infty$, $\mu[A_t] > 0$ with $A_t \equiv \{x| \, |f(x)| > t\}$, and $\|f\|_p \geq t(\mu[A_t])^{1/p}$ by definition. Consider the limit inferior in the cases where $\mu[A_t] < \infty$ or $\mu[A_t] = \infty$.

Our goal is to prove that the $L_p(X)$-spaces are Banach spaces, but we have a problem that may not have been apparent initially. As stated, $\|f\|_p$ is not actually a norm because it violates item 4 of Definition 9.5 that $\|x\| = 0$ if and only if $x = 0$. By item 2 of Proposition 3.29, if $1 \leq p < \infty$ and $\|f\|_p = 0$, we can only assert that $f = 0$ μ-a.e. The same conclusion follows from $\|f\|_\infty = 0$ by definition.

Remark 9.20 (Each $f \in L_p(X)$ is a class of functions) *The most common approach to circumventing this "problem" is to identify a function f of an $L_p(X)$-space with an* ***equivalence class*** *of functions, so that by definition:*

$$f \equiv \{f_0 | f_0 = f, \ \mu\text{-a.e.}\}.$$

In other words, we will not distinguish between different representatives of this equivalence class. Any statement about given f must be interpreted as a statement that applies equally well to any function in the equivalence class of f.

With this convention, $\|f\|_p = 0$ if and only if $f = 0$ since the latter statement simply means that $f_0 \equiv 0 \in f$, where this last f denotes the class.

Looking at the last sentence, the notion that $\|f\|_p = 0$ is well-defined whether f denotes a function or an equivalence class. If this is true for a given f of the class, it is true for all f in that class by item 4 of Proposition 3.29. On the other hand, the statement $f = 0$ potentially has two different meanings depending on the context. Interpreted as a statement about a function, in calculus say, this implies $f \equiv 0$. However, when understood as a statement about the equivalence class in measure theory, this only implies that $0 \in f$ and so $f = 0$, μ-a.e.

While potentially confusing initially, it must be emphasized that in virtually any context in measure theory, there is no meaningful distinction between determining that $f = 0$, or that $f = 0$, μ-a.e. In measure theory, one has very little control over measurable functions at the pointwise level, and in general can only define or identify such functions relative to sets of non-zero measure.

Remark 9.21 (On equivalence classes) *The notion of equivalence classes is formalized in mathematics by first defining an* ***equivalence relation*** *on a given collection, where above we start with the collection $\mathcal{C}$ of all μ-measurable functions on X. An equivalence relation, almost universally denoted $\sim$ and sometimes by R, is then a binary relation on $\mathcal{C}$ which satisfies three properties:*

1. ***Reflexivity:*** *$f \sim f$ for all $f \in \mathcal{C}$.*

2. ***Symmetry:*** *$f \sim g$ implies $g \sim f$ for all $f, g \in \mathcal{C}$.*

3. ***Transitivity:*** *$f \sim g$ and $g \sim h$ implies $f \sim h$ for all $f, g, h \in \mathcal{C}$.*

It is an exercise to check that the binary relation:

$$f \sim g \Leftrightarrow f = g, \ \mu\text{-a.e.},$$

is an equivalence relation on $\mathcal{C}$.

An equivalence relation then induces equivalence classes on the original collection $\mathcal{C}$. This new collection of classes $\{\mathcal{C}_\alpha\}$ is denoted $\mathcal{C}/\tilde{}$ or $\mathcal{C}/R$, and read $\mathcal{C}$ ***modulo*** *$\tilde{}$, or $\mathcal{C}$* ***modulo*** *R, is the defined by:*

$$f, g \in \mathcal{C}_\alpha \Longleftrightarrow f \sim g.$$

Example 9.22 (Vitali's nonmeasurable set) *The reader may be interested in revisiting Proposition I.2.31, which implemented* ***Giuseppe Vitali's*** *(1875–1932) construction of a Lebesgue nonmeasurable set. It begins by defining an equivalence relation on $[0, 1]$ and creating equivalence classes, and then uses the axiom of choice to select one element from each class, creating his famous nonmeasurable set.*

We now prove that for any measure space $(X, \sigma(X), \mu)$, the corresponding $L_p(X)$-space is a Banach space. This proof will be split into two parts, the first establishing that $L_p(X)$ is a normed linear space over $\mathbb{R}$, the second step proving completeness as a normed space.

Proposition 9.23 ($L_p(X)$ is a normed linear space) *Given a measure space $(X, \sigma(X), \mu)$, the space $L_p(X)$ is a normed linear space over $\mathbb{R}$ for $1 \leq p \leq \infty$.*

Proof. *It is an exercise to show that these spaces satisfy the vector space and norm requirements other than item 1.a. of Definition 9.2, that $L_p(X)$ is closed under addition, and item 6, that the $L_p(X)$-norm satisfies the triangle inequality. Both results are proved by Minkowski's inequality of Proposition 9.29.* ■

Minkowski's inequality, or **the Minkowski inequality**, is named for **Hermann Minkowski** (1864–1909). For this proof, we require Hölder's inequality, a result introduced in Proposition IV.4.54 in the context of moment inequalities. Recall that expectations of random variables and μ-integration of measurable functions was formally related in Section IV.4.1.2, so it will be no surprise that this result extends beyond probability spaces to general measure spaces. While it is an exercise to check that the proof of this result did not require that the underlying measure space was a probability space, we provide a proof for completeness.

Hölder's inequality was derived by **Otto Hölder** (1859–1937) in 1889, citing the original 1888 derivation by **Leonard James Rogers** (1862–1933). For the current proof of this inequality, we utilize **Young's inequality,** derived by **W. H. Young** (1863–1942) in 1912.

Proposition 9.24 (Young's inequality) *Given p, q with $1 < p, q < \infty$ and $\frac{1}{p} + \frac{1}{q} = 1$, then for all $a, b > 0$:*

$$ab \leq \frac{a^p}{p} + \frac{b^q}{q}. \tag{9.9}$$

Proof. *This was proved as Proposition IV.4.51, but we include this short proof for completeness.*

The function $g(x) = \ln x$ is concave on $(0, \infty)$ by Proposition IV.4.40, since $g'(x) = 1/x$ is a decreasing function. Thus by definition of concave, if $x, y \in (0, \infty)$:

$$\ln(tx + (1-t)y) \geq t\ln(x) + (1-t)\ln(y), \quad \text{for } t \in [0, 1]. \tag{9.10}$$

Applying (9.10) with $t = 1/p$, $x = a^p$ and $y = b^q$:

$$\begin{aligned} \ln(ab) &\equiv (\ln a^p)/p + (\ln b^q)/q \\ &\leq \ln\left(\frac{a^p}{p} + \frac{b^q}{q}\right), \end{aligned}$$

and (9.9) follows by exponentiation. ■

Proposition 9.25 (Hölder's inequality) *Given p, q with $1 < p, q < \infty$ and $1/p + 1/q = 1$, assume that $f \in L_p(X)$ and $g \in L_q(X)$.*

Then $fg \in L_1(X)$ and:

$$\|fg\|_1 \leq \|f\|_p \|g\|_q. \tag{9.11}$$

For $p = 1$, if $f \in L_p(X)$ and $g \in L_\infty(X)$, then $fg \in L_1(X) < \infty$ and:

$$\|fg\|_1 \leq \|f\|_1 \|g\|_\infty. \tag{9.12}$$

Proof. *For $p > 1$, if either or both $\|f\|_p = 0$ and $\|g\|_q = 0$, then either or both $f = 0$ and $g = 0$ outside a set of μ-measure 0 by item 2 of Proposition 3.29. Thus $fg = 0$ outside such a set, $\|fg\|_1 = 0$ by item 4 of Proposition 3.29, and (9.11) follows in these cases.*

Thus assuming that $\|f\|_p \neq 0$ and $\|g\|_q \neq 0$, apply Young's inequality with $a = |f| / \|f\|_p$ and $b = |g| / \|g\|_q$:

$$\frac{|fg|}{\|f\|_p \|g\|_q} \leq \frac{1}{p}\frac{|f|^p}{\|f\|_p^p} + \frac{1}{q}\frac{|g|^q}{\|g\|_q^q}.$$

The existence of $\int_X |f|^p \, d\mu$ and $\int_X |g|^q \, d\mu$ therefore assures the existence of $\int_X |fg| \, d\mu$ by item 5 of Proposition 3.29. Integrating, the right-hand side reduces to 1 for conjugate indexes, and (9.11) follows.

If $p = 1$ and $q = \infty$, then from $|fg| \leq |f| \, \|g\|_\infty$ μ-a.e. by (9.7), we obtain by integration:

$$\|fg\|_1 \leq \|f\|_1 \, \|g\|_\infty \, .$$

■

Recalling Exercise 9.17, the following exercise proves that for finite measure spaces, the $L_p(X)$-spaces are nested.

Exercise 9.26 (On $\{L_p(X)\}_{p \geq 1}$ for $\mu(X) < \infty$) *Using Hölder's inequality, prove that if $(X, \sigma(X), \mu)$ is a finite measure space, then for $1 \leq p \leq p' \leq \infty$:*

$$L_{p'}(X) \subset L_p(X).$$

Hint: Confirm that $L_\infty(X) \subset L_p(X)$ for all $p \geq 1$ (In fact this extends to $0 < p < 1$). If $f \in L_{p'}(X)$ for $p' < \infty$, use Hölder to bound the integral of $|f|^p$ by the integral of $|f|^{p'}$.

The **Cauchy-Schwarz inequality** was originally proved by **Augustin-Louis Cauchy** (1759–1857) in 1821 in the context of the n-dimensional real Euclidean space $\mathbb{R}^n$ and generalized 25 years later to all inner product spaces by **Hermann Schwarz** (1843–1921). Though derived much earlier, the Cauchy-Schwarz inequality is now a simple corollary of Hölder's inequality. For an independent proof, see Proposition IV.4.47 and Corollary IV.4.48.

Proposition 9.27 (Cauchy-Schwarz inequality) *If $f, g \in L_2(X)$, then $fg \in L_1(X)$ and:*

$$\|fg\|_1 \leq \|f\|_2 \, \|g\|_2 \, . \tag{9.13}$$

Proof. *Immediate from Hölder's inequality.* ■

Remark 9.28 (On $p = 2$; Conjugate (p, q) pairs) *Are conjugate pairs (p, q), and in particular $p = q = 2$, really the special cases within the $L_p(X)$-space hierarchy as implied Hölder's inequality?*

This result seems to imply that $p = 2$ is the only case for which fg is integrable when both $f, g \in L_p(X)$, and that fg is integrable for $f \in L_p(X)$ and $g \in L_q(X)$ only for conjugate pairs (p, q). But that is not what the above proved. Without contradicting the above results, it is certainly possible that $fg \in L_1(X)$ for $f, g \in L_p(X)$ and $p \neq 2$, or, for $f \in L_p(X)$ and $g \in L_q(X)$ for a non-conjugate pair (p, q), since this result does not prove otherwise.

However, simple examples resolve this question.

Let $X = [0, \infty)$ with Lebesgue measure and define $f(x) = x^a$, $g(x) = x^b$ on $[1, \infty)$ and 0 otherwise. Then $f \in L_p(X)$ if $a < -1/p$, and similarly $g \in L_q(X)$ for $b < -1/q$ and $fg \in L_1(X)$ for $a + b < -1$. Given such f, g obtains $a + b < -(1/p + 1/q)$, and thus $fg \in L_1(X)$ if and only if $1/p + 1/q \geq 1$.

Now let $\bar{f}(x) = x^a$, $\bar{g}(x) = x^b$ on $[0, 1]$ and 0 otherwise. Then $\bar{f} \in L_p(X)$ if $a > -1/p$, and similarly $\bar{g} \in L_q(X)$ for $b > -1/q$ and $\bar{f}\bar{g} \in L_1(X)$ for $a + b > -1$. Given such $\bar{f}$, $\bar{g}$ obtains $a + b > -(1/p + 1/q)$, and thus $\bar{f}\bar{g} \in L_1(X)$ if and only if $1/p + 1/q \leq 1$.

This example illustrates that for general $f \in L_p(X)$ and $g \in L_q(X)$, it is not possible for $fg \in L_1(X)$ to hold unless (p, q) is conjugate, while Hölder's inequality proves that it is always true that $fg \in L_1(X)$ in such cases. Thus for integrability of fg, having $f \in L_p(X)$ and $g \in L_q(X)$ for conjugate (p, q) is necessary and sufficient.

Letting $p = q$ in the above examples proves that for integrability of fg, where $f, g \in L_p(X)$, it is necessary and sufficient that $p = 2$. See Section 9.5.

With Hölder's inequality, we are now able to prove Minkowski's inequality, and thereby complete the proof of Proposition 9.23 that $L_p(X)$ is a normed linear space over $\mathbb{R}$ for $1 \leq p \leq \infty$.

Proposition 9.29 (Minkowski's inequality) *If $1 \leq p \leq \infty$ and $f, g \in L_p(X)$, then $f+g \in L_p(X)$ and:*

$$\|f+g\|_p \leq \|f\|_p + \|g\|_p . \tag{9.14}$$

Proof. *As $f+g$ is always measurable by Proposition 2.13, $f+g \in L_p(X)$ follows from a proof of (9.14).*

If $p=1$, this norm inequality follows from the triangle inequality, that states $|f+g| \leq |f|+|g|$. For $p=\infty$, this result again follows from the triangle inequality and (9.7):

$$\begin{aligned} |f+g| &\leq |f|+|g| \\ &\leq \|f\|_\infty + \|g\|_\infty , \mu\text{-a.e.} \end{aligned}$$

Thus $||f||_\infty + ||g||_\infty$ is an essential upper bound for $|f+g|$. Since $||f+g||_\infty$ is the smallest such upper bound, the proof is complete.

Now assume that $1<p<\infty$. Since $h(x)=x^p$ is convex (Definition IV.4.39) on $[0,\infty)$ for $p>1$ by Proposition IV.4.40, $|(f+g)/2|^p \leq (|f|^p+|g|^p)/2$ and thus by integration:

$$\|f+g\|_p \leq 2^{(p-1)/p}\left[\|f\|_p^p + \|g\|_p^p\right]^{1/p}.$$

While this upper bound assures that $f+g \in L_p(X)$, it falls short of the triangle inequality in (9.14). The final step requires Hölder's inequality, where we assume that $\|f+g\|_p \neq 0$, since (9.14) is apparently valid if $\|f+g\|_p = 0$.

By the triangle inequality, $|f+g| \leq |f|+|g|$ and so:

$$\int_X |f+g|^p \, d\mu \leq \int_X |f|\,|f+g|^{p-1} \, d\mu + \int_X |g|\,|f+g|^{p-1} \, d\mu. \tag{1}$$

Now $f, g, f+g \in L_p(X)$, and $|f+g|^{p-1} \in L_q(X)$ with q conjugate to p since $(p-1)q=p$, and so:

$$\left\| |f+g|^{p-1} \right\|_q = \left(\|f+g\|_p \right)^{p/q} < \infty.$$

Thus by (9.11):

$$\begin{aligned} \textstyle\int_X |f|\,|f+g|^{p-1} \, d\mu &\leq \|f\|_p \left\| |f+g|^{p-1} \right\|_q , \\ \textstyle\int_X |g|\,|f+g|^{p-1} \, d\mu &\leq \|g\|_p \left\| |f+g|^{p-1} \right\|_q . \end{aligned} \tag{2}$$

Combining results of (1) *and* (2)*:*

$$\|f+g\|_p^p \leq \left(\|f\|_p + \|g\|_p \right) \left(\|f+g\|_p \right)^{p/q},$$

and (9.14) follows by division since $p - p/q = 1$. ■

The final step for the proof that $L_p(X)$ is a Banach space is to address completeness. The needed result is called the **Riesz-Fischer theorem,** named for **Frigyes Riesz** (1880–1956) and **Ernst Fischer** (1875–1954), who independently derived it for the space L_2 in 1907.

Proposition 9.30 (Riesz-Fischer theorem - $L_p(X)$ is a Banach space) *Given a measure space $(X, \sigma(X), \mu)$, the space $L_p(X)$ is a complete normed linear space over $\mathbb{R}$ for $1 \leq p \leq \infty$, and hence is a Banach space.*

Proof.

1. $\mathbf{p = \infty}$***:***

Let $\{f_n\}_{n=1}^{\infty}$ be a Cauchy sequence in $L_\infty(X)$ and choose an increasing sequence $\{n_k\}_{k=1}^{\infty}$ so that $\|f_n - f_m\|_\infty < 2^{-k}$ for $n, m \geq n_k$. Then $\|f_{n_{k+1}} - f_{n_k}\|_\infty < 2^{-k}$ and recalling Remark 9.18, there exists $A_k \in \sigma(X)$ with $\mu(A_k) = 0$ and $|f_{n_{k+1}}(x) - f_{n_k}(x)| < 2^{-k}$ for $x \notin A_k$. In other words, such $\{A_k\}_{k=1}^{\infty}$ exists for any function sequence from these equivalence classes. If $A \equiv \bigcup_k A_k$, then $\mu(A) = 0$, and $|f_{n_{k+1}}(x) - f_{n_k}(x)| < 2^{-k}$ for all k for $x \notin A$.

Hence for each $x \notin A$, $\{f_{n_k}(x)\}_{k=1}^{\infty}$ is a Cauchy sequence in $\mathbb{R}$. By completeness, define $f(x) = \lim_{k\to\infty} f_{n_k}(x)$ for $x \notin A$, and $f(x) = 0$ for $x \in A$. We prove that $f(x)$ so defined is μ-measurable, that $f \in L_\infty(X)$, and $f_n \to f$ in $L_\infty(X)$.

Define $\widetilde{f}_{n_k}(x) \equiv f_{n_k}(x)$ for $x \notin A$ and $\widetilde{f}_{n_k}(x) = 0$ for $x \in A$. Then $\widetilde{f}_{n_k}(x) = f_{n_k}(x)\chi_{\tilde{A}}(x)$ is μ-measurable since A is measurable, $f(x) = \lim_{k\to\infty} \widetilde{f}_{n_k}(x)$ for all x, and $f(x)$ is μ-measurable by Proposition 2.19.

To see that $f \in L_\infty(X)$, Minkowski's inequality obtains that:

$$\|f\|_\infty \leq \|f_n - f\|_\infty + \|f_n\|_\infty .$$

Now $\|f_n\|_\infty \leq K$ by Proposition 9.10, and the result will follow from $\|f_n - f\|_\infty \to 0$.

For this, if $n \geq n_k$ and $x \notin A$:

$$\begin{aligned} |f_n(x) - f(x)| &\leq |f_n(x) - f_{n_k}(x)| + |f_{n_k}(x) - f(x)| \\ &\leq 2^{-k} + 2^{-k+1}. \end{aligned}$$

The first bound follows from the definition of n_k. For the second, $f(x) = \lim_{k\to\infty} f_{n_k}(x)$ for $x \notin A$ and the triangle inequality obtain that for all m:

$$\left|f_{n_k}(x) - f_{n_{k+m}}(x)\right| \leq \sum\nolimits_{j=0}^{m-1} \left|f_{n_{k+j+1}} - f_{n_{k+j}}\right| < \sum\nolimits_{j=0}^{m-1} 2^{-(k+j)} < 2^{-k+1}.$$

Thus $f_n \to f$ outside A, a set of measure 0, and so $\|f_n - f\|_\infty \to 0$.

The last detail, which we will not repeat again, is to prove that given another function sequence $\{\tilde{f}_n(x)\}_{n=1}^{\infty}$ from the associated equivalence classes, and corresponding $\{\tilde{A}_k\}_{k=1}^{\infty}$-collection, the resulting function $\tilde{f}$ will satisfy $f = \tilde{f}$, μ-a.e. This assures that $\tilde{f}$ is in the same equivalence class as the above constructed f. But this follows by Minkowki's inequality:

$$\left\|\tilde{f} - f\right\|_\infty \leq \left\|\tilde{f}_n - \tilde{f}\right\|_\infty + \|f_n - f\|_\infty + \left\|\tilde{f}_n - f_n\right\|_\infty ,$$

because $f_n = \tilde{f}_n$ μ-a.e. assures that $\left\|\tilde{f}_n - f_n\right\|_\infty = 0$ for all n.

2. $\mathbf{1 \leq p < \infty}$***:***

Let $\{f_n\}_{n=1}^{\infty}$ be an absolutely summable sequence in $L_p(X)$, meaning $\sum_{n=1}^{\infty} \|f_n\|_p = M < \infty$. By Proposition 9.12, completeness will follow by showing that $\{f_n\}_{n=1}^{\infty}$ is summable, meaning that there exists $f \in L_p(X)$ with $\sum_{j=1}^{n} f_j \to f$ in $L_p(X)$.

To this end, define the partial summation sequence $g_n(x) = \sum_{j=1}^{n} |f_j(x)|$. By Minkowski's inequality, $g_n \in L_p(X)$, $\|g_n\|_p \leq M$, and hence:

$$\int_X g_n^p d\mu \leq M^p. \tag{1}$$

Thus for each x, $\{g_n(x)\}_{n=1}^{\infty}$ is an increasing sequence of extended real numbers in $\overline{\mathbb{R}}$ (Definition I.3.1), so define $g(x) = \lim_{n\to\infty} g_n(x)$. Then $g(x)$ is measurable by Proposition 2.19 as a limit of measurable functions, and since $g_n(x) \geq 0$ we conclude by (3.13) of Fatou's lemma and (1) that $g \in L_p(X)$ and $\|g\|_p \leq M$.

This implies by Proposition 3.40 that $g(x) \equiv \sum_{j=1}^{\infty} |f_j(x)| < \infty$ μ*-a.e. Thus there exists* $A \in \sigma(X)$ *with* $\mu(A) = 0$ *so that for* $x \notin A$, $\sum_{j=1}^{\infty} f_j(x)$ *is absolutely convergent to a real number and hence is convergent. For* $x \notin A$, *define* $s(x) = \lim_{n\to\infty} s_n(x)$ *with* $s_n(x) = \sum_{j=1}^{n} f_j(x)$, *and* $s(x) = 0$ *on* A. *As above, defining* $\widetilde{f}_j(x) = f_j(x)\chi_{\tilde{A}}(x)$ *obtains that* $\widetilde{f}_j(x)$ *is* μ*-measurable and* $s(x) = \sum_{j=1}^{\infty} \widetilde{f}_j(x)$ *for all* x.

Hence, $s(x)$ *is measurable and* $s \in L_p(X)$ *since* $|s| \leq g \in L_p(X)$. *As* $|s_n| \leq g$:

$$|s(x) - s_n(x)| \leq 2\,|g(x)|,$$

and so $|s(x) - s_n(x)|^p$ *is dominated by the integrable function* $2^p\,|g(x)|^p$. *By Lebesgue's dominated convergence theorem,* $|s(x) - s_n(x)|^p \to 0$ *obtains:*

$$\int_X |s(x) - s_n(x)|^p \, d\mu \to 0,$$

and so $\|s - s_n\|_p \to 0$.

Then by Minkowski's inequality:

$$\left\| s - \sum_{j=1}^{n} f_j \right\|_p \leq \|s - s_n\|_p + \left\| s_n - \sum_{j=1}^{n} f_j \right\|_p.$$

This last term is 0 *by item 4 of Proposition 3.29 since* $s_n = \sum_{j=1}^{n} f_j$ μ*-a.e., and thus* $\{f_n\}_{n=1}^{\infty}$ *is summable to* $s \in L_p(X)$. ■

There is an important corollary to the Riesz-Fischer theorem that connects the notion of L_p-convergence to convergence μ-a.e.

Corollary 9.31 (Riesz-Fischer theorem) *If* $\{f_n\}_{n=1}^{\infty}$ *is a Cauchy sequence in* $L_p(X)$ *with limit* f:

1. *If* $1 \leq p < \infty$, *then there is a subsequence* $\{f_{n_k}\}_{k=1}^{\infty}$ *so that* $f_{n_k} \to f$, μ*-a.e.*
2. *If* $p = \infty$, *then* $f_n \to f$ *uniformly on* $X - A$ *with* $\mu(A) = 0$.

Proof. *If* $p < \infty$, *let* $\{f_n\}_{n=1}^{\infty}$ *be a Cauchy sequence in* $L_p(X)$ *and choose an increasing sequence* $\{n_k\}_{k=1}^{\infty}$ *so that* $\|f_n - f_m\|_p < 2^{-k}$ *for* $n, m \geq n_k$, *and thus* $\left\|f_{n_{k+1}} - f_{n_k}\right\|_p < 2^{-k}$. *Define* $g_j(x) = \sum_{k=1}^{j} \left|f_{n_{k+1}}(x) - f_{n_k}(x)\right|$ *and* $g(x) = \sum_{k=1}^{\infty} \left|f_{n_{k+1}}(x) - f_{n_k}(x)\right|$. *By construction and the Minkowski inequality,* $\|g_j\|_p \leq 1$ *for all* j, *and since* $p < \infty$ *we can apply Fatou's lemma of Proposition 3.19 to* $\{g_j^p\}_{j=1}^{\infty}$ *to conclude that* $\|g\|_p \leq 1$.

Hence, $g(x) < \infty$ μ*-a.e. by Proposition 3.40, and so the series:*

$$f_{n_1}(x) + \sum_{k=1}^{\infty} \left(f_{n_{k+1}}(x) - f_{n_k}(x)\right),$$

converges absolutely μ*-a.e. Define the function* $f(x)$ *to equal this summation when convergent and to equal* 0 *on the non-convergence set of* μ*-measure* 0. *Then* $f(x)$ *is measurable as in the above proof, and since:*

$$f_{n_j}(x) = f_{n_1}(x) + \sum_{k=1}^{j-1} \left(f_{n_{k+1}}(x) - f_{n_k}(x)\right),$$

it follows that $f_{n_j} \to f$ μ*-a.e.*

If $p = \infty$, *then* $\|f_n - f\|_\infty \to 0$ *implies that for each* $\epsilon = 1/m$, *there exists* N_m *and* $A_m \subset X$ *with* $\mu(A_m) = 0$, *so that* $\sup|f_n(x) - f(x)| \leq 1/m$ *for* $x \notin A_m$ *and all* $n \geq N_m$. *Letting* $A = \bigcup_{n=1}^{\infty} A_n$, *then* $\mu(A) = 0$ *and we claim that* $f_n \to f$ *uniformly on* $X - A$.

Let $\epsilon > 0$ *be given. Then if* $1/m < \epsilon$, *the above obtained that* $\sup|f_n(x) - f(x)| \leq 1/m < \epsilon$ *for* $x \in X - A_m$ *and all* $n \geq N_m$. *Since* $A_m \subset A$ *implies that* $X - A \subset X - A_m$, *the proof is complete.* ■

9.3 Approximating $L_p(X)$-Functions

The following result proves that when $1 \leq p \leq \infty$, every $f \in L_p(X)$ is the $L_p(X)$-limit of a sequence of simple functions, and when X is σ-finite and $p < \infty$, each such function can be defined to be zero outside X sets of finite μ-measure. Example 9.33 demonstrates that the second part of this result does not hold in $L_\infty(X)$.

When X has a topology and $\sigma(X)$ contains the open sets and hence contains the Borel sigma algebra $\mathcal{B}(X)$ of Example 1.2, each $f \in L_p(X)$ is also the $L_p(X)$-limit of a sequence of continuous functions on X. These continuous functions again can be defined to be zero outside sets of finite μ-measure when $p < \infty$ and X is σ-finite. The proof of this relies on a generalization of Lusin's theorem of Proposition I.4.10. We do not prove these results, but reference **Doob** (1994) or **Rudin** (1974).

For the following result, $\mathcal{S}_0$ is in fact dense in $L_1(X)$ without the σ-finite (Definition 1.21) assumption on $(X, \sigma(X), \mu)$, since then Corollary 2.32 can be applied in the last step based on the integrability of $f(x)$. Details are left as an exercise.

Proposition 9.32 (Simple function approximations in $L_p(X)$) *Given a measure space $(X, \sigma(X), \mu)$, let $\mathcal{S}$ denote the space of simple functions on X as in Definition 2.26. Then $\mathcal{S}$ is dense in $L_p(X)$ for $1 \leq p \leq \infty$. In other words, for any $f \in L_p(X)$ there is a sequence $\{\varphi_n\}_{n=1}^\infty \subset \mathcal{S}$ so that $\varphi_n \to f$ pointwise and:*

$$\|f - \varphi_n\|_p \to 0.$$

If $(X, \sigma(X), \mu)$ is σ-finite, then $\mathcal{S}_0 \subset \mathcal{S}$ is dense in $L_p(X)$ for $1 \leq p < \infty$, where $\mathcal{S}_0$ is defined:

$$\mathcal{S}_0 = \{\varphi \in \mathcal{S} | \mu\left[\{x \in X | \varphi(x) \neq 0\}\right] < \infty\}.$$

In other words, $\mathcal{S}_0$ is the collection of simple functions that are 0 outside sets of finite measure.

Proof. *If $f \in L_p(X)$ is nonnegative, $f \geq 0$, define increasing $\{\varphi_n\}_{n=1}^\infty$ as in Proposition 2.28 so $\{\varphi_n\}_{n=1}^\infty \subset \mathcal{S}$ and $\varphi_n \to f$ pointwise. Since $0 \leq \varphi_n \leq f$, it follows that $\varphi_n \in L_p(X)$ for all n. By construction $|f - \varphi_n|^p \leq |f|^p$ for $1 \leq p < \infty$. Lebesgue's dominated convergence theorem can then be applied to conclude that as $n \to \infty$:*

$$\int_X |f(x) - \varphi_n(x)|^p \, d\mu \to 0,$$

and hence $\|f - \varphi_n\|_p \to 0$.

When $p = \infty$, the definition of $\|f\|_\infty < \infty$ and Remark 9.18 obtain that $|f| \leq M$ for $x \notin A \in \sigma(X)$ with $\mu(A) = 0$. By the construction in Proposition 2.28, if $n > M$ then $\sup |f - \varphi_n| \leq 2^{-n}$ for $x \notin A$, and thus $\|f - \varphi_n\|_\infty \to 0$.

For general f write $f(x) = f^+(x) - f^-(x)$, where $f^+(x)$ and $f^-(x)$ are nonnegative measurable functions as in Definition 3.37. Then, there are sequences of simple functions $\{\varphi_n^+(x)\}$ and $\{\varphi_n^-(x)\}$ so that $\varphi_n^+(x) \to f^+(x)$ and $\varphi_n^-(x) \to f^-(x)$ pointwise. Defining $\varphi_n(x) = \varphi_n^+(x) - \varphi_n^-(x)$ obtains $\varphi_n(x) \to f(x)$ pointwise. Also, since $f - \varphi_n = (f^+ - \varphi_n^+) - (f^- - \varphi_n^-)$, the Minkowski's inequality can be applied:

$$\|f - \varphi_n\|_p \leq \left\|f^+ - \varphi_n^+\right\|_p + \left\|f^- - \varphi_n^-\right\|_p,$$

and the conclusion for nonnegative f then proves that $\|f - \varphi_n\|_p \to 0$.

If X is σ-finite and $p < \infty$, the simple functions in the above proof can be chosen from $\mathcal{S}_0$ by Corollary 2.32. ■

Example 9.33 *To see that the second statement in Proposition 9.32 is not true in $L_\infty(X)$, let $X = \mathbb{R}$ with Lebesgue measure, which is a σ-finite measure space. If $f \equiv 1$ and g is any function that equals 0 outside a set of finite measure, then $\|f - g\|_\infty \geq 1$.*

9.4 Bounded Linear Functionals on $L_p(X)$-Spaces

Bounded linear transformations $T : (V_1, \|\cdot\|_1) \to (V_2, \|\cdot\|_2)$ between normed vector spaces were introduced in Definition III.4.55. The topic of this section represents a special case of linear transformations, called linear functionals, and these are characterized as transformations for which the range space V_2 is replaced by the field of scalars underlying the domain space V_1.

Focusing on real normed spaces, we have the following:

Definition 9.34 (Linear functional; bounded LF) *A **linear functional** T on a real normed linear space $(V, \|\cdot\|)$ is a mapping $T : (V, \|\cdot\|) \to (\mathbb{R}, |\cdot|)$, so that for all $x, y \in V$, $\alpha, \beta \in \mathbb{R}$:*

$$T(\alpha x + \beta y) = \alpha T(x) + \beta T(y).$$

*A **bounded linear functional** is a linear functional with the additional property that there is a constant $M < \infty$ so that for all $x \in V$:*

$$|T(x)| \leq M \|x\|.$$

By the linearity property of the above definition, a linear functional is bounded if and only if $|T(x)| \leq M$ for all $x \in V$ with $\|x\| = 1$.

Definition 9.35 (Norm of a bounded LF) *The **norm of a bounded linear functional** $T : V \to \mathbb{R}$, denoted $\|T\|$, is the minimum value of all such M in Definition 9.34. Equivalently:*

$$\|T\| = \sup_{x \neq 0} \left\{ \frac{|T(x)|}{\|x\|} \right\}. \tag{9.15}$$

By linearity, this can be equivalently defined:

$$\|T\| = \sup_{\|x\|=1} \{|T(x)|\}.$$

Remark 9.36 *Note that for a bounded linear functional T, that for all x:*

$$|T(x)| \leq \|T\| \|x\|. \tag{9.16}$$

This follows from (9.15) which states that $\|T\| \geq \frac{|T(x)|}{\|x\|}$ for all $x \neq 0$, noting that (9.16) is apparently satisfied for $x = 0$.

Properties of bounded linear transformations, and thus bounded linear functionals, were derived in Proposition III.4.58, which we partially restate in the context of linear functionals for completeness.

Proposition 9.37 (On boundedness; continuity) *Given a linear functional $T : (V, \|\cdot\|) \to (\mathbb{R}, |\cdot|)$, the following are equivalent:*

1. *T is bounded.*

2. *T is continuous: Given* $x \in V$, *for any* $\epsilon > 0$ *there is a* δ *so that:*

$$|Tx - Ty| < \epsilon \text{ if } \|x - y\| < \delta.$$

3. *T is sequentially continuous: Given* $x \in V$, *if* $x_n \to x$ *in* V *then* $Tx_n \to Tx$ *in* $\mathbb{R}$. *In other words, if* $\|x_n - x\| \to 0$ *then* $|Tx_n - Tx| \to 0$.

Example 9.38 ($L_q(X)$ and bounded LFs on $L_p(X)$ for $1 \leq p \leq \infty$) *Given an* $L_p(X)$-*space with* $1 \leq p \leq \infty$, *let* $g \in L_q(X)$, *where* q *is conjugate to* p, *meaning* $\frac{1}{p} + \frac{1}{q} = 1$ *and where we invoke the convention that* $\frac{1}{\infty} = 0$.

Define T *on* $L_p(X)$ *by:*

$$T(f) = \int_X f g d\mu.$$

Once we prove that this integral is well-defined, note that T *is linear by items 1 and 2 of Proposition 3.42.*

If $1 < p < \infty$, *then by the triangle inequality and then Hölder's inequality of Proposition 9.25:*

$$|T(f)| \leq \|fg\|_1 \leq \|g\|_q \|f\|_p .$$

So T *is a bounded linear functional on* $L_p(X)$ *with* $\|T\| \leq \|g\|_q$.

If $p = 1$ *and* $g \in L_\infty(X)$, *then the definition of* $T(f)$ *is unchanged by redefining* g *to be* 0 *on the set of* μ-*measure* 0 *where* $|g| > \|g\|_\infty$. *Hence by the triangle inequality,*

$$|T(f)| \leq \|g\|_\infty \|f\|_1 ,$$

and so again $\|T\| \leq \|g\|_\infty$. *The result for* $p = \infty$ *follows by symmetry relative to* $p = 1$.

In summary, it follows that for all p *with* $1 \leq p \leq \infty$, *and* T *defined by* $g \in L_q(X)$, *that* $\|T\| \leq \|g\|_q$.

It is in fact true that in all such cases:

$$\|T\| = \|g\|_q , \tag{9.17}$$

though an extra assumption is needed when $p = 1$. *Considering the cases:*

- $1 < p < \infty$: *Let* $f = |g|^{q/p} sgn[g]$ *where* $sgn[g] = +1$ *if* $g \geq 0$, $sgn[g] = -1$ *if* $g < 0$. *Then* $f \in L_p(X)$ *and* $T(f) = \|g\|_q \|f\|_p$. *This implies* $\|T\| \geq \|g\|_q$.

- $p = \infty$: *If* $g \in L_1(X)$, *then choosing* $f \equiv c \in L_\infty(X)$ *for any constant* $c \neq 0$ *yields* $\|T\| \geq \|g\|_1$.

- $p = 1$: *If* $g \in L_\infty(X)$, *let* $A_\epsilon \equiv \{x | \, |g| > \|g\|_\infty - \epsilon\}$. *For arbitrary* $\epsilon > 0$, A_ϵ *is* μ-*measurable with non-zero measure by definition of* $\|g\|_\infty$, *though* A_ϵ *need not have finite measure. If* μ *is* σ-*finite, meaning* $X = \bigcup_{i=1}^\infty X_i$ *with* $\mu(X_i) < \infty$, *and let* $f_n = \chi_{B_n}$ *where* $B_n \equiv A_\epsilon \bigcap \bigcup_{i=1}^n X_i$ *with* n *large enough so that* $\mu[B_n] > 0$. *Then* $f_n \in L_1(X)$, $T(f_n) \geq (\|g\|_\infty - \epsilon) \|f_n\|_1$, *and the result* $\|T\| \geq \|g\|_\infty$ *follows since* ϵ *is arbitrary.*

In summary, given an $L_p(X)$ *space and* $1 < p \leq \infty$, *every function in the conjugate space* $L_q(X)$ *induces a bounded linear functional on* $L_p(X)$ *with norm equal to the* $L_q(X)$-*norm of that function. When* $p = 1$, *this result is again true but requires* σ-*finiteness of the measure space.*

An important question that arises with general normed linear spaces is:

Can all bounded linear functionals be characterized in a useful way?

In the case of the $L_p(X)$ spaces, this question could potentially be restated given the above discussion:

Are all bounded linear functionals on $L_p(X)$ given by some $g \in L_q(X)$?

The answer is in the affirmative for $1 \leq p < \infty$, but not so for $L_\infty(X)$. This famous result is the **Riesz Representation theorem,** named for **Frigyes Riesz** (1880–1956), who also developed many results of this type for various normed linear spaces, all of which are referred to by his name.

While not needed for $1 < p < \infty$ as noted above, we assume σ-finiteness of the measure space. This is the most applicable result, and this assumption simplifies the proof, which admittedly, is already quite long.

Remark 9.39 (On T and $L_p(X)$ equivalence classes) *Recalling Remark 9.20, the statement that $f \in L_p(X)$ is a statement about equivalence classes:*

$$f \equiv \{f_0 | f_0 = f, \ \mu\text{-a.e.}\}.$$

By item 4 of Proposition 3.29, if $f \in L_p(X)$, then all such $f_0 \in L_p(X)$ and $\|f\|_p = \|f_0\|_p$. But this raises a potential question about linear functionals T on $L_p(X)$.

In short, in order for T to be a linear transformation on $L_p(X)$, it must be the case that $T(f) = T(f_0)$ for all f_0 in the class of f. By item 3 of Proposition 3.42, this statement is true for linear transformations of Example 9.38 that are defined in terms of $g \in L_q(X)$. For a general bounded linear functional, the same result follows from (9.16):

$$|T(f) - T(f_0)| \leq \|T\| \, \|f - f_0\|_p = 0.$$

Proposition 9.40 (Riesz representation theorem) *Given a σ-finite measure space $(X, \sigma(X), \mu)$, let T be a bounded linear functional on $L_p(X)$ for $1 \leq p < \infty$. Then there exists $g \in L_q(X)$ so that:*

$$T(f) = \int_X fg d\mu, \tag{9.18}$$

and:

$$\|T\| = \|g\|_q .$$

Further, if $g' \in L_q(X)$ and satisfies (9.18), then $g = g'$, μ-a.e.

Proof.

1. ***Uniqueness:*** *Assume such g and g' exist. Taking $f(x) = \chi_B(x)$ for $B \in \sigma(X)$ with $\mu(B) < \infty$, then by Definition 3.7:*

$$\int_B (g - g') \, d\mu = 0.$$

By σ-finiteness, there exists a countable collection $\{B_j\}_{j=1}^\infty \subset \sigma(X)$ with $\mu(B_j) < \infty$ for all j, and $X = \bigcup_{j=1}^\infty B_j$. Letting $B^{(n)} = \bigcup_{j=1}^n B_j$, then $\mu\left(B^{(n)}\right) < \infty$ for all n and by the above result, if $E \in \sigma(X)$ then with $E^{(n)} \equiv E \bigcap B^{(n)}$:

$$0 = \int_{E^{(n)}} (g - g') \, d\mu = \int_E (g - g') \, \chi_{B^{(n)}} d\mu.$$

Since $(g-g')\chi_{B^{(n)}} \to g-g'$ and is monotonically increasing, Lebesgue's monotone convergence theorem obtains for all $E \in \sigma(X)$:

$$\int_E (g-g')\,d\mu = 0.$$

Hence $g = g'$, μ-a.e. by item 8 of Proposition 3.42.

2. ***Existence for Finite*** $(X, \sigma(X), \mu)$: *Define a set function ν on $\sigma(X)$ by:*

$$\nu(A) = T(\chi_A),$$

which is well-defined since $\chi_A \in L_p(X)$ for all $1 \leq p < \infty$ by finiteness of μ. To prove that ν is a signed measure by Definition 8.5, first note that $\nu(\emptyset) = T(0)$, where 0 denotes the function $g(x) = 0$. Then $\nu(\emptyset) = 0$ since $T(0) = T(0) + T(0)$ by linearity of T.

Now let $\{A_j\}_{j=1}^{\infty} \subset \sigma(X)$ be disjoint, $A^{(n)} \equiv \bigcup_{j=1}^{n} A_j$ and $A \equiv \bigcup_{j=1}^{\infty} A_j$. Then $\chi_{A^{(n)}} = \sum_{j=1}^{n} \chi_{A_j}$, and by linearity of T :

$$\nu\left(\bigcup_{j=1}^{n} A_j\right) = \sum\nolimits_{j=1}^{n} \nu(A_j). \tag{1}$$

As $\chi_{A^{(n)}} \to \chi_A$ in all given $L_p(X)$ by Lebesgue's monotone convergence theorem, it follows that $T(\chi_{A^{(n)}}) \to T(\chi_A)$ by (9.16). Hence $\nu\left(\bigcup_{j=1}^{n} A_j\right) \to \nu\left(\bigcup_{j=1}^{\infty} A_j\right)$, and by (1) this proves that ν is countably additive, and thus a signed measure on $\sigma(X)$.

By the Jordan decomposition theorem of Proposition 8.13, there are mutually singular measures ν^+ and ν^- defined on $\sigma(X)$ with $\nu = \nu^+ - \nu^-$. Now if $A \in \sigma(X)$ with $\mu(A) = 0$, then since $\chi_A = 0$ μ-a.e. it follows that $\nu(A) = T(\chi_A) = T(0) = 0$. This obtains $\nu^+(A) = \nu^-(A)$ for all such A. Then by monotonicity and mutual singularity, $\nu^+(A \bigcap A^+) = \nu^-(A \bigcap A^+) = 0$ for the positive set A^+, and similarly, $0 = \nu^+(A \bigcap A^-) = \nu^-(A \bigcap A^-)$ for the negative set A^-. Thus by finite additivity, both $\nu^+ \ll \mu$ and $\nu^- \ll \mu$. The Radon-Nikodým theorem obtains μ-measurable functions g^+ and g^- so that for all $A \in \sigma(X)$, $\nu^{\pm}(A) = \int_A g^{\pm} d\mu$.

Defining $g = g^+ - g^-$ obtains by Definition 3.7 that for all $A \in \sigma(X)$:

$$T(\chi_A) = \int_A g\,d\mu \equiv \int \chi_A g\,d\mu.$$

Thus by linearity of T and of integrals, this proves that for all simple functions φ :

$$T(\varphi) = \int \varphi g\,d\mu. \tag{2}$$

If it can be proved that (2) is satisfied for all $f \in L_q(X)$ in place of φ and that $g \in L_q(X)$, then then $\|T\| = \|g\|_q$ will follow from Example 9.38.

a. $1 < p < \infty$: *Let $G_n = \{|g| < n\}$, and define:*

$$f_n = (sgn\ g)\,|g|^{q/p} \chi_{G_n}.$$

Then $f_n \in L_p(X)$ by boundedness of f_n and finiteness of μ. By Proposition 9.32, there exists a simple function sequence $\{\varphi_j\}_{j=1}^{\infty}$ so that $\varphi_j \to f_n$ pointwise, and $\|f_n - \varphi_j\|_p \to 0$ as $j \to \infty$. Hence, $T(\varphi_j) \to T(f_n)$ by (9.16), and by (2) and the bounded convergence theorem of Proposition 3.50:

$$T(\varphi_j) = \int \varphi_j g\,d\mu \to \int f_n g\,d\mu.$$

Combining obtains for all n:

$$T(f_n) = \int |g|^{1+q/p} \chi_{G_n} d\mu = \int |g|^q \chi_{G_n} d\mu, \tag{3}$$

since $1 + q/p = q$.

By (9.16):

$$|T(f_n)| \leq \|T\| \, \|f_n\|_p = \|T\| \left(\int |g|^q \chi_{G_n} d\mu \right)^{1/p},$$

and combining with (3) *obtains:*

$$\int |g|^q \chi_{G_n} d\mu \leq \|T\| \left(\int |g|^q \chi_{G_n} d\mu \right)^{1/p}. \tag{4}$$

So for all n:

$$\left\| g\chi_{G_n} \right\|_q \leq \|T\|.$$

Letting $n \to \infty$ and applying the monotone convergence theorem obtains that $g \in L_q(X)$.

Finally, to prove that (2) *is satisfied for all $f \in L_p(X)$ in place of φ, let such f be given, and $\{\varphi_j\}_{j=1}^{\infty}$ the sequence in Proposition 9.32 with $\left\| f - \varphi_j \right\|_p \to 0$ as $j \to \infty$. Then again $T(\varphi_j) \to T(f)$ by (9.16), and by* (2), *the triangle inequality, and then Hölder's inequality:*

$$\left| \int f g d\mu - T(\varphi_j) \right| \leq \int \left| (f - \varphi_j) \, g \right| d\mu \leq \left\| f - \varphi_j \right\|_p \|g\|_q \to 0.$$

Hence, $T(f) = \int f g d\mu$ and (2) *is satisfied for all $f \in L_p(X)$.*

b. $\mathbf{p = 1}$: *With G_n as in part 2.a., and $N > 0$, define:*

$$f_n = (sgn\ g) \, |g|^N \chi_{G_n}.$$

As $f_n \in L_1(X)$ by boundedness of f_n and finiteness of μ, with the same steps as above, (4) *now becomes:*

$$\int |g|^{1+N} \chi_{G_n} d\mu \leq \|T\| \int |g|^N \chi_{G_n} d\mu.$$

Using Hölder's inequality with indexes $p' = (1 + N)/N$ and $q' = 1 + N$ obtains:

$$\int |g|^N \chi_{G_n} d\mu \leq \left(\int |g|^{1+N} d\mu \right)^{N/(1+N)} \left(\int \chi_{G_n} d\mu \right)^{1/(1+N)},$$

and combining with the above bound and simplifying obtains:

$$\left(\int |g|^{1+N} \chi_{G_n} d\mu \right)^{1/(1+N)} \leq \|T\| \left(\mu(X) \right)^{1/(1+N)}.$$

Letting $N \to \infty$, it follows by (9.8) of Exercise 9.19 that:

$$\left\| g\chi_{G_n} \right\|_\infty \leq \|T\|,$$

and letting $n \to \infty$ proves that $g \in L_\infty(X)$.

That (2) *is satisfied for all $f \in L_1(X)$ in place of φ follows as in item 2.a.*

3. ***Existence for σ-finite*** $(X, \sigma(X), \mu)$ *: By Definition 1.21, there exists a countable collection $\{B_j\}_{j=1}^{\infty} \subset \sigma(X)$ with $\mu(B_j) < \infty$ for all j, and $X = \bigcup_{j=1}^{\infty} B_j$. For each j, $(B_j, \sigma(X), \mu_j)$ is a finite measure space where $\mu_j(A) \equiv \mu(A \bigcap B_j)$ for all $A \in \sigma(X)$. Further, if $f_j \in L_p(B_j, \mu_j)$ is identified with $\bar{f}_j \in L_p(X, \mu)$ by defining $\bar{f}_j = f_j$ on B_j and $\bar{f}_j = 0$ otherwise, we can identify $L_p(B_j, \mu_j) \subset L_p(X, \mu)$.*

For $1 \leq p < \infty$, T induces a bounded linear functional on $L_p(B_j, \mu_j)$ by this identification, so by part 2 there exists $g_j \in L_q(B_j, \mu_j)$ with:

$$T(f_j) = \int_{B_j} f_j g_j d\mu_j,$$

and:

$$\|g_j\|_{L_q(B_j, \mu_j)} \leq \|T\|.$$

Define the measurable function $g = \sum_{j=1}^{\infty} g_j \chi_{B_j}$ so that $g = g_j$ on B_j. If $f \in L_p(X, \mu)$, then since $f\chi_{B_j} \in L_p(B_j, \mu_j)$ and $\mu = \mu_j$ on B_j, this obtains:

$$T(f\chi_{B_j}) = \int f\chi_{B_j} g d\mu, \quad \left\|g\chi_{B_j}\right\|_q \leq \|T\|. \tag{5}$$

By linearity, it follows that for all $f \in L_p(X, \mu)$ and all M:

$$T\left(f \sum\nolimits_{j=1}^{M} \chi_{B_j}\right) = \int f \left(\sum\nolimits_{j=1}^{M} \chi_{B_j} g\right) d\mu.$$

Rewriting $\sum_{j=1}^{M} \chi_{B_j} = \chi_{B^{(M)}}$ with $B^{(M)} \equiv \bigcup_{j=1}^{M} B_j$ obtains for all $f \in L_p(X, \mu)$:

$$T\left(f\chi_{B^{(M)}}\right) = \int fg\chi_{B^{(M)}} d\mu. \tag{6}$$

We claim that for all M:

$$\|g\chi_{B^{(M)}}\|_q \leq \|T\|. \tag{7}$$

To see this, note that the above derivation of (5) could also have been implemented on the finite measure space $\left(B^{(M)}, \sigma(X), \mu^{(M)}\right)$, where $\mu^{(M)}(A) \equiv \mu\left(A \bigcap B^{(M)}\right)$ for all $A \in \sigma(X)$. It then follows from part 2 as above that there exists $g^{(M)} \in L_q(B^{(M)}, \mu^{(M)})$ so that:

$$T(f\chi_{B^{(M)}}) = \int f g^{(M)} \chi_{B^{(M)}} d\mu,$$

and:

$$\left\|g^{(M)}\right\|_q \leq \|T\|.$$

Then by uniqueness, $g^{(M)} = \sum_{j=1}^{M} g_j$ μ-a.e. on $B^{(M)}$, meaning $g^{(M)} = g\chi_{B^{(M)}}$ μ-a.e., and the claim in (7) is proved.

Letting $M \to \infty$ in (7) for $1 < p$, it follows from Lebesgue's monotone convergence theorem that $g \in L_q(X, \mu)$ with $\|g\|_q \leq \|T\|$. For $p = 1$, $\|g\chi_{B^{(M)}}\|_\infty \to \|g\|_\infty$ by definition.

Then for $f \in L_p(X, \mu)$, it is an exercise to check that $f\chi_{B^{(M)}} \to f$ in $L_p(X, \mu)$ as $M \to \infty$, and so $T\left(f\chi_{B^{(M)}}\right) \to T(f)$ by continuity of T. In addition, by the triangle and Hölder's inequalities:

$$\left|\int fg\chi_{B^{(M)}} d\mu - \int fg d\mu\right| \leq \|g\|_q \left\|f\chi_{B^{(M)}} - f\right\|_p,$$

and so by (6):

$$T(f\chi_{B(M)}) = \int fg\chi_{B(M)}d\mu \to \int fgd\mu. \tag{8}$$

Combining (8) with $T(f\chi_{B(M)}) \to T(f)$ *obtains that* $T(f) = \int fgd\mu$ *for all* $f \in L_p(X,\mu)$, *for* $g \in L_q(X,\mu)$ *with* $\|g\|_q \leq \|T\|$.

■

Remark 9.41 (Dual space) *Given a vector space* V *over a field* $\mathcal{F}$, *the collection of linear functionals on* V:

$$\{T : V \to \mathcal{F}\},$$

is called the **dual space for** V, *and denoted* V^*. *Then* V^* *is also a vector space over* $\mathcal{F}$, *where we define the linear functional* $\alpha T_1 + \beta T_2$ *pointwise.*

For example, if $V = \mathbb{R}^n$ *then also* $V^* = \mathbb{R}^n$, *where this second equality is interpreted as,* V^* **can be identified with** $\mathbb{R}^n$. *To see this, let* T *be given and for* $e_j \in \mathbb{R}^n$ *defined by* 1 *in the* j*th component and* 0*s elsewhere, let* $a_j = T(e_j)$. *Then* T *can be identified with the point* $(a_1, ..., a_n) \in \mathbb{R}^n$ *in the sense that:*

$$Tx = \sum\nolimits_{j=1}^{n} a_j x_j,$$

as follows from $x = \sum_{j=1}^{n} a_j e_j$ *and linearity.*

When V *is a normed vector space, then it is typical to define the dual space* V^* *as the collection of bounded linear functionals, recalling Definition 9.34. This space is also called the* **continuous dual space for** V, *since by Proposition 9.37,* T *is bounded if and only if* T *is continuous. Then* V^* *is not only a vector space but is a normed vector space by defining* $\|\cdot\|$ *as the norm of* T *from Definition 9.35.*

When the field $\mathcal{F}$ *underlying* V *is a Banach space, for example, when* $\mathcal{F} = \mathbb{R}$, *then* V^* *is not only a normed vector space but is a Banach space. If* $\{T_n\}_{n=1}^{\infty} \subset V^*$ *is a Cauchy sequence, then there exists* $T \in V^*$ *so that* $\|T - T_n\| \to 0$. *The functional* T *is defined pointwise, since if* $\{T_n\}_{n=1}^{\infty} \subset V^*$ *is Cauchy, then so too is* $\{T_n x\}_{n=1}^{\infty} \subset \mathcal{F}$ *for all* $x \in V$. *Since* $\mathcal{F}$ *is by assumption complete, there exists* $y \in \mathcal{F}$ *with* $T_n x \to y$. *Now define* $Tx = y$, *and left to prove is that* T *so defined is linear and bounded.*

The prior result showed that for $1 \leq p < \infty$, *not only is* $L_p(X)^*$ *a Banach space, but it is one that we recognize:*

$$L_p(X)^* = L_q(X),$$

where as above, this equality is meant in the sense that $L_p(X)^*$ **can be identified with** $L_q(X)$. *Specifically,* $T \in L_p(X)^*$ *is identified with* $g \in L_q(X)$ *by (9.18).*

9.5 Hilbert Space: A Special Case of $p = 2$

That $p = 2$ is a special case within the $L_p(X)$-space hierarchy is initially appreciated with Hölder's inequality and Remark 9.28. Namely, $p = 2$ is the only case for which fg is integrable when both $f, g \in L_p(X)$. In this special case, Hölder's inequality is the **Cauchy-Schwarz inequality**:

$$\|fg\|_1 \leq \|f\|_2 \|g\|_2,$$

and this inequality alone implies that every $g \in L_2(X)$ induces a bounded linear functional on $L_2(X)$.

That $p = 2$ is a special case within the $L_p(X)$-space hierarchy is also appreciated with the Riesz Representation theorem. Again, $L_2(X)$ is the only $L_p(X)$ space for which all bounded linear functionals are given by (9.18) with functions g within this same space. In the notation of Remark 9.41:

$$L_2(X)^* = L_2(X).$$

The Banach space $L_2(X)$ also has more structure than the other $L_p(X)$ spaces, and as a result it is a special type of Banach space that is called a **Hilbert Space.** This space is named for **David Hilbert** (1862–1943), who investigated solutions to integral equations in this space in the early 1900s. In short, a Hilbert space has all the structure of the Euclidean space $\mathbb{R}^n$ or $\mathbb{C}^n$. Recalling Example 9.13, this means that a Hilbert space is first and foremost a Banach space, but with the additional property that it possesses an **inner product** which is compatible with the given norm.

Definition 9.42 (Hilbert space) *A* ***real Hilbert space*** *H is a real Banach space $(H, \|\cdot\|)$ on which is defined an* **inner product**, *denoted (x, y), which produces a mapping $H \times H \to \mathbb{R}$ with the properties that for $x_j, y \in H$ and $\alpha, \beta \in \mathbb{R}$:*

1. $(\alpha x_1 + \beta x_2, y) = \alpha(x_1, y) + \beta(x_2, y)$,
2. $(x, y) = (y, x)$,
3. $(x, x) = \|x\|^2$.

For a ***complex Hilbert space,*** *the inner product is a mapping $H \times H \to \mathbb{C}$, and item 2 is changed to:*

2′. $(x, y) = \overline{(y, x)}$,

where $\overline{(y, x)}$ denotes the complex conjugate of (y, x). Thus if $(y, x) = a + bi$ then $\overline{(y, x)} = a - bi$.

Remark 9.43 *Combining items 1 and 2, we have that in a real Hilbert space that for $\alpha, \beta \in \mathbb{R}$:*

4. $(x, \alpha y_1 + \beta y_2) = \alpha(x, y_1) + \beta(x, y_2)$,

while combining items 1 and 2′ for a complex Hilbert space obtains for $\alpha, \beta \in \mathbb{C}$:

4′. $(x, \alpha y_1 + \beta y_2) = \overline{\alpha}(x, y_1) + \overline{\beta}(x, y_2)$.

Example 9.44 ($\mathbb{R}^n$, $\mathbb{C}^n$ and $L_2(X)$) *As noted in the introduction, $\mathbb{R}^n$ and $\mathbb{C}^n$ are the original examples of real and complex Hilbert spaces, respectively.*

By Example 9.13, for $x, y \in \mathbb{R}^n$:

$$(x, y) = \sum\nolimits_{i=1}^{n} x_i y_i,$$

while for $x, y \in \mathbb{C}^n$:

$$(x, y) = \sum\nolimits_{i=1}^{n} x_i \bar{y}_i.$$

It is an exercise to verify that these are indeed inner products on the respective spaces by the above definition.

On $L_2(X)$, define an inner product on $f, g \in L_2(X)$ by:

$$(f, g) = \int_X f g d\mu,$$

for real-valued functions, or:

$$(f, g) = \int_X f\overline{g}d\mu,$$

for complex-valued functions (recalling Section 7.1). Again, that (f, g) *so defined is an inner product on the respective space can be verified as an exercise.*

Proposition 9.45 ((x, y) **is continuous)** *If* H *is a Hilbert space, then the inner product is a continuous function from* $H \times H$ *to* $\mathbb{R}$ *or* $\mathbb{C}$.
Proof. *In real or complex Hilbert spaces:*

$$(x_1, y_1) - (x_2, y_2) = (x_1 - x_2, y_1) + (x_2 - x_1, y_1 - y_2) + (x_1, y_1 - y_2),$$

and so by the triangle and Cauchy-Schwarz inequalities:

$$|(x_1, y_1) - (x_2, y_2)| \leq \|x_1 - x_2\| \, \|y_1\| + \|x_2 - x_1\| \, \|y_1 - y_2\| + \|x_1\| \, \|y_1 - y_2\| .$$

Hence, $|(x_1, y_1) - (x_2, y_2)| \to 0$ *as* $\|x_1 - x_2\| \to 0$ *and* $\|y_1 - y_2\| \to 0$. ■

The significance of the additional structure in a Hilbert space is that with an inner product, one can introduce the notion of **orthogonality.**

Definition 9.46 (Orthogonality) *Given a Hilbert space* H, x *is defined to be* ***orthogonal*** *to* y, *denoted* $x \perp y$, *if* $(x, y) = 0$.

This is precisely the notion of orthogonality in $\mathbb{R}^n$ or $\mathbb{C}^n$ with (x, y) defined above, and this notion can be used in the same way in this general setting.

Example 9.47 (Fourier series) *Let* $\{x_n\}_{n=1}^{\infty} \subset H$ *be pairwise orthogonal,* $(x_n, x_m) = 0$ *for* $n \neq m$, *and have unit norm,* $(x_n, x_n) = \|x_n\|^2 = 1$. *Such a collection is called an* ***orthonormal system.***

Given $x \in H$, *assume there exists constants* $\{\alpha_n\}_{n=1}^{\infty} \subset \mathbb{R}/\mathbb{C}$ *so that* $x = \sum_{n=1}^{\infty} \alpha_n x_n$ *in* H *in the sense that as* $N \to \infty$:

$$\left\| \sum_{n=1}^{N} \alpha_n x_n - x \right\| \to 0.$$

Then for any m, *by continuity of inner products:*

$$\left(\sum_{n=1}^{N} \alpha_n x_n - x, x_m \right) \to 0.$$

But by linearity of inner products and orthonormality of $\{x_n\}_{n=1}^{\infty}$, *if* $N \geq m$:

$$\left(\sum_{n=1}^{N} \alpha_n x_n - x, x_m \right) = \alpha_m - (x, x_m),$$

and thus it follows that for all n:

$$\alpha_n = (x, x_n).$$

The coefficients $\{(x, x_n)\}_{n=1}^{\infty}$ *are called the* ***Fourier coefficients of*** x, *and the* ***Fourier series expansion of*** x *is given by:*

$$x = \sum_{n=1}^{\infty} (x, x_n) x_n.$$

These are named for ***Jean-Baptiste Joseph Fourier*** *(1768–1830) who developed these ideas in the context* ***Fourier series****, or trigonometric series expansions of integrable functions.*

For example, consider the complex Hilbert space $H \equiv L_2([0,2\pi])$, of complex-valued, square integrable functions defined on the interval $[0,2\pi]$, with inner product defined by:

$$(f,g) = \int_0^{2\pi} f(x)\overline{g(x)}dx.$$

Then $\{e^{inx}/\sqrt{2\pi}\}_{n=-\infty}^{\infty} \subset L_2([0,2\pi])$ is an orthonormal system, as can be verified as an exercise.

Hence, if $f \in L_2([0,2\pi])$ has a Fourier expansion, $f(x) = \frac{1}{\sqrt{2\pi}}\sum_{n=-\infty}^{\infty} \alpha_n e^{inx}$, it must be the case that:

$$\alpha_n = (f, e^{inx}/\sqrt{2\pi}) = \frac{1}{\sqrt{2\pi}}\int_0^{2\pi} f(x)e^{-inx}dx.$$

The Fourier coefficients $\{\alpha_n\}_{n=-\infty}^{\infty}$ are often denoted $\left\{\widehat{f}(n)\right\}_{n=-\infty}^{\infty}$, and so:

$$f = \frac{1}{\sqrt{2\pi}}\sum_{n=-\infty}^{\infty} \widehat{f}(n)e^{inx}. \tag{9.19}$$

That this is a trigonometric series expansion of f follows from (7.4).

An important question is, when does $f \in L_2([0,2\pi])$ have such a Fourier expansion? If $f \in L_1([0,2\pi]) \cap L_2([0,2\pi])$, then α_n is well-defined with $|\alpha_n| \leq \|f\|_1/\sqrt{2\pi}$, but the general case is more subtle.

It turns out, though we will not prove this, that every function $f \in L_2([0,2\pi])$ has this expansion in the sense that as $N \to \infty$:

$$\frac{1}{\sqrt{2\pi}}\sum_{n=-N}^{N} \widehat{f}(n)e^{inx} \to f.$$

Here convergence is defined in $L_2([0,2\pi])$:

$$\left\|\frac{1}{\sqrt{2\pi}}\sum_{n=-N}^{N} \widehat{f}(n)e^{inx} - f\right\|_2 \to 0.$$

By continuity of the inner product from Proposition 9.45, we can therefore conclude that

$$\left(\frac{1}{\sqrt{2\pi}}\sum_{n=-N}^{N} \widehat{f}(n)e^{inx}, \frac{1}{\sqrt{2\pi}}\sum_{n=-N}^{N} \widehat{f}(n)e^{inx}\right) \to \|f\|_2^2.$$

But by orthonormality, this inner product can be evaluated:

$$\left(\frac{1}{\sqrt{2\pi}}\sum_{n=-N}^{N} \widehat{f}(n)e^{inx}, \frac{1}{\sqrt{2\pi}}\sum_{n=-N}^{N} \widehat{f}(n)e^{inx}\right) = \sum_{n=-N}^{N} \left|\widehat{f}(n)\right|^2,$$

and so every $f \in L_2([0,2\pi])$ has a Fourier series expansion, where as $N \to \infty$:

$$\sum_{n=-N}^{N} \left|\widehat{f}(n)\right|^2 \to \|f\|_2^2.$$

In other words, every $f \in L_2([0,2\pi])$ has a Fourier series expansion for which the Fourier coefficients are square summable, meaning as a series in the Hilbert space $l_2(\mathbb{C})$, recalling Example 9.13 but defined with respect to vectors:

$$x = (..., x_{-2}, x_{-1}, x_0, x_1, x_2, ...).$$

Another important question is, when is a given series $\{\alpha_n\}_{n=-\infty}^{\infty}$ equal to the Fourier coefficients of a function $f \in L_2([0, 2\pi])$?

It turns out that if $\{\alpha_n\}_{n=-\infty}^{\infty} \in l_2(\mathbb{C})$, so:

$$\sum\nolimits_{n=-\infty}^{\infty} |\alpha_n|^2 < \infty,$$

then there exists $f \in L_2([0, 2\pi])$ with $\widehat{f}(n) = \alpha_n$. This is the original version of the ***Riesz-Fischer theorem,*** *since the essential ingredient for its proof is to first prove that $L_2([0, 2\pi])$ is complete. Hence, it is also common to refer to this theorem as in Proposition 9.30, with the emphasis on completeness of the $L_p(X)$-spaces.*

Summarizing, these results assert that there is a one-to-one correspondence between $L_2([0, 2\pi])$, the space of square integrable, complex-valued functions on $[0, 2\pi]$, and the space $l_2(\mathbb{C})$ of square summable double-sided series. Moreover, this correspondence is norm-preserving in that:

$$\|f\|_{L_2} = \left\| \left\{ \widehat{f}(n) \right\}_{n=-\infty}^{\infty} \right\|_{l_2}. \tag{9.20}$$

The identity in (9.20) is known as ***Parseval's identity*** *and named for* ***Marc-Antoine Parseval*** *(1755–1836).*

For more on this topic and generalizations, see ***Stein and Weiss*** *(1971).*

References

I have listed below a number of textbook references for the mathematics and finance presented in this series of books. All provide both theoretical and applied materials in their respective areas that are beyond those developed here and are worth pursuing by those interested in gaining a greater depth or breadth of knowledge. This list is by no means complete and is intended only as a guide to further study. In addition, various published research papers have been identified in some chapters where these results were discussed.

The reader will no doubt observe that the mathematics references are somewhat older than the finance references and upon web searching will find that some older texts have been updated to newer editions, sometimes with additional authors. Since I own and use the editions below, I decided to present these editions rather than reference the newer editions which I have not reviewed. As many of these older texts are considered "classics," they are also likely to be found in university and other libraries.

That said, there are undoubtedly many very good new texts by both new and established authors with similar titles that are also worth investigating. One that I will at the risk of immodesty recommend for more introductory materials on mathematics, probability theory and finance is:

1. Reitano, Robert, R. *Introduction to Quantitative Finance: A Math Tool Kit.* Cambridge, MA: The MIT Press, 2010.

Linear Algebra, Topology, Measure and Integration

2. Doob, J. L. *Measure Theory.* New York, NY: Springer-Verlag, 1994.

3. Dugundji, James. *Topology.* Boston, MA: Allyn and Bacon, 1970.

4. Edwards, Jr., C. H. *Advanced Calculus of Several Variables.* New York, NY: Academic Press, 1973.

5. Gemignani, M. C. *Elementary Topology.* Reading, MA: Addison-Wesley Publishing, 1967.

6. Halmos, Paul R. *Measure Theory.* New York, NY: D. Van Nostrand, 1950.

7. Hewitt, Edwin, and Karl Stromberg. *Real and Abstract Analysis.* New York, NY: Springer-Verlag, 1965.

8. Royden, H. L. *Real Analysis,* 2nd Edition. New York, NY: The MacMillan Company, 1971.

9. Rudin, Walter. *Principals of Mathematical Analysis,* 3rd Edition. New York, NY: McGraw-Hill, 1976.

10. Rudin, Walter. *Real and Complex Analysis,* 2nd Edition. New York, NY: McGraw-Hill, 1974.

11. Shilov, G. E., and B. L. Gurevich. *Integral, Measure & Derivative: A Unified Approach.* New York, NY: Dover Publications, 1977.

12. Stein, Elias M., and Guido Weiss. *Introduction to Fourier Analysis on Euclidean Spaces.* Princeton, NJ: Princeton University Press, 1971.

13. Strang, Gilbert. *Introduction to Linear Algebra,* 4th Edition. Wellesley, MA: Cambridge Press, 2009.

Probability Theory & Stochastic Processes

14. Billingsley, Patrick. *Probability and Measure,* 3rd Edition. New York, NY: John Wiley & Sons, 1995.

15. Chung, K. L., and R. J. Williams. *Introduction to Stochastic Integration.* Boston, MA: Birkhäuser, 1983.

16. Davidson, James. *Stochastic Limit Theory.* New York, NY: Oxford University Press, 1997.

17. de Haan, Laurens, and Ana Ferreira. *Extreme Value Theory, An Introduction.* New York, NY: Springer Science, 2006.

18. Durrett, Richard. *Probability: Theory and Examples,* 2nd Edition. Belmont, CA: Wadsworth Publishing, 1996.

19. Durrett, Richard. *Stochastic Calculus, A Practical Introduction.* Boca Raton, FL: CRC Press, 1996.

20. Feller, William. *An Introduction to Probability Theory and Its Applications,* Volume I. New York, NY: John Wiley & Sons, 1968.

21. Feller, William. *An Introduction to Probability Theory and Its Applications,* Volume II, 2nd Edition. New York, NY: John Wiley & Sons, 1971.

22. Friedman, Avner. *Stochastic Differential Equations and Applications, Volume 1 and 2.* New York, NY: Academic Press, 1975.

23. Ikeda, Nobuyuki, and Shinzo Watanabe. *Stochastic Differential Equations and Diffusion Processes.* Tokyo: Kodansha Scientific, 1981.

24. Karatzas, Ioannis, and Steven E. Shreve. *Brownian Motion and Stochastic Calculus.* New York, NY: Springer-Verlag, 1988.

25. Kloeden, Peter E., and Eckhard Platen. *Numerical Solution of Stochastic Differential Equations.* New York, NY: Springer-Verlag, 1992.

26. Lowther, George, *Almost Sure, A Maths Blog on Stochastic Calculus,* https://almostsure.wordpress.com/stochastic-calculus/

27. Lukacs, Eugene. *Characteristic Functions.* New York, NY: Hafner Publishing, 1960.

28. Nelson, Roger B. *An Introduction to Copulas,* 2nd Edition. New York, NY: Springer Science, 2006.

29. Øksendal, Bernt. *Stochastic Differential Equations, An Introduction with Applications,* 5th Edition. New York, NY: Springer-Verlag, 1998.

30. Protter, Phillip. *Stochastic Integration and Differential Equations, A New Approach.* New York, NY: Springer-Verlag, 1992.

31. Revuz, Daniel, and Marc Yor. *Continuous Martingales and Brownian Motion,* 3rd Edition. New York, NY: Springer-Verlag, 1991.

32. Rogers, L. C. G., and D. Williams. *Diffusions, Markov Processes and Martingales,* Volume 1, Foundations, 2nd Edition. Cambridge, UK: Cambridge University Press, 2000.

33. Rogers, L. C. G., and D. Williams. *Diffusions, Markov Processes and Martingales,* Volume 2, Itô Calculus, 2nd Edition. Cambridge, UK: Cambridge University Press, 2000.

34. Sato, Ken-Iti. *Lévy Processes and Infinitely Divisible Distributions.* Cambridge, UK: Cambridge University Press, 1999.

35. Schilling, René L. and Lothar Partzsch. *Brownian Motion: An Introduction to Stochastic Processes,* 2nd Edition. Berlin/Boston: Walter de Gruyter GmbH, 2014.

36. Schuss, Zeev, *Theory and Applications of Stochastic Differential Equations.* New York, NY: John Wiley and Sons, 1980.

Finance Applications

37. Etheridge, Alison. *A Course in Financial Calculus.* Cambridge, UK: Cambridge University Press, 2002.

38. Embrechts, Paul, Claudia Klüppelberg, and Thomas Mikosch. *Modelling Extremal Events for Insurance and Finance.* New York, NY: Springer-Verlag, 1997.

39. Hunt, P. J., and J. E. Kennedy. *Financial Derivatives in Theory and Practice,* Revised Edition. Chichester, UK: John Wiley & Sons, 2004.

40. McLeish, Don L. *Monte Carlo Simulation and Finance.* New York, NY: John Wiley, 2005.

41. McNeil, Alexander J., Rüdiger Frey, and Paul Embrechts. *Quantitative Risk Management: Concepts, Techniques, and Tools.* Princeton, NJ.:Princeton University Press, 2005.

Research Paper for Book V

42. Horst, H. J. Ter. "Riemann-Stieltjes and Lebesgue-Stieltjes Integrability." The American Mathematical Monthly 91, 9, 1984.

Index

Printed in the United States
by Baker & Taylor Publisher Services